PACEMAKER®

# Biology

## TEACHER'S ANSWER EDITION

GLOBE FEARON
Pearson Learning Group

# Contents

| | |
|---|---|
| About the Program | **T3** |
| Correlations to National Science Education Standards | **T4** |
| Materials For Lab Activities in the Student Edition | **T5** |
| Using the Student Edition | **T6** |
| Using the Workbook | **T11** |
| Using the Teacher's Answer Edition | **T12** |
| Using the Classroom Resource Binder | **T13** |
| Using the SciLinks® Web Site | **T14** |
| Using the ESL/ELL Teacher's Guide | **T15** |
| Related Resources | **T16** |

*Pacemaker® Biology*, Third Edition

We would like to thank the following educators, who provided valuable comments and suggestions during the development of this book.

**CONSULTANTS**

*Content Consultant:* **Richard Lowell**, Ramapo College of New Jersey, Mahwah, NJ
*Education Consultants:* **Richard Grybos**, Director of Special Education Instruction, Rochester City Schools, Rochester, NY **Christine Mason**, Center for Exceptional Children, Curriculum Assessment, Arlington, VA
*ESL/ELL Consultant:* **Elizabeth Jimenez**, GEMAS Consulting, Pomona, CA

**REVIEWERS**

*Content Reviewers:* **Sharon Danielson**, New York State Department of Health, Albany, NY **Sukamol Jacobson**, University of California, San Diego, CA **Rusty Lansford**, California Institute of Technology, Pasadena, CA **Helen McBride**, California Institute of Technology, Pasadena, CA

*Teacher Reviewers:* **Miriam Gage**, Newman Smith High School, Carrollton, TX **Jude Ann Morino**, Pequannock Township High School, Pequannock, NJ **Celiamarie Narro**, Ysleta Independent School District, El Paso, TX **Mohini Robinson**, Granada High School, Livermore, CA

**PROJECT STAFF**

*Art and Design:* Evelyn Bauer, Jenifer Hixson, Dan Trush, Jennifer Visco *Editorial:* Stephanie Cahill, Martha Feehan, Shirley White *Manufacturing:* Mark Cirillo *Marketing:* Clare Harrison, Anna Mazzoccoli *Production:* Irene Belinsky, Karen Edmonds, Jill Kuhfuss, Cynthia Lynch, Phyllis Rosinsky, Susan Tamm *Publishing Operations:* Carolyn Coyle, Tom Daning, Richetta Lobban

ISBN 0-13-024045-1

Printed in the United States of America

5 6 7 8 9 10     09 08 07

Globe Fearon
Pearson Learning Group

**1-800-321-3106**
**www.pearsonlearning.com**

The complete program that gives you everything you need to help all your students succeed in today's biology classroom

### The Student Edition and the Workbook

- Make biology interesting and understandable for students with individual needs and diverse learning styles
- Help students meet state academic content standards and boost their performance on state tests
- Prepare students with consistent vocabulary development and a preview of key learning objectives at the beginning of each chapter
- Keep chapter size and concept load manageable, empowering students to progress successfully through the text at their own pace
- Provide opportunities for discovery through lab activities and enrichment exercises that encourage students to apply newly learned concepts
- Offer extensive opportunities for practice and assessment
- Optimize comprehension through a controlled reading level
- Teach age-appropriate, high-school level biology content that students with differing ability levels will find motivating and challenging
- Integrate content with other important curriculum areas

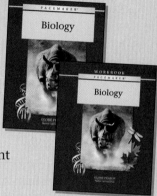

### The Teacher's Answer Edition and the Classroom Resource Binder

- Provide complete support for the mixed-ability classroom, with practice worksheets, motivational exercises, enrichment activities, lab activities, and more
- Help you target instruction to individual needs
- Offer frequent opportunities to assess understanding
- Help establish real-life science connections
- Simplify preparation with point-of-use answers and teaching strategies

### The ESL/ELL Teacher's Guide

- Speeds lesson planning with customized content that parallels the Student Edition
- Provides specific strategies and tips using proven CALLA and SDAIE methodologies
- Features vocabulary-building activities to teach and reinforce scientific terms, idioms, and expressions
- Taps students' prior knowledge and relates it to scientific concepts
- Offers three levels of questions for varying degrees of language fluency
- Uses graphic organizers to facilitate comprehension of science concepts

---

**How Pacemaker Biology *fosters successful self-paced learning:***

- **Appealing full-color design** holds students' interest.
- **High-interest photos, illustrations, charts, diagrams,** and **graphic organizers** support visual learning.
- **Upfront vocabulary instruction, controlled reading level,** and **manageable, consistent format** support comprehension and build strong reading skills.
- **Extensive practice opportunities** reinforce key skills.

*Pacemaker® Biology,* Third Edition was developed to meet national standards. The following chart correlates each chapter to the National Science Education Standards for grades 9–12.

## National Science Education Standards for Grades 9–12

### CHAPTERS

| | 1 | 2 | 3 | 4 | 5 | 6 | 7 | 8 | 9 | 10 | 11 | 12 | 13 | 14 | 15 | 16 | 17 | 18 | 19 | 20 | 21 | 22 | 23 | 24 | 25 |
|---|---|---|---|---|---|---|---|---|---|---|---|---|---|---|---|---|---|---|---|---|---|---|---|---|---|
| **Unifying Concepts and Process** | | | | | | | | | | | | | | | | | | | | | | | | | |
| Systems, order, and organization | • | • | • | • | • | • | • | • | • | • | • | • | • | • | • | • | • | • | • | • | • | • | • | • | • |
| Evidence, models, and explanation | • | • | • | • | • | • | • | • | • | • | • | • | • | • | • | • | • | • | • | • | • | • | • | • | • |
| Change, constancy, and measurement | • | • | • | • | • | • | • | • | • | • | • | • | • | • | • | • | • | • | • | • | • | • | • | • | • |
| Evolution and equilibrium | | | | | | | | | • | | | | | | | | | | | | | | | | • |
| Form and function | • | • | • | • | • | • | • | • | • | • | • | • | • | • | • | • | • | • | • | • | • | • | • | • | • |
| **Science as Inquiry** | | | | | | | | | | | | | | | | | | | | | | | | | |
| Abilities necessary to do scientific inquiry | • | | • | • | • | • | • | • | • | • | • | • | • | • | • | • | • | • | • | • | • | • | • | • | • |
| Understandings about scientific inquiry | • | • | • | • | • | • | • | • | • | • | • | • | • | • | • | • | • | • | • | • | • | • | • | • | • |
| **Life Science** | | | | | | | | | | | | | | | | | | | | | | | | | |
| The cell | | • | • | • | | | | | | | | | | | | | | | • | | | | | | |
| Molecular basis of heredity | | • | • | | • | | | | | • | | | | | | | | | | | | | | | |
| Biological evolution | | | | | | | | • | | | | | | | | | | | | | | | | | • |
| Interdependence of organisms | | | | • | | • | • | | | • | • | • | • | | | | | | | • | • | | | | |
| Matter, energy, and organization in living systems | • | • | • | • | • | • | • | • | • | • | • | • | • | • | • | • | • | • | • | • | • | • | • | | • |
| Behavior of organisms | • | • | | • | • | • | • | • | • | • | • | • | • | • | | • | | • | • | • | • | • | • | | • |
| **Science and Technology** | | | | | | | | | | | | | | | | | | | | | | | | | |
| Abilities of technological design | • | • | • | • | • | • | • | • | • | • | • | • | • | • | • | • | • | • | • | • | • | • | • | • | • |
| Understanding science and technology | • | • | • | • | • | • | • | • | • | • | • | • | • | • | • | • | • | • | • | • | • | • | • | • | • |
| **Science in Personal and Social Perspectives** | | | | | | | | | | | | | | | | | | | | | | | | | |
| Personal and community health | | • | | • | • | | | | | • | • | | • | • | • | • | • | • | • | • | | | • | • | |
| Population growth | | | | • | | | | | | | | | | | | | | | | • | • | | • | • | |
| Natural resources | | | | | | • | • | | • | | | | | | | | | | | | | | | • | |
| Environmental quality | | | | • | • | | | | | • | | | | | | | | | | • | | | • | • | • |
| Natural and human-induced hazards | | | | • | | | | • | | • | | | • | • | | | • | • | | • | | | • | • | • |
| Science and technology in local, national, and global challenges | • | • | • | • | • | • | • | • | • | • | • | • | • | • | • | • | • | • | • | • | • | • | • | • | • |
| **History and Nature of Science** | | | | | | | | | | | | | | | | | | | | | | | | | |
| Science as a human endeavor | • | • | • | • | • | • | • | • | • | • | • | • | • | • | • | • | • | • | • | • | • | • | • | • | • |
| Nature of scientific knowledge | • | • | • | • | • | • | • | • | • | | • | | • | | • | • | • | • | • | • | | • | • | • | |
| Historical perspectives | | • | • | • | • | • | | • | | • | | | • | | • | | | • | | | | • | | • | • |

# Lab activities use accessible materials.

## Materials for Lab Activities in the Student Edition
### per class of 30

| Items | Quantity | Chapter | Items | Quantity | Chapter |
|---|---|---|---|---|---|
| Acidity test kit | 1 kit | 23 | Microscope | 10 | 2, 3, 9, 11, 15 |
| Apple juice | 1 bottle | 19 | | | |
| Baby-food jar with lid | 30 | 4 | Microscope depression slides | several boxes | 11 |
| Balloons | 2 bags | 12 | Microscope slides | several boxes | 2 |
| Beakers, various sizes | 30 | 6, 12 | Milk | 1 gallon | 19 |
| Clay soil | 1 gallon | 7 | Mirror, small | 20 | 18 |
| Cone-bearing plants | 10 samples | 8 | Mosses | 10 samples | 8 |
| Construction paper | 4 packages | 11, 25 | Onion | 3 | 3 |
| Cooking oil | 1 bottle | 24 | Orange juice | 1 gallon | 19 |
| Corn syrup | 1 small bottle | 2 | Paper plate | 1 package | 10 |
| Cotton ball | 1 bag | 6 | Penlight | 20 | 18 |
| Cover slip | several boxes | 2, 3 | Philodendron plant | 10 samples | 10 |
| Crackers, unsalted | 1 box | 16 | Planaria, live | 1 sample | 11 |
| Dish detergent, liquid | 1 bottle | 24 | Plants, various house plants | 10 samples | 21 |
| Disks, two different colors | 20 of each color | 4 | Plastic container | 30 | 24 |
| Dropper | 30 | 3 | Plastic cups, various sizes | 2 packages | 2, 10 |
| Fabric, green | 3–4 yards | 25 | Plastic soda bottle, 2 liter | 20 | 21, 23 |
| Ferns | 10 samples | 8 | Plastic spoons | 1 box | 4 |
| Filter paper | 1 package | 4 | Potting soil | 3 bags | 7, 10 |
| Flowering plants | 10 samples | 8 | Prepared slides of dicot and monocot stems, human blood, and protists | 15 of each | 6, 9, 15 |
| Funnels | 10 | 7 | | | |
| Graduated cylinder, various sizes | 10 | 7, 19 | Rubber band | 1 box | 12 |
| Graph paper | several packages | 10, 20 | Rubber tubing | 4 feet | 12 |
| Gravel | 1 large bag | 21 | Rubbing alcohol | 1 bottle | 4 |
| Hand lens | 30 | 11 | Ruler | 30 | 10 |
| Index cards | 5 packages | 13 | Safety gloves, disposable | several boxes | 16 |
| Indophenol | 1 small bottle | 19 | Safety goggles | 30 | 2, 3, 4, 10, 11, 12, 24 |
| Insects (optional) | 10 samples | 21 | | | |
| Iodine | 1 small bottle | 3 | Sand | 2 bags | 7, 21 |
| Jade plant | 90 leaves | 10 | Scissors | 30 | 4, 10 |
| Jigsaw puzzle | 30 | 22 | Spinach leaves | 3 packages | 4 |
| Lab apron | 30 | 2, 16 | Stones | 1/2 gallon | 6 |
| Leaves | 1 gallon bucket | 6 | Stopwatch | 30 | 7, 22, 25 |
| Lemon juice | 1 small bottle | 19 | Tablespoon | 10 | 2 |
| Lens paper | several boxes | 3 | Test tube rack | 10 | 19 |
| Marble | 1 bag | 12 | Test tubes | 60 | 19 |
| Marker | 30 | 5 | Vermiculite | 2 bags | 7 |
| Methylene blue stain | 1 small bottle | 2 | Vinegar eels, live | 1 sample | 11 |
| | | | Water collected from a lake or pond | 1 gallon | 6 |
| | | | Yeast | 3 packages | 2 |

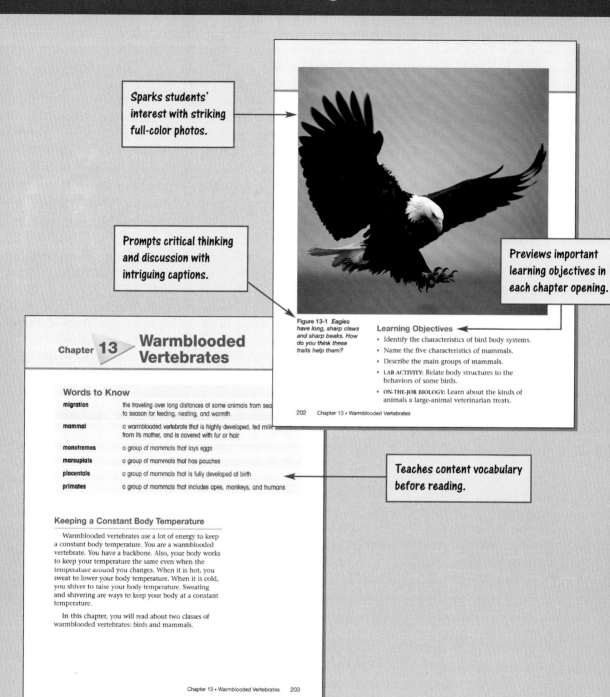

Sparks students' interest with striking full-color photos.

Prompts critical thinking and discussion with intriguing captions.

Previews important learning objectives in each chapter opening.

Figure 13-1 *Eagles have long, sharp claws and sharp beaks. How do you think these traits help them?*

**Learning Objectives**

- Identify the characteristics of bird body systems.
- Name the five characteristics of mammals.
- Describe the main groups of mammals.
- **LAB ACTIVITY:** Relate body structures to the behaviors of some birds.
- **ON-THE-JOB BIOLOGY:** Learn about the kinds of animals a large-animal veterinarian treats.

202    Chapter 13 • Warmblooded Vertebrates

Chapter **13**    **Warmblooded Vertebrates**

**Words to Know**

| | |
|---|---|
| migration | the traveling over long distances of some animals from season to season for feeding, nesting, and warmth |
| mammal | a warmblooded vertebrate that is highly developed, fed milk from its mother, and is covered with fur or hair |
| monotremes | a group of mammals that lays eggs |
| marsupials | a group of mammals that has pouches |
| placentals | a group of mammals that is fully developed at birth |
| primates | a group of mammals that includes apes, monkeys, and humans |

Teaches content vocabulary before reading.

**Keeping a Constant Body Temperature**

Warmblooded vertebrates use a lot of energy to keep a constant body temperature. You are a warmblooded vertebrate. You have a backbone. Also, your body works to keep your temperature the same even when the temperature around you changes. When it is hot, you sweat to lower your body temperature. When it is cold, you shiver to raise your body temperature. Sweating and shivering are ways to keep your body at a constant temperature.

In this chapter, you will read about two classes of warmblooded vertebrates: birds and mammals.

Chapter 13 • Warmblooded Vertebrates    203

*Pacemaker® Biology Student Edition*, pages 202–203

# The Student Edition makes biology understandable through readable text, consistent reading support, and clear visuals.

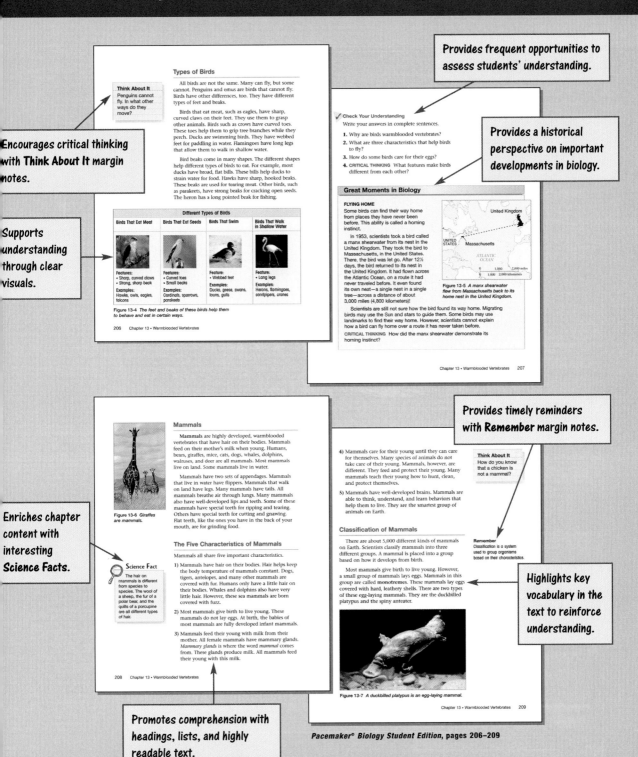

**Provides frequent opportunities to assess students' understanding.**

**Encourages critical thinking with Think About It margin notes.**

**Supports understanding through clear visuals.**

**Provides a historical perspective on important developments in biology.**

**Provides timely reminders with Remember margin notes.**

**Enriches chapter content with interesting Science Facts.**

**Highlights key vocabulary in the text to reinforce understanding.**

**Promotes comprehension with headings, lists, and highly readable text.**

*Pacemaker® Biology Student Edition, pages 206–209*

Uses full-color visuals to increase appeal and clarify concepts.

Provides new perspectives on content in **Science Facts**.

Organizes content with clear headings and subheadings.

The second group of mammals has pouches. These mammals are called **marsupials**. They give birth to live young that are not very well developed. After they are born, the young must crawl into a pouch in their mother's body. Inside the pouch, these young mammals drink their mother's milk. There, they finish developing. Kangaroos, koalas, and opossums are mammals with pouches.

The third group of mammals is fully developed at birth. These mammals are called **placentals**. After they are born, the parents of these young mammals take care of them until the young mammals are ready to take care of themselves. Most mammals belong to this group. They include deer, mice, elephants, raccoons, dolphins, and humans.

Figure 13-8 *Koalas are mammals with pouches.*

### Groups of Placentals

There are many mammals that belong to the placentals group. These mammals can be further divided into smaller groups. Placental mammals can be placed into smaller groups based on how and what they eat, how they move about, or where they live.

Rodents
The largest group of mammals is the rodent group. There are more species of rodents than any other mammal group. Rats, mice, squirrels, and beavers are rodents. They have sharp teeth for gnawing and cutting. Some rodents reproduce very quickly.

Flying Mammals
The only true flying mammal is the bat. The wings of bats are made from their long finger bones. Skin stretches across these bones to make wings. Bats usually sleep hanging upside down in caves during the day. They become active at night.

**Science Fact**
One kind of bat can be harmful to other mammals. Vampire bats feed on blood. They wait until a large animal is sleeping. Then, they bite the animal and feed on the blood that flows. Vampire bats can carry diseases from one animal to another. Vampire bats are usually found in Central and South America.

210   Chapter 13 • Warmblooded Vertebrates

Meat-Eating Mammals
Meat-eating mammals have special body structures to help them capture and eat other animals. Some of these structures include special teeth and sharp claws. Wolves, cats, dogs, bears, seals, and walruses are meat eaters.

Trunk-Nosed Mammals
The trunk-nosed mammals are all elephants. An elephant's trunk is really a nose and upper lip. An elephant uses its trunk to drink, to lift objects, and to scratch. An elephant's trunk contains many muscles.

**Science F**
Today, there are three kinds o elephants on Ear two types of Afric elephants and the Asian elephant.

### On the Cutting Edge

**TRACKING ANIMALS**
Elephants are in danger. Many have been killed for their tusks. People make jewelry and other objects from elephant tusks. In order to save elephants, scientists record their movements in different parts of the world.

Tiny radio transmitters are attached to collars that the elephants wear. Information about the animals, such as their movements and their location, is sent to a satellite that orbits Earth. The information is then sent to a ground station. Scientists use computers to read the information. Knowing where the elephants are and where they travel can help scientists set up places of protection for them.

Figure 13-9 *Elephants are tracked using radio transmitters, satellites, and computers.*

Many different kinds of animals, including birds and fish, are tracked using satellites. Scientists and researchers examine the information to find out more about the tracked animals.

CRITICAL THINKING  Why do you think satellite-tracking collars are useful?

Chapter 13 • Warmblooded Vertebrates   211

Relates content to current issues and technologies in **On the Cutting Edge** features.

*Pacemaker® Biology Student Edition*, pages 210–211

# The Student Edition develops science inquiry skills and establishes connections between biology and life.

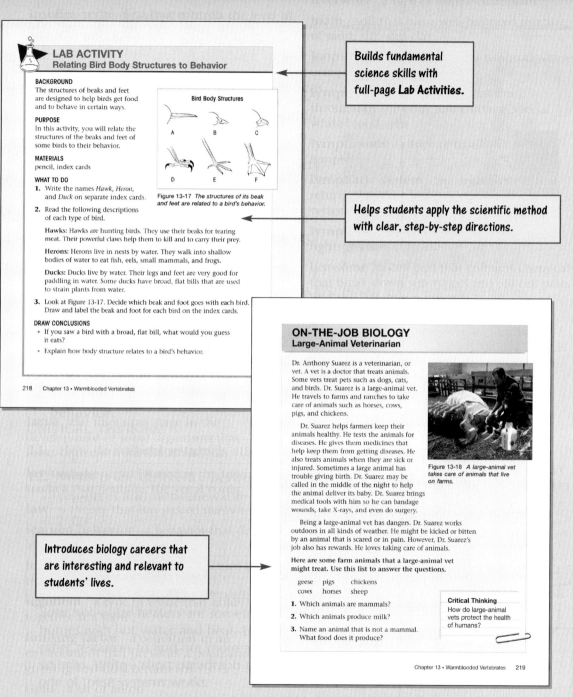

## LAB ACTIVITY
### Relating Bird Body Structures to Behavior

**BACKGROUND**
The structures of beaks and feet are designed to help birds get food and to behave in certain ways.

**PURPOSE**
In this activity, you will relate the structures of the beaks and feet of some birds to their behavior.

**MATERIALS**
pencil, index cards

**WHAT TO DO**
1. Write the names *Hawk*, *Heron*, and *Duck* on separate index cards.
2. Read the following descriptions of each type of bird.

   **Hawks:** Hawks are hunting birds. They use their beaks for tearing meat. Their powerful claws help them to kill and to carry their prey.

   **Herons:** Herons live in nests by water. They walk into shallow bodies of water to eat fish, eels, small mammals, and frogs.

   **Ducks:** Ducks live by water. Their legs and feet are very good for paddling in water. Some ducks have broad, flat bills that are used to strain plants from water.

3. Look at Figure 13-17. Decide which beak and foot goes with each bird. Draw and label the beak and foot for each bird on the index cards.

**DRAW CONCLUSIONS**
* If you saw a bird with a broad, flat bill, what would you guess it eats?
* Explain how body structure relates to a bird's behavior.

**Bird Body Structures**

Figure 13-17 *The structures of its beak and feet are related to a bird's behavior.*

218    Chapter 13 • Warmblooded Vertebrates

> Builds fundamental science skills with full-page **Lab Activities**.

> Helps students apply the scientific method with clear, step-by-step directions.

## ON-THE-JOB BIOLOGY
### Large-Animal Veterinarian

Dr. Anthony Suarez is a veterinarian, or vet. A vet is a doctor that treats animals. Some vets treat pets such as dogs, cats, and birds. Dr. Suarez is a large-animal vet. He travels to farms and ranches to take care of animals such as horses, cows, pigs, and chickens.

Dr. Suarez helps farmers keep their animals healthy. He tests the animals for diseases. He gives them medicines that help keep them from getting diseases. He also treats animals when they are sick or injured. Sometimes a large animal has trouble giving birth. Dr. Suarez may be called in the middle of the night to help the animal deliver its baby. Dr. Suarez brings medical tools with him so he can bandage wounds, take X-rays, and even do surgery.

Being a large-animal vet has dangers. Dr. Suarez works outdoors in all kinds of weather. He might be kicked or bitten by an animal that is scared or in pain. However, Dr. Suarez's job also has rewards. He loves taking care of animals.

Figure 13-18 *A large-animal vet takes care of animals that live on farms.*

Here are some farm animals that a large-animal vet might treat. Use this list to answer the questions.

geese    pigs    chickens
cows    horses    sheep

1. Which animals are mammals?
2. Which animals produce milk?
3. Name an animal that is not a mammal. What food does it produce?

**Critical Thinking**
How do large-animal vets protect the health of humans?

Chapter 13 • Warmblooded Vertebrates    219

> Introduces biology careers that are interesting and relevant to students' lives.

*Pacemaker® Biology Student Edition*, pages 218–219

# The Student Edition includes extensive review to improve comprehension and test-taking skills.

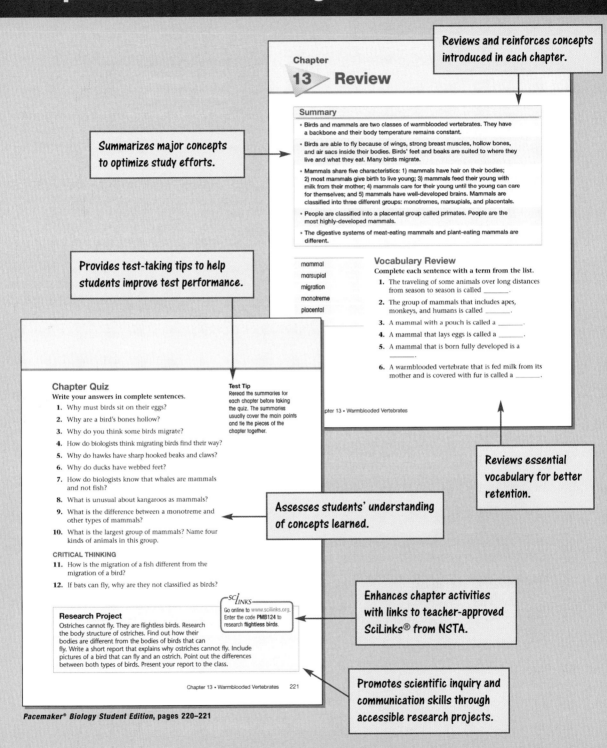

**Reviews and reinforces concepts introduced in each chapter.**

**Summarizes major concepts to optimize study efforts.**

**Provides test-taking tips to help students improve test performance.**

**Reviews essential vocabulary for better retention.**

**Assesses students' understanding of concepts learned.**

**Enhances chapter activities with links to teacher-approved SciLinks® from NSTA.**

**Promotes scientific inquiry and communication skills through accessible research projects.**

## Chapter 13 Review

### Summary

- Birds and mammals are two classes of warmblooded vertebrates. They have a backbone and their body temperature remains constant.
- Birds are able to fly because of wings, strong breast muscles, hollow bones, and air sacs inside their bodies. Birds' feet and beaks are suited to where they live and what they eat. Many birds migrate.
- Mammals share five characteristics: 1) mammals have hair on their bodies; 2) most mammals give birth to live young; 3) mammals feed their young with milk from their mother; 4) mammals care for their young until the young can care for themselves; and 5) mammals have well-developed brains. Mammals are classified into three different groups: monotremes, marsupials, and placentals.
- People are classified into a placental group called primates. People are the most highly-developed mammals.
- The digestive systems of meat-eating mammals and plant-eating mammals are different.

mammal
marsupial
migration
monotreme
placental

### Vocabulary Review

**Complete each sentence with a term from the list.**

1. The traveling of some animals over long distances from season to season is called _____.
2. The group of mammals that includes apes, monkeys, and humans is called _____.
3. A mammal with a pouch is called a _____.
4. A mammal that lays eggs is called a _____.
5. A mammal that is born fully developed is a _____.
6. A warmblooded vertebrate that is fed milk from its mother and is covered with fur is called a _____.

Chapter 13 • Warmblooded Vertebrates

### Chapter Quiz

**Write your answers in complete sentences.**

1. Why must birds sit on their eggs?
2. Why are a bird's bones hollow?
3. Why do you think some birds migrate?
4. How do biologists think migrating birds find their way?
5. Why do hawks have sharp hooked beaks and claws?
6. Why do ducks have webbed feet?
7. How do biologists know that whales are mammals and not fish?
8. What is unusual about kangaroos as mammals?
9. What is the difference between a monotreme and other types of mammals?
10. What is the largest group of mammals? Name four kinds of animals in this group.

**CRITICAL THINKING**

11. How is the migration of a fish different from the migration of a bird?
12. If bats can fly, why are they not classified as birds?

**Test Tip**
Reread the summaries for each chapter before taking the quiz. The summaries usually cover the main points and tie the pieces of the chapter together.

**SciLINKS**
Go online to www.scilinks.org.
Enter the code PMB124 to research **flightless birds**.

### Research Project

Ostriches cannot fly. They are flightless birds. Research the body structure of ostriches. Find out how their bodies are different from the bodies of birds that can fly. Write a short report that explains why ostriches cannot fly. Include pictures of a bird that can fly and an ostrich. Point out the differences between both types of birds. Present your report to the class.

Chapter 13 • Warmblooded Vertebrates    221

*Pacemaker® Biology Student Edition*, pages 220–221

# The Workbook promotes critical thinking and application of science process skills.

**Provides exercises that parallel the Student Edition.**

**Gives students the opportunity to apply newly learned concepts.**

**Builds critical-thinking skills through carefully structured exercises.**

**Reinforces content understanding while building comprehension skills.**

**Helps students learn science process skills such as interpreting visuals and analyzing data.**

**Integrates technology, everyday applications, and careers with science skills to make content relevant.**

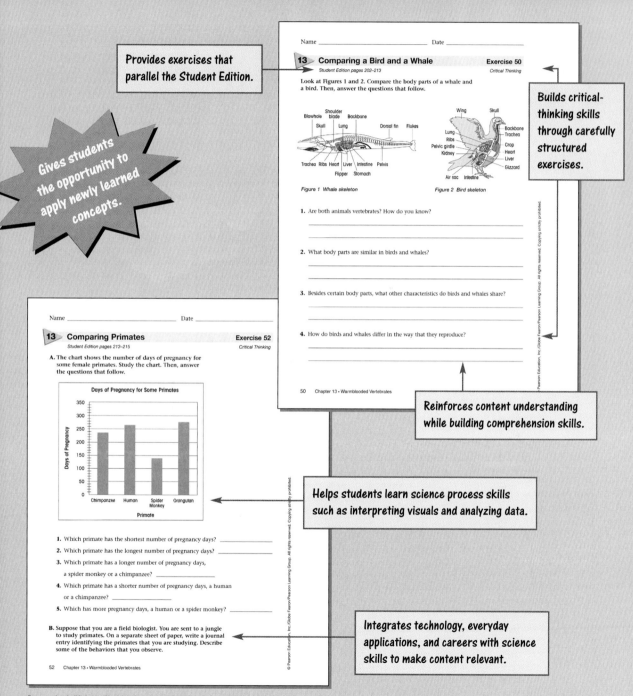

Name _____ Date _____

**13** ▸ **Comparing a Bird and a Whale**          Exercise 50
Student Edition pages 202–213                    Critical Thinking

Look at Figures 1 and 2. Compare the body parts of a whale and a bird. Then, answer the questions that follow.

Figure 1  Whale skeleton

Figure 2  Bird skeleton

1. Are both animals vertebrates? How do you know?
   _____
   _____

2. What body parts are similar in birds and whales?
   _____
   _____

3. Besides certain body parts, what other characteristics do birds and whales share?
   _____
   _____

4. How do birds and whales differ in the way that they reproduce?
   _____
   _____

50    Chapter 13 • Warmblooded Vertebrates

Name _____ Date _____

**13** ▸ **Comparing Primates**          Exercise 52
Student Edition pages 213–215                    Critical Thinking

**A.** The chart shows the number of days of pregnancy for some female primates. Study the chart. Then, answer the questions that follow.

**Days of Pregnancy for Some Primates**

1. Which primate has the shortest number of pregnancy days? _____
2. Which primate has the longest number of pregnancy days? _____
3. Which primate has a longer number of pregnancy days, a spider monkey or a chimpanzee? _____
4. Which primate has a shorter number of pregnancy days, a human or a chimpanzee? _____
5. Which has more pregnancy days, a human or a spider monkey? _____

**B.** Suppose that you are a field biologist. You are sent to a jungle to study primates. On a separate sheet of paper, write a journal entry identifying the primates that you are studying. Describe some of the behaviors that you observe.

52    Chapter 13 • Warmblooded Vertebrates

*Pacemaker® Biology Workbook,* **pages 50, 52**

# ▶ The Teacher's Answer Edition includes point-of-use answers and teaching tips for flexible classroom support.

**Gives information for ESL instruction.**

**ESL Note** To help students understand warmblooded animals, distribute thermometers and have students take their own temperatures. Remind students that 98.6°F (about 37°C) is the normal body temperature for humans. No matter what the temperature of the air is around them, their bodies work to keep their body temperatures at around 98.6°F.

Figure 13-1 *Eagles have long, sharp claws and sharp beaks. How do you think these traits help them?*

Possible answer: Sharp claws help eagles capture small animals. Sharp beaks help them to tear meat.

**Eases classroom management by offering answers at point of use.**

### Learning Objectives

- Identify the characteristics of bird body systems.
- Name the five characteristics of mammals.
- Describe the main groups of mammals.
- **LAB ACTIVITY:** Relate body structures to the behaviors of some birds.
- **ON-THE-JOB BIOLOGY:** Learn about the kinds of animals a large-animal veterinarian treats.

202     Chapter 13 • Warmblooded Vertebrates

Chapter **13** ▷ **Warmblooded Vertebrates**

### Words to Know

| | |
|---|---|
| **migration** | the traveling over long distances of some animals from season to season for feeding, nesting, and warmth |
| **mammal** | a warmblooded vertebrate that is highly developed, fed milk from its mother, and is covered with fur or hair |
| **monotremes** | a group of mammals that lays eggs |
| **marsupials** | a group of mammals that has pouches |
| **placentals** | a group of mammals that is fully developed at birth |
| **primates** | a group of mammals that includes apes, monkeys, and humans |

### Keeping a Constant Body Temperature

Warmblooded vertebrates use a lot of energy to keep a constant body temperature. You are a warmblooded vertebrate. You have a backbone. Also, your body works to keep your temperature the same even when the temperature around you changes. When it is hot, you sweat to lower your body temperature. When it is cold, you shiver to raise your body temperature. Sweating and shivering are ways to keep your body at a constant temperature.

In this chapter, you will read about two classes of warmblooded vertebrates: birds and mammals.

**Linking Prior Knowledge**
Students should recall information that they learned about invertebrates (Chapter 11) and coldblooded vertebrates (Chapter 12).

**Linking Prior Knowledge cross-references related content from prior chapters.**

Chapter 13 • Warmblooded Vertebrates     203

*Pacemaker® Biology Teacher's Answer Edition, pages 202–203*

# The Classroom Resource Binder provides more than 200 reproducibles for ready-made practice and assessment.

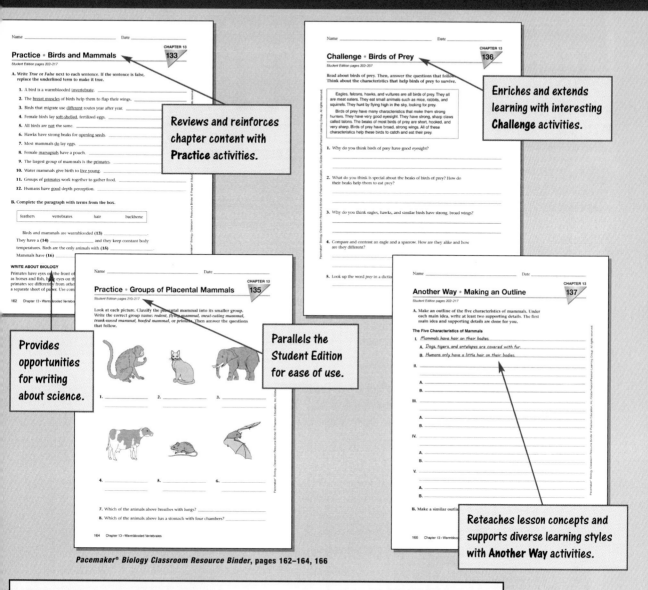

**Reviews and reinforces chapter content with Practice activities.**

**Enriches and extends learning with interesting Challenge activities.**

**Provides opportunities for writing about science.**

**Parallels the Student Edition for ease of use.**

**Reteaches lesson concepts and supports diverse learning styles with Another Way activities.**

Pacemaker® Biology Classroom Resource Binder, pages 162–164, 166

## Classroom Resource Binder

- Encourages hands-on learning with additional Lab Activities
- Offers students opportunities to practice their skills using the Scientific Skills Handbook
- Builds standardized test-taking skills with cumulative Unit Tests
- Provides two parallel chapter tests for flexible use; tests can be used as pretests and posttests or for students with particular format needs

*SCi*
*LINKS*

*Pacemaker® Biology* has partnered with SciLinks® and NSTA to connect students' textbooks and the World Wide Web. Each SciLinks reference contains a code and key term that students use to access teacher-approved Web sites. Using the SciLinks service available from NSTA can enhance the chapter activities. Go to www.scilinks.org/register.asp to register your class. When students use SciLinks for the first time, they will need to register as well. Registration requires the first name, last initial, and zip code of each student. This information is kept strictly confidential. Students will also be asked to create a unique user name and password.

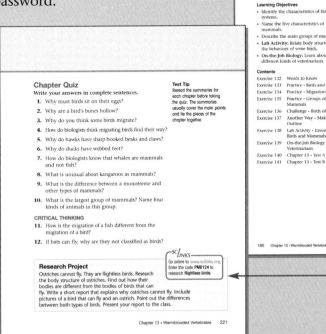

*Sample pages from the Student Edition and the Classroom Resource Binder*

Connect your students' textbooks to the World Wide Web using SciLinks. Look for this logo in the *Pacemaker® Biology Student Edition* and *Classroom Resource Binder* to enhance your students' science activities online.

# The ESL/ELL Teacher's Guide enhances content instruction for English-language learners.

Taps students' prior knowledge and relates it to biology concepts.

Provides three levels of questions for varying degrees of language fluency.

Parallels the Student Edition to ease lesson planning.

Emphasizes teaching and reinforcement of scientific terms, idioms, and expressions.

Helps English-language learners succeed on the same tests as their English-speaking peers.

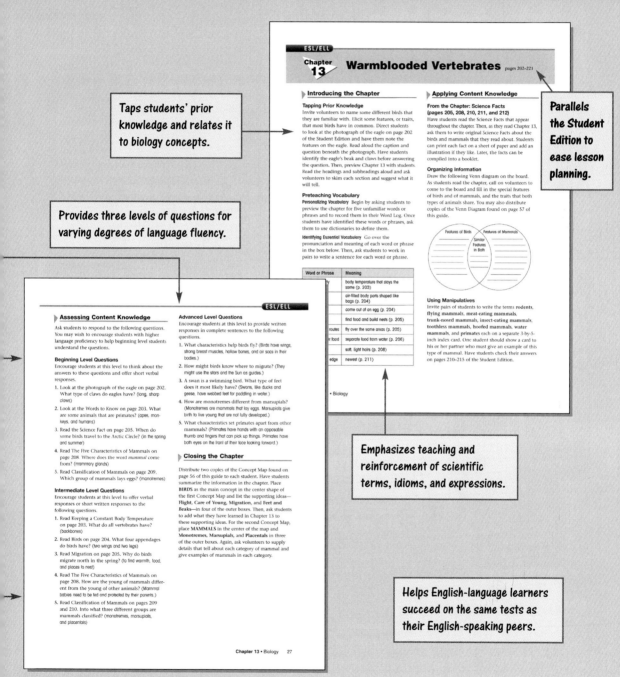

**ESL/ELL**

## Chapter 13 — Warmblooded Vertebrates  pages 202–221

### Introducing the Chapter

**Tapping Prior Knowledge**
Invite volunteers to name some different birds that they are familiar with. Elicit some features, or traits, that most birds have in common. Direct students to look at the photograph of the eagle on page 202 of the Student Edition and have them note the features on the eagle. Read aloud the caption and question beneath the photograph. Have students identify the eagle's beak and claws before answering the question. Then, preview Chapter 13 with students. Read the headings and subheadings aloud and ask volunteers to skim each section and suggest what it will tell.

**Preteaching Vocabulary**
**Personalizing Vocabulary** Begin by asking students to preview the chapter for five unfamiliar words or phrases and to record them in their Word Log. Once students have identified these words or phrases, ask them to use dictionaries to define them.

**Identifying Essential Vocabulary** Go over the pronunciation and meaning of each word or phrase in the box below. Then, ask students to work in pairs to write a sentence for each word or phrase.

| Word or Phrase | Meaning |
| --- | --- |
| | body temperature that stays the same (p. 203) |
| | air-filled body parts shaped like bags (p. 204) |
| | come out of an egg (p. 204) |
| routes | find food and build nests (p. 205) |
| | fly over the same areas (p. 205) |
| food | separate food from water (p. 206) |
| | soft, light hairs (p. 208) |
| edge | newest (p. 211) |

• Biology

### Applying Content Knowledge

**From the Chapter: Science Facts**
**(pages 205, 208, 210, 211, and 212)**
Have students read the Science Facts that appear throughout the chapter. Then, as they read Chapter 13, ask them to write original Science Facts about the birds and mammals that they read about. Students can print each fact on a sheet of paper and add an illustration if they like. Later, the facts can be compiled into a booklet.

**Organizing Information**
Draw the following Venn diagram on the board. As students read the chapter, call on volunteers to come to the board and fill in the special features of birds and of mammals, and the traits that both types of animals share. You may also distribute copies of the Venn Diagram found on page 57 of this guide.

Features of Birds — Similar Features in Both — Features of Mammals

**Using Manipulatives**
Invite pairs of students to write the terms **rodents, flying mammals, meat-eating mammals, trunk-nosed mammals, insect-eating mammals, toothless mammals, hoofed mammals, water mammals,** and **primates** each on a separate 3-by-5-inch index card. One student should show a card to his or her partner who must give an example of this type of mammal. Have students check their answers on pages 210–215 of the Student Edition.

**ESL/ELL**

### Assessing Content Knowledge

Ask students to respond to the following questions. You may wish to encourage students with higher language proficiency to help beginning level students understand the questions.

**Beginning Level Questions**
Encourage students at this level to think about the answers to these questions and offer short verbal responses.

1. Look at the photograph of the eagle on page 202. What type of claws do eagles have? (long, sharp claws)

2. Look at the Words to Know on page 203. What are some animals that are primates? (apes, monkeys, and humans)

3. Read the Science Fact on page 205. When do some birds travel to the Arctic Circle? (in the spring and summer)

4. Read The Five Characteristics of Mammals on page 208. Where does the word *mammal* come from? (mammary glands)

5. Read Classification of Mammals on page 209. Which group of mammals lays eggs? (monotremes)

**Intermediate Level Questions**
Encourage students at this level to offer verbal responses or short written responses to the following questions.

1. Read Keeping a Constant Body Temperature on page 203. What do all vertebrates have? (backbones)

2. Read Birds on page 204. What four appendages do birds have? (two wings and two legs)

3. Read Migration on page 205. Why do birds migrate north in the spring? (to find warmth, food, and places to nest)

4. Read The Five Characteristics of Mammals on page 208. How are the young of mammals different from the young of other animals? (Mammal babies need to be fed and protected by their parents.)

5. Read Classification of Mammals on pages 209 and 210. Into what three different groups are mammals classified? (monotremes, marsupials, and placentals)

**Advanced Level Questions**
Encourage students at this level to provide written responses in complete sentences to the following questions.

1. What characteristics help birds fly? (Birds have wings, strong breast muscles, hollow bones, and air sacs in their bodies.)

2. How might birds know where to migrate? (They might use the stars and the Sun as guides.)

3. A swan is a swimming bird. What type of feet does it most likely have? (Swans, like ducks and geese, have webbed feet for paddling in water.)

4. How are monotremes different from marsupials? (Monotremes are mammals that lay eggs. Marsupials give birth to live young that are not fully developed.)

5. What characteristics set primates apart from other mammals? (Primates have hands with an opposable thumb and fingers that can pick up things. Primates have both eyes on the front of their face looking forward.)

### Closing the Chapter

Distribute two copies of the Concept Map found on page 56 of this guide to each student. Have students summarize the information in the chapter. Place **BIRDS** as the main concept in the center shape of the first Concept Map and list the supporting ideas—**Flight, Care of Young, Migration,** and **Feet and Beaks**—in four of the outer boxes. Then, ask students to add what they have learned in Chapter 13 to these supporting ideas. For the second Concept Map, place **MAMMALS** in the center of the map and **Monotremes, Marsupials,** and **Placentals** in three of the outer boxes. Again, ask volunteers to supply details that tell about each category of mammal and give examples of mammals in each category.

Chapter 13 • Biology    27

*Pacemaker® Biology ESL/ELL Teacher's Guide, pages 26–27*

### Pacemaker® General Science

*Pacemaker® General Science* introduces the basic concepts and principles of life, physical, and earth sciences, using content that fully aligns with the National Science Education Standards. With expanded real-world activities, the addition of relevant lab activities, test preparation, and more comprehensive review, this program is designed so that students of all abilities can master basic science concepts.

### Pacemaker® Health

This text supports students in learning the essential health concepts that prepare them to make sensible, educated choices about their own health care. *Pacemaker® Health* also helps students grasp the importance of good health in their own lives.

### Globe Biology

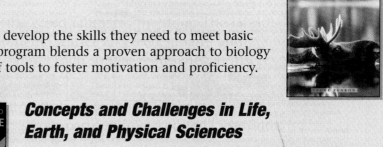

*Globe Biology* helps students develop the skills they need to meet basic science requirements. This program blends a proven approach to biology instruction with a variety of tools to foster motivation and proficiency.

### Concepts and Challenges in Life, Earth, and Physical Sciences

This program makes key concepts accessible, provides reinforcement, and helps students build confidence. By integrating science with technology and with other curriculum areas, *Concepts and Challenges* helps students understand how science impacts everyday life and the world around them.

### Science Workshop Series

The Science Workshop Series is a complete, up-to-date, and flexible science program you can use to easily supplement your science curriculum. This program helps students develop competency in the areas of biology, earth science, chemistry, and physical science.

PACEMAKER®

# Biology

GLOBE FEARON

Pearson Learning Group

## *Pacemaker® Biology*, Third Edition

We would like to thank the following educators, who provided valuable comments and suggestions during the development of this book.

## Consultants

*Content Consultant:* **Richard Lowell**, Ramapo College of New Jersey, Mahwah, NJ

*Education Consultants:* **Richard Grybos**, Director of Special Education Instruction, Rochester City Schools, Rochester, NY **Christine Mason**, Center for Exceptional Children, Curriculum Assessment, Arlington, VA

*ESL/ELL Consultant:* **Elizabeth Jimenez**, GEMAS Consulting, Pomona, CA

## Reviewers

*Content Reviewers:* **Sharon Danielson**, New York State Department of Health, Albany, NY **Sukamol Jacobson**, University of California, San Diego, CA **Rusty Lansford**, California Institute of Technology, Pasadena, CA **Helen McBride**, California Institute of Technology, Pasadena, CA

*Teacher Reviewers:* **Miriam Gage**, Newman Smith High School, Carrollton, TX **Jude Ann Morino**, Pequannock Township High School, Pequannock, NJ **Celiamarie Narro**, Ysleta Independent School District, El Paso, TX **Mohini Robinson**, Granada High School, Livermore, CA

## Project Staff

*Art and Design:* Evelyn Bauer, Jenifer Hixson, Dan Trush, Jennifer Visco *Editorial:* Stephanie Cahill, Martha Feehan, Shirley White *Manufacturing:* Mark Cirillo *Marketing:* Clare Harrison, Anna Mazzoccoli *Production:* Irene Belinsky, Karen Edmonds, Jill Kuhfuss, Cynthia Lynch, Phyllis Rosinsky, Susan Tamm *Publishing Operations:* Carolyn Coyle, Tom Daning, Richetta Lobban

**About the Cover:** Biology is the study of living things and the principles that determine how living things function. In this course, you will learn about a wide variety of living things. The rhinoceros and the dragonfly on the cover are both living things. Although they look very different, they share many things in common. They are both animals and they both must find food to survive. The maple leaves show how living things can change. As the temperature becomes colder in the fall, the leaves stop producing the green pigment called chlorophyll. Then, the leaves change to yellow, orange, or red and eventually fall from the tree. The spiral-shaped structure is a model of DNA. This molecule is found inside the cells of all living things. DNA contains hereditary information that controls the traits of an organism.

ISBN 0-13-024044-3

Printed in the United States of America
7 8 9 10     09 08 07

Globe Fearon
Pearson Learning Group

1-800-321-3106
www.pearsonlearning.com

# Contents

**A Note to the Student**    **x**

| | | |
|---|---|---|
| **UNIT ONE** | **SCIENCE COMES TO LIFE** | **1** |
| **Chapter 1** | **The Science of Living Things** | **2** |

*Modern Leaders in Biology*: Kevin Andrewin,
Research Scientist   7
Lab Activity: Thinking Like a Scientist   12
Biology in Your Life: Making a Diagnosis   13
Chapter Review   14

| | | |
|---|---|---|
| **Chapter 2** | **Characteristics of Living Things** | **16** |

*Great Moments in Biology*: Disproving Spontaneous
Generation   23
Lab Activity: Investigating Yeast Reproduction   26
Biology in Your Life: Choosing a Dog   27
Chapter Review   28

Unit 1 Review   30

| | | |
|---|---|---|
| **UNIT TWO** | **CELLS AND SIMPLE ORGANISMS** | **31** |
| **Chapter 3** | **Cell Structure and Function** | **32** |

*Great Moments in Biology*: The Cell Theory   35
*A Closer Look*: Cancer Cells   45
Lab Activity: Identifying Parts of a Plant Cell   46
On-the-Job Biology: Cytotechnologist   47
Chapter Review   48

| | | |
|---|---|---|
| **Chapter 4** | **Cells and Energy** | **50** |

*A Closer Look*: Photosynthesis and Life   59
Lab Activity: Getting Chlorophyll From Spinach   60
Biology in Your Life: Counting Calories   61
Chapter Review   62

| | | |
|---|---|---|
| **Chapter 5** | **Heredity and Genetics** | **64** |
| | *Great Moments in Biology*: The Human Genome Project | 83 |
| ▶ | Lab Activity: Modeling Genetics | 84 |
| ▶ | On-the-Job Biology: Rose Breeder | 85 |
| | Chapter Review | 86 |
| **Chapter 6** | **Simple Organisms** | **88** |
| | *On the Cutting Edge*: Modern Uses of Bacteria | 96 |
| ▶ | Lab Activity: Identifying Protists | 104 |
| ▶ | Biology in Your Life: Living With Simple Organisms | 105 |
| | Chapter Review | 106 |
| | Unit 2 Review | 108 |
| **UNIT THREE** | **THE PLANT KINGDOM** | **109** |
| **Chapter 7** | **The Study of Plants** | **110** |
| | *Modern Leaders in Biology*: Michael Kasperbauer, Plant Researcher | 115 |
| | *On the Cutting Edge*: Plants as Healers | 117 |
| ▶ | Lab Activity: Investigating Properties of Soils | 118 |
| ▶ | On-the-Job Biology: Farmer | 119 |
| | Chapter Review | 120 |
| **Chapter 8** | **Types of Plants** | **122** |
| | *A Closer Look*: Plants as Fuel | 127 |
| ▶ | Lab Activity: Classifying Plants by Characteristics | 132 |
| ▶ | Biology in Your Life: Studying Cone-Bearing Plants | 133 |
| | Chapter Review | 134 |

| Chapter 9 | **Plants at Work** | **136** |
|---|---|---|
| | *On the Cutting Edge*: Hydroponics | 142 |
| ▶ | Lab Activity: Identifying a Monocot and a Dicot | 146 |
| ▶ | On-the-Job Biology: Horticulturist | 147 |
| | Chapter Review | 148 |

| Chapter 10 | **Reproduction in Seed Plants** | **150** |
|---|---|---|
| | *On the Cutting Edge*: T-Bud Grafting | 157 |
| ▶ | Lab Activity: Growing Plants From Cuttings | 158 |
| ▶ | Biology in Your Life: Attracting Pollinators to a Garden | 159 |
| | Chapter Review | 160 |

| | Unit 3 Review | 162 |
|---|---|---|

| **UNIT FOUR** | **THE ANIMAL KINGDOM** | **163** |
|---|---|---|

| Chapter 11 | **Invertebrates** | **164** |
|---|---|---|
| | *A Closer Look*: Parasitic Worms | 170 |
| | *A Closer Look*: Regeneration | 179 |
| ▶ | Lab Activity: Comparing Worms | 180 |
| ▶ | On-the-Job Biology: Clammer | 181 |
| | Chapter Review | 182 |

| Chapter 12 | **Coldblooded Vertebrates** | **184** |
|---|---|---|
| | *A Closer Look*: The Pine Barrens Tree Frog | 194 |
| ▶ | Lab Activity: Investigating Swim Bladders | 198 |
| ▶ | Biology in Your Life: Keeping Fish as Pets | 199 |
| | Chapter Review | 200 |

**Chapter 13**   **Warmblooded Vertebrates**                                           **202**

*Great Moments in Biology*: Flying Home                                        207
*On the Cutting Edge*: Tracking Animals                                        211
*Modern Leaders in Biology*: Jane Goodall,
   Primate Biologist                                           214
▶ Lab Activity: Relating Bird Body Structures to Behavior        218
▶ On-the-Job Biology: Large-Animal Veterinarian                  219
Chapter Review                                                                220

**Unit 4 Review**                                                             222

**UNIT FIVE**   **THE HUMAN BODY**                                             **223**

**Chapter 14**   **Support and Movement**                                      **224**

*A Closer Look*: Muscular Dystrophy                                            233
▶ Lab Activity: Identifying Joints                                            234
▶ Biology in Your Life: Treating a Pulled Muscle                 235
Chapter Review                                                                236

**Chapter 15**   **Circulation, Respiration, and Excretion**                   **238**

*Great Moments in Biology*: Blood Circulates in the Body         242
▶ Lab Activity: Observing Blood Cells                                         252
▶ On-the-Job Biology: Personal Trainer                           253
Chapter Review                                                                254

**Chapter 16**   **Digestion**                                                 **256**

*A Closer Look*: Lactose Intolerance                                          261
▶ Lab Activity: Investigating the Digestion of Starch             262
▶ Biology in Your Life: Preventing Tooth Decay                   263
Chapter Review                                                                264

**Chapter 17**  **Regulating the Body**  **266**

*On the Cutting Edge*: Treating Spinal Cord Injuries — 274
▶ Lab Activity: Testing Coordination — 278
▶ On-the-Job Biology: Nuclear Medicine Technologist — 279
Chapter Review — 280

**Chapter 18**  **The Sense Organs**  **282**

*Great Moments in Biology*: The Invention of Braille — 287
*On the Cutting Edge*: Advances in Eye Surgery — 289
*A Closer Look*: Sound Waves — 291
▶ Lab Activity: Investigating the Action of the Pupil — 292
▶ Biology in Your Life: Studying Optical Illusions — 293
Chapter Review — 294

**Chapter 19**  **Human Health**  **296**

*A Closer Look*: The Dangers of Driving Under
    the Influence — 305
▶ Lab Activity: Testing Foods for Vitamin C — 306
▶ On-the-Job Biology: Dietitian — 307
Chapter Review — 308

**Chapter 20**  **Reproduction and Development**  **310**

*On the Cutting Edge*: Prenatal Care — 317
▶ Lab Activity: Graphing Human Development — 318
▶ Biology in Your Life: Keeping Healthy and Safe — 319
Chapter Review — 320

**Unit 5 Review** — 322

| UNIT SIX | LIVING TOGETHER ON EARTH | 323 |
|---|---|---|
| **Chapter 21** | **Ecology** | **324** |
| | *A Closer Look*: Human Population Growth | 336 |
| ▶ | Lab Activity: Modeling a Land Biome | 342 |
| ▶ | On-the-Job Biology: Ecologist | 343 |
| | Chapter Review | 344 |
| **Chapter 22** | **Animal Behavior** | **346** |
| | *Modern Leaders in Biology*: Edward Osborne Wilson, Biologist | 349 |
| | *A Closer Look*: Pheromones | 354 |
| ▶ | Lab Activity: Learning to Solve a Puzzle | 358 |
| ▶ | On-the-Job Biology: Animal Trainer | 359 |
| | Chapter Review | 360 |
| **Chapter 23** | **Cycles in Nature** | **362** |
| | *A Closer Look*: Acid Rain | 365 |
| | *On the Cutting Edge*: Monitoring Global Warming | 368 |
| ▶ | Lab Activity: Testing the Acidity of Rainwater | 370 |
| ▶ | Biology in Your Life: Dealing with Drought | 371 |
| | Chapter Review | 372 |
| **Chapter 24** | **Natural Resources** | **374** |
| | *Great Moments in Biology*: Rachel Carson's *Silent Spring* | 384 |
| ▶ | Lab Activity: Cleaning Up an Oil Spill | 388 |
| ▶ | Biology in Your Life: Driving Hybrid Electric Cars | 389 |
| | Chapter Review | 390 |

**Chapter 25**   **Evolution**                                                    **392**

*Modern Leaders in Biology*: Stephen Jay Gould,
   Evolutionary Biologist                                           399
▶   Lab Activity: Modeling Natural Selection                        406
▶   On-the-Job Biology: Paleontologist                              407
   Chapter Review                                                   408

**Unit 6 Review**                                                   410

**Appendix A:**  Science Terms                                      **411**

**Appendix B:**  The Microscope                                     **412**

**Appendix C:**  The Six Kingdoms                                   **414**

**Appendix D:**  Some Important Vitamins and Minerals               **416**

**Glossary**                                                        **417**

**Index**                                                           **431**

# A Note to the Student

Biology is the science of living things. By studying biology, you will learn about yourself and the environment around you. Knowing something about biology is in your own interest. As you read this, rain forests are being cut down in South America. Somewhere oil is being spilled into the ocean. These things may be happening thousands of miles away. However, they affect you and every other living thing on Earth that needs to breathe air and drink water. As we move through the twenty-first century, some important choices will have to be made. What can we do to protect our environment? How can we cure diseases and feed the growing human population? Only people who understand the basic concepts of biology will be able to make wise decisions.

While reading this book, look for the notes in the margins of the pages. These notes are there to make you stop and think. Sometimes they provide interesting facts about material you are learning. Sometimes they give examples of biology in everyday life. Other times they remind you of something you already know.

You will also find several study aids in the book. At the beginning of every chapter, you will find **Learning Objectives**. Take a moment to study these goals. They will help you focus on the important points covered in the chapter. **Words to Know** will give you a look at some biology vocabulary you will find in your reading. At the end of each chapter, a **Summary** will give you a quick review of what you have just learned.

In order to help you with **Research Projects**, there are boxed instructions for using the SciLinks® Web site in each chapter of this book. This site is designed to help you find more information for reports and other research projects. Whenever you see this box, log on to www.scilinks.org, and use the key terms and codes provided to start your search. Your teacher will help you register.

We hope you enjoy reading about biology. Everyone who put this book together worked hard to make it interesting as well as useful. The rest is up to you. We wish you well in your studies.

Chapter 1 **The Science of Living Things**

Chapter 2 **Characteristics of Living Things**

Figure U1-1 *The diver, the fish, the coral, and the moray eel shown above are living things.*

### Biology Journal

Earth is home to a great variety of living things. Scientists have identified over one million different kinds of living things. In this unit, you will learn about the methods scientists use to study living things. You will also learn about the main characteristics that all living things share. In your journal, write a list of questions that you have about living things. As you read each chapter, go back and answer your questions. You can also use the Internet or other references to help answer your questions.

**Biology Journal** The journal activity can be an alternative assessment, a portfolio project, or an enrichment exercise. Have students write at least five questions. As they read the chapters, have them answer the questions.

Figure 1-1 *The science of biology is the study of living things. Biologists may study plants, animals, or other living things. What types of things in this picture do you think a biologist would want to study?*

Biologists might study the many different types of plants, including the grasses and small trees. Biologists might also study the moose.

## Learning Objectives

- Define *biology*.
- Contrast theory and fact.
- Explain the purpose of an experiment.
- Describe the steps in the scientific method.
- Identify the fields of biology.
- Name some fields and jobs in biology.
- **LAB ACTIVITY:** Create scientific questions.
- **BIOLOGY IN YOUR LIFE:** Describe how doctors use the scientific method.

# The Science of Living Things

## Words to Know

| | |
|---|---|
| **biology** | the study of living things |
| **experiment** | a special kind of test to get information |
| **theory** | an idea that explains something based on repeated scientific observations and experiments |
| **law** | a theory that has passed many tests |
| **scientific method** | a set of steps used to find answers to questions |
| **hypothesis** | a possible answer to a scientific question |
| **control** | the part of an experiment in which conditions are kept constant; something used to make comparisons to |
| **botany** | the study of plants |
| **zoology** | the study of animals |
| **genetics** | the study of how the characteristics of a living thing are passed along from generation to generation |
| **microbiology** | the study of living things too small to be seen with the naked eye |
| **ecology** | the study of how all living things relate to one another and their world |

## Biology: The Study of Living Things

Stars, mountains, drops of water, and the human brain are all subjects of science. Science is the study of the whole universe.

In this book, you will study one field of science. You will study living things. The study of living things is called **biology**.

**Linking Prior Knowledge**
Students should recall several inventions or discoveries that were the result of science.

## Discovering Life

The world you live in is full of life. You can probably list at least 20 living things. Foxes, snakes, birds, and mice are all living things. Animals are not the only things that are alive. Trees, grass, and the moss growing on a rock are living things. Humans are living things, too.

Living things can be as big as an elephant. They can also be smaller than a speck of dust. Living things exist on land, in the air, and in the deepest parts of the ocean. They can be found almost everywhere on Earth.

In this chapter, you will learn about the five fields of biology. You will learn how biologists go about their work. You will also read about some jobs in biology.

Moss

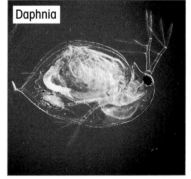
Daphnia

Fox

Figure 1-2 *Living things come in all shapes and sizes. Each of the photos above shows a living thing.*

## Asking Questions

Scientists try to understand the world around them by asking questions. Have you ever wondered about nature? Have you ever been curious about the world around you? Asking questions is the beginning of all scientific studies. Think about the following questions.

- How do spiders make webs?
- What are viruses?
- How do birds fly?
- How does the human heart work?
- Why do bears hibernate?
- What makes trees grow so big?

Figure 1-3 *Spiders use silk to create complex webs.*

### Everyday Science

Science is about asking questions. What kinds of questions have you asked about the world around you? Make a list of three science questions you have thought about. Then, use an encyclopedia or other references to try and find answers to your questions.

**Figure 1-4** *This gardener is trying to find answers to her questions about different kinds of soil.*

**Think About It**

Besides soil type, what other parts of a garden could you experiment with?

Possible answer: You could experiment with the amount of light, the amount of water, and the type of fertilizer.

# Experimenting to Find Answers

When people have questions, they want to find answers. One way to find answers is to do an **experiment**. An experiment is a test to get information. People often experiment without even knowing it. For example, someone may test different kinds of soil. She could see which works best in her garden. She might use one type of soil in one part of the garden. Then, use a different kind of soil in another part. She would wait to see which soil worked better. This is a form of experimenting.

Scientists use observations and experiments to test their ideas. They collect information from experiments. This helps them find answers to their questions.

### Theories and Laws

Some scientific ideas are based on information from experiments and observations. However, they have not yet been proven. This kind of idea is called a **theory**. A theory is a statement of an idea or principle that has been tested by many scientists over a period of time. An example is: *The Theory of Natural Selection*. A theory is not the same as a fact. A fact is something that has already been proven.

Scientists test theories to see if they can be shown to be false. Some theories have been tested many times and are not proven false. Then they may become a scientific **law**. A law is usually accepted by most scientists. A description of how gravity works is an example of a scientific law. It is called: *The Law of Universal Gravitation*.

Science is an on-going process. This means scientists test and retest their ideas all the time. Sometimes the results of new experiments can go against the results of past experiments. Theories and laws may change if new information is found.

✓ **Check Your Understanding**

Answer the questions in complete sentences. Write your answers on a separate sheet of paper.

**1.** What is biology?

**2.** Name four things a biologist might study.

**3.** What is an experiment?

**4.** Why do scientists need to do experiments?

**5.** What is the difference between a theory and a fact?

**Check Your Understanding**

1. the study of living things

2. Possible Answers: trees, insects, fish, humans.

3. a special kind of test to get information

4. Scientists do experiments to test their ideas and get answers to their questions.

5. A theory is something that has not yet been proven true. A fact has been proven true.

## Modern Leaders in Biology

**KEVIN ANDREWIN, Research Assistant**

Kevin Andrewin is a research assistant from Belize (buh-LEEZ). He has been interested in manatees his whole life. Kevin observes the behavior of the manatees. He studies their migration. Migration is the movement of an animal to a new location. Many animals migrate during certain seasons in order to find food.

Kevin attaches special radio equipment to the manatees. This allows his team to follow the movements of the manatees. Manatees are sometimes hurt or killed accidentally by boats. The information Kevin finds out about the movement of the manatees could help to protect them from harm.

**CRITICAL THINKING** What types of questions about the manatee do you think Kevin wants to answer?

Possible answer: He may want to know when the manatees migrate or where they go.

Figure 1-5 *Kevin Andrewin is a research assistant. He studies manatees.*

# The Scientific Method

Scientists today often follow a set of steps to find answers to their questions. These steps are called the **scientific method**.

Suppose you have an aquarium. You notice the fish in your aquarium have been swimming very slowly. You wonder what could be causing this strange behavior. You can use the following steps to find answers to your question.

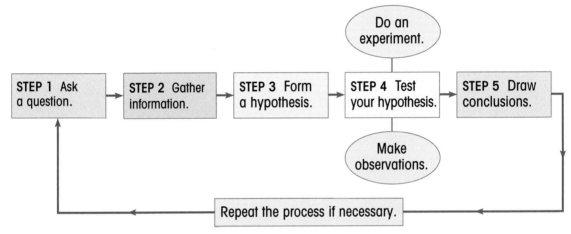

Figure 1-6 *Scientific method flowchart*

Possible answers: research for school projects, reports, or science fairs

**STEP 1  Ask a question.** Your question might be "What is causing my fish to swim so slowly?"

**STEP 2  Gather information.** You might begin by talking to someone else who has an aquarium. That person may have information about aquariums that you did not know. Next, you might talk to someone who works at a pet store. You could read articles in books or magazines about aquarium fish. You could also use the Internet to find information.

**STEP 3** Suggest a good answer, or hypothesis, for your question. You then look at the aquarium tank again. You notice the thermometer says the tank is 70 degrees Fahrenheit (70°F), or 18 degrees Celsius (18°C). You wonder if the temperature of the water could be the cause of the problem. A possible answer to a scientific question is called a **hypothesis**. Your hypothesis is "the water temperature affects the way the fish behave."

**STEP 4** Test your answer with observations or experiments. To test your hypothesis, you decide to do an experiment. You set up two fish tanks of the same size. You put half of your fish in the new tank. You leave the other half in the original tank. You write down the temperature of the water in both tanks. Then, you raise the temperature of the water in one of the tanks three degrees each day.

**Safety Alert**

Make sure when doing experiments that you use proper safety equipment, such as gloves and safety goggles.

Figure 1-7 *Take the temperature of the fish tank every three days.*

Every three days, you compare the fish in one tank with the fish in the other. The tank that had the same temperature every day is called the **control**. A control is something you already know a lot about. It is the part of an experiment in which conditions are kept the same. The control is used to make comparisons.

**Science Fact**

In an experiment, the condition that is being tested is called the *variable*. A good experiment will have only one variable. This makes it easier to analyze your results.

**STEP 5 Draw conclusions.** You should think about the results of your experiment. You then decide if the results support your hypothesis. If you were still unsure, you would do another experiment. Many times scientists repeat experiments to make sure their results are accurate.

After scientists complete an experiment, they often write about it in a report. This is often called a lab report. A lab report is a way to communicate the results and conclusions of an experiment to others. Lab reports sometimes include tables, charts, and graphs. Figure 1-8 is an example of a table you would use to record your observations.

| Effects of Temperature on Aquarium Fish | | | |
|---|---|---|---|
| Date | Tank A | Tank B | Observations |
| 2/3 | 70°F | 65°F | |
| 2/6 | 70°F | 67°F | |
| 2/9 | 70°F | 70°F | |
| 2/12 | 70°F | 75°F | |
| 2/15 | 70°F | 80°F | |

**Figure 1-8** *The table above can be used to record your observations.*

**mycology** – study of fungi
**paleobotany** – study of prehistoric plants
**anatomy** – study of the parts of the body
**physiology** – study of how the body functions

## Science Fact

There are many more specific fields of biology. Some examples are mycology, paleobotany, anatomy, and physiology. Use a dictionary to find out what these terms mean.

## Different Fields of Biology

"Living things" is a huge topic. Most biologists choose one field in which to specialize. Five important areas of biology are botany, zoology, genetics, microbiology, and ecology.

**Botany** is the study of plants. Biologists who study botany may work with plants from a specific part of the world, such as the rain forest. **Zoology** is the study of animals. These biologists study many different animals, such as insects, birds, and fish.

**Genetics** is the study of how characteristics are passed from one generation to the next. Characteristics are qualities that describe something or make it unique. Eye color is an example of a characteristic. **Microbiology** is the study of living things that are too small to be seen with the naked eye. *Micro-* means "small." **Ecology** is the study of how living things interact with one another and with the nonliving world.

**Think About It**

The word part *micro* means "small." What other words have you heard of that contain "micro"?

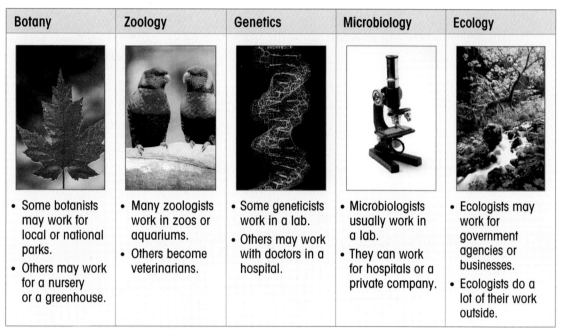

| Botany | Zoology | Genetics | Microbiology | Ecology |
|---|---|---|---|---|
| • Some botanists may work for local or national parks.<br>• Others may work for a nursery or a greenhouse. | • Many zoologists work in zoos or aquariums.<br>• Others become veterinarians. | • Some geneticists work in a lab.<br>• Others may work with doctors in a hospital. | • Microbiologists usually work in a lab.<br>• They can work for hospitals or a private company. | • Ecologists may work for government agencies or businesses.<br>• Ecologists do a lot of their work outside. |

Figure 1-9 *Biologists usually specialize in one area.*

✓ **Check Your Understanding**

Answer the questions in complete sentences.

**1.** What is the scientific method?

**2.** Why is it important to have a control in an experiment?

**3.** What are the five major fields of biology?

**Check Your Understanding**
**1.** a set of steps used to answer a scientific question
**2.** so that you have something to compare your results to
**3.** botany, zoology, genetics, microbiology, and ecology

# LAB ACTIVITY
## Thinking Like a Scientist

## BACKGROUND
Science is about asking questions. Biologists ask questions about living things. Then they test their hypotheses with experiments or observations.

## PURPOSE
In this activity, you will develop scientific questions.

## MATERIALS
paper, pencil

## WHAT TO DO
1. Suppose you are a biologist interested in beetles. Before you begin researching, you have to come up with some good questions about beetles.

2. For each of the following subjects, write two questions that could be answered using the scientific method. Use complete sentences.

   a. beetles and temperature

   b. beetles and food

   c. beetles and movement

   d. beetles and humans

Possible answers: **a.** In what temperatures can beetles survive? In what temperature should young beetles be kept? **b.** What kind of food do beetles eat? How do beetles find food? **c.** Does size of the wing affect the speed a beetle can fly? Do beetles migrate? **d.** How can beetles be helpful to humans? How does a beetle's eyesight compare to that of a human?

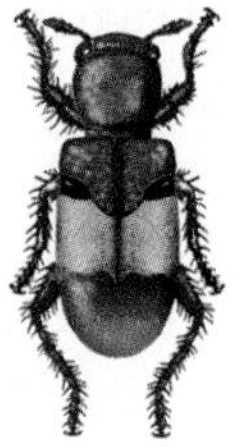

**Common name:** Checkered Beetle

**Family:** Cleridae

**Habitat:** can be found on trunks of dying trees or trees recently killed; can also be found on flowers and foliage

**Diet:** feed on the larvae of other insects

**Other features:** The checkered beetle has a narrow body shape. It can be black, with red, yellow, and blue markings. They are helpful in keeping down the population of wood-eating insects.

Figure 1-10 *Biologists often use field guides such as this one.*

## DRAW CONCLUSIONS
• An experiment is supposed to produce useful information about a question. What kinds of questions do you think make the best starting point for an experiment? Questions that ask about only one topic would work the best in an experiment.

• After scientists complete experiments, they often write about their work in a report. What kinds of things would you include in a report on your beetle investigation? A lab report might include information about the types of beetles studied, the purpose of the experiments, the materials used, and the results.

# BIOLOGY IN YOUR LIFE
## Making a Diagnosis

You wake up one morning with a sore throat. You wonder what's wrong. Should you take medicine? If so, what kind and how much?

When you ask yourself these questions, you are thinking like a scientist. When doctors observe your symptoms and do tests, they are using the scientific method.

Your doctor will ask you questions about your symptoms and check for others. Say your doctor thinks you have strep throat. This is her hypothesis. To test it, she takes a sample from your throat. The sample may be tested for signs of the bacteria that cause strep throat. If the test is positive, she concludes that you have strep throat.

Figure 1-11 *Fever is a common sign that the body is fighting an infection.*

**Match each statement with a step of the scientific method. Write your answers on a separate sheet of paper.**

1. The tests and observations support my hypothesis. I have strep throat.
   Draw conclusions.
2. My throat is very red. It is painful to swallow. A mucous sample shows signs of strep bacteria. Test your hypothesis with observations or experiments.
3. Why does my throat hurt so much?
   Ask a question.
4. I think I have strep throat. Form a hypothesis.
5. A health book says some sore throats are caused by bacteria known as strep bacteria. Symptoms of strep include severe sore throat, bright red throat, swollen glands, and a fever higher than 101°F.
   Gather information.

Critical Thinking You cannot tell if your body temperature is elevated unless you know what it is normally.

**Critical Thinking**
The normal body temperature for humans is about 98.6°F. For some people, though, it is higher, and for others it is lower. Most people only take their temperature when they feel sick. Why is it a good idea to take your temperature when you are well?

# Chapter

## 1 ▷ Review

## Summary

- The study of living things is called biology.

- All science starts with questions. Answers to these questions are tested with experiments.

- Some scientific ideas have not yet been proven. These ideas are called theories. Scientists test theories to find out how well they explain things. Theories are different from facts.

- Scientists today use the scientific method to study problems and answer questions. A hypothesis is a possible answer to a scientific question.

- Most biologists specialize in a specific field. Some of the fields of biology are botany, zoology, genetics, microbiology, and ecology.

biology

botany

control

experiment

hypothesis

law

scientific method

theory

## Vocabulary Review

**Write the word from the list that matches each definition.**

1. a theory that has passed many tests  law

2. the part of an experiment in which conditions are kept constant; something used to make comparisons to  control

3. a set of steps used to find answers to questions  scientific metho

4. a special kind of test to get information  experiment

5. the study of living things  biology

6. an idea that explains something based on scientific information, but has not yet been proven  theory

7. a possible answer to a scientific question  hypothesis

8. the study of plants  botany

## Chapter Quiz

**Answer the questions in complete sentences. Write your answers on a separate sheet of paper.**

1. What is the study of living things called?  biology

2. What is a test to answer scientific questions called?  an experiment

3. What is a theory?  an idea that explains something based on scientific information but has not yet been proven

4. What are the five steps in the scientific method?

5. Suppose you wanted to gather information about different kinds of birds. What are two places you could go to find this kind of information?

6. The study of the eating habits of grizzly bears is in which field of biology?  zoology

7. What does a botanist study?  plants

8. What are three places you could work if you were educated in biology?  in a hospital, private company, aquarium, or zoo

**CRITICAL THINKING**

9. Why do you think it is important to have a control as part of an experiment?

10. How could an everyday problem be solved using the scientific method? Give an example in your answer.

**Test Tip**
It is a good idea to rewrite the definitions of vocabulary words to prepare for a test.

4. ask a question, gather information, suggest a good answer to your question, test your idea, draw conclusions
5. library references such as books on birds or encyclopedias, or the Internet

9. Possible answer: A control helps you to analyze your results. It is used so that you have something to compare to your observations.
10. Possible answer: "Why won't my car start?" You could pick one possibility, such as an old battery, and see if that is the problem. To test this, you would install a new battery and see if the car works.

SC*L*INKS

Go online to www.scilinks.org. Enter the code **PMB100** to research **biology careers**.

## Research Project

Choose a job that is related to biology. Research this job using library references or the Internet. On a separate sheet of paper, write a summary about the job. Then, write two things about the job you would like and two things you would not like.

See the *Classroom Resource Binder* for a scoring rubric for the Research Project.

**Figure 2-1** *All living things have the ability to reproduce. Many living things take care of their young. In what ways do you think this elephant takes care of its young?*

Possible answer: by providing food, teaching it useful habits, and protecting it from predators

## Learning Objectives

- Explain the meaning of *organism*.
- Name seven important characteristics of living things.
- Define *element* and give examples.
- Name four kinds of molecules found in living things.
- **LAB ACTIVITY:** Investigate yeast reproduction.
- **BIOLOGY IN YOUR LIFE:** List some things to consider when choosing a dog.

## Words to Know

| | |
|---|---|
| **organism** | any living thing |
| **energy** | the ability to do work and cause change |
| **stimulus** | a change in the environment that causes a response—plural *stimuli* |
| **offspring** | a new organism produced by a living thing |
| **reproduction** | the way organisms make more of their own kind |
| **cell** | the basic unit of structure and function in all living things |
| **unicellular** | having one cell |
| **multicellular** | having more than one cell |
| **homeostasis** | (hoh-mee-oh-STAY-sihs) the process by which organisms keep their internal conditions relatively stable |
| **element** | a pure substance made of the same kinds of atoms |
| **atom** | smallest part of an element that can be identified as that element |
| **molecule** | the smallest part of a compound |
| **compound** | a combination of two or more elements |

## Being Alive

Think of all the ways you are different from a rock. A rock cannot breathe. It does not grow. You could probably think of many other differences between you and a rock. The biggest difference between you and a rock is that you are alive. In this chapter, you will learn what it means to be alive.

**Linking Prior Knowledge**
Students should recall five things that are alive.

## Common Features of Living Things

Anything that is alive is called an **organism**. Bushes, turtles, whales, grasses, trees, and microscopic living things are all organisms. You are an organism, too.

It may seem as if a grizzly bear and a microscopic organism have nothing in common. However, they are both alive. That means they have some very important features in common. The following list gives seven basic characteristics of most living things.

1) Most living things move in some way.

2) Living things use energy.

3) Living things grow and develop.

4) Living things respond to the environment.

5) Living things reproduce.

6) Living things are made of cells.

7) Living things keep stable conditions inside their bodies.

### Everyday Science

Humans share the same characteristics as other living things. In what ways do you move, use energy, grow and develop, and respond to the environment?

Possible answer: running, walking, biking, getting taller, reacting to other students

**Remember**
Something that is microscopic is so small it cannot be seen with your eyes alone. A microscope is used to look at these organisms, such as the stentor in Figure 2-2.

stentor (100x)

Figure 2-2 *A grizzly bear can grow to be 900 pounds. A stentor is a microscopic organism. Both are living things.*

## Movement

You can kick a rock across a street. However, a rock cannot roll across a street on its own. Many living things can move in some way. The ability to move helps these living things to survive. Organisms may move to find food. They also move to get away from danger.

Plants can also move. However, they move more slowly than animals. Plants often move toward sunlight. If you have a houseplant, take a good look at it. Chances are the leaves have turned toward the window.

Figure 2-3 *Movement allows organisms, like this orangutan, to find food.*

## Using Energy

You are using up energy even as you read this page. **Energy** is the ability to do work and cause change. All organisms need energy to survive. Walking, talking, playing sports, even reading all require energy. You get energy from the foods you eat. Other animals, like the cheetah shown below, also eat to get energy.

Figure 2-4 *Cheetahs use energy to run at great speeds. Some cheetahs have been known to run up to 70 miles per hour.*

### Science Fact

Energy comes in different forms. The Sun gives off many different kinds of energy including light energy. The energy found in food is called chemical energy. X-rays and radio waves are also forms of energy.

Plants also do work and use energy. Most plants do not eat food though. They absorb light energy from the Sun. Plants use this energy to make their own food. Then, plants use the energy in the food to carry out activities. Plants use energy to take up water and minerals from the soil. Plants also use energy to grow leaves and flowers.

## Growth and Development

All living things grow at some time in their life. In most organisms, the fastest growth occurs in the first part of their life. You have grown a great deal since you were born. For example, an oak tree starts out as an acorn. After a long period of time, it can become a large, adult tree.

### Science Fact

The range of growth of each living thing depends on what kind of organism it is. People usually reach a height of about 5 to 6 feet. Oak trees can grow up to 200 feet tall!

Figure 2-5 *Over many years, an acorn can become an adult oak tree.*

Besides growing, many living things also develop as they get older. Development is a change in the shape or structure of their bodies over a period of time. As a caterpillar develops, it changes into a butterfly. Humans also develop as they grow. After a period of time, many living things stop growing.

**Figure 2-6** *A caterpillar develops into a butterfly over time.*

## ✓ Check Your Understanding

Answer the questions in complete sentences. Write your answers on a separate sheet of paper.

**1.** What is another term for "living thing"?

**2.** What is energy?

**3.** Where do you get energy?

**4.** Where do plants get energy?

**5.** What is a change in body shape called?

**Check Your Understanding**
1. organism
2. the ability to do work
3. food
4. from the Sun
5. development

## Responding to the Environment

The environment is all the things around you. Your home, your school, the sky, the soil, and many organisms are all part of your environment. All organisms respond to the things in their environment. To respond means to act in return or in answer to something. For example, a cat will respond to a flea on its back by scratching.

Organisms often respond to changes in their environment. The change that occurs is called the **stimulus**. The way an organism reacts to the stimulus is the response.

A stimulus can be almost anything in the organism's environment. Chemicals in the environment, temperature, and amount of light can also cause an organism to respond. A plant will respond to a source of light by growing toward the light.

Possible answer: It allows you to get food, keep warm, and avoid danger.

**Think About It**
Like all organisms, humans respond to their environment. How do you think the ability to respond to your surroundings helps you to survive?

Sometimes an organism helps itself to survive by responding to a stimulus. For example, if an octopus senses danger, it may respond by releasing ink. The ink clouds the water and allows the octopus to escape. A frilled lizard will respond to predators by expanding flaps of skin around its neck. Other animals respond by changing colors, releasing chemicals, or making noise.

Figure 2-7 *How do you think the response shown here helps the frilled lizard to survive?*

It may scare away predators.

## Reproduction

All organisms have some form of **offspring**. Offspring are organisms produced from parent organisms of the same kind. The offspring of a human is another human. The offspring of a dog is another dog. The offspring of a plant is another plant.

Offspring are produced by a process called **reproduction**. All living things reproduce. Some simple kinds of organisms reproduce by splitting in two. In more complex organisms, two members of the same species are needed for reproduction. Usually, one of these organisms is a female and one is a male.

Figure 2-8 *These baby owls are the offspring of a male and female owl.*

### DISPROVING SPONTANEOUS GENERATION

Spontaneous generation is the theory that living things can come from nonliving things. For centuries, most people believed this. In 1668, Italian doctor Francesco Redi challenged this idea.

Redi performed experiments to show that living things come only from other living things. In one experiment, he compared jars containing meat with mesh covers against jars that had no covers. Redi then allowed the meat to sit in the jars for several days. He found maggots only in the jars that were uncovered. Redi concluded that maggots grew only where flies could lay their eggs. The meat itself did not produce the maggots.

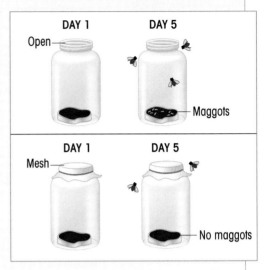

Figure 2-9 *This diagram shows one of Redi's experiments.*

Redi's experiment helped to disprove the theory of spontaneous generation. It also introduced a new way of performing experiments. More than 300 years later, this form of experimenting is still used.

CRITICAL THINKING Why were maggots or fly eggs not found in all of the jars? Flies, eggs, and maggots would not be found in all the jars because the adult flies could not enter the jars that were covered.

## All Living Things Are Made of Cells

**Cells** are the building blocks of life. They are the basic units of structure and function in living things. Most cells are too small to be seen without a microscope.

All living things have at least one cell. An organism with only one cell is called **unicellular**. An organism with more than one cell is called **multicellular**. Humans are multicellular organisms. The human body may contain more than 60 trillion (60,000,000,000,000) cells.

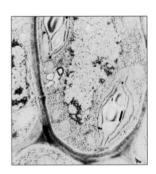

Figure 2-10 *This cell comes from the leaf of a plant. It has been magnified 6,000 times!*

## Science Fact

Viruses are small structures that can be puzzling. They are not made of cells and do not take in food or make waste. However, they can reproduce by invading another organism. Scientists do not agree whether to consider viruses living or non-living.

In multicellular organisms, each cell has a specific purpose, or job. These are called specialized cells. For example, bone cells have a different purpose than skin cells. In plants, leaf cells have a different purpose than root cells.

A few kinds of organisms, such as bacteria, are unicellular. These organisms carry out all of the important life functions using only one cell. They do not have specialized cells.

## Maintaining Stable Conditions

Organisms change all of the time. They also respond to stimuli in their environment. However, organisms must keep stable conditions inside their bodies. This is called **homeostasis**. For example, a plant may have to deal with periods of little rainfall. In order to survive, the plant has to keep the amount of water in its cells within a certain range. Keeping a balance of water is an important part of homeostasis for all organisms.

## The Chemistry of Living Things

When you eat, your body uses nutrients in the food to survive. The nutrients provide your body with energy.

### Atoms and Elements

**Atoms** are small particles of elements that still can be identified as that element. All forms of matter are made up of atoms. Atoms group together to form elements. An **element** is a pure substance made of the same kinds of atoms. Nutrients are made up of elements. The air you breathe is also made of elements. Oxygen (O), carbon (C), hydrogen (H), and nitrogen (N) are the most common elements in living things.

### Elements in Living Things

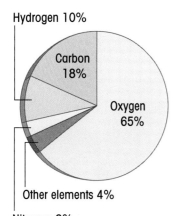

Hydrogen 10%

Carbon 18%

Oxygen 65%

Other elements 4%

Nitrogen 3%

Figure 2-11 *These common elements are found in all living things.*

## Molecules

When two or more atoms combine, a **molecule** is formed. When similar molecules group together, a **compound** is formed. Everything on Earth is made up of some kind of element or compound.

Water is made up of molecules. A molecule of water is made up of the elements hydrogen and oxygen. It contains two atoms of hydrogen and one atom of oxygen.

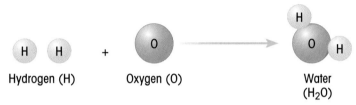

Hydrogen (H)          Oxygen (O)          Water
                                          (H₂O)

**Figure 2-12** *Water is made up of the elements oxygen and hydrogen.*

## Compounds in Living Things

The main types of compounds in living things are carbohydrates, lipids, proteins, and nucleic acids.

- *Carbohydrates* are made up of carbon, hydrogen, and oxygen. They provide energy for living things.

- *Lipids* are fats and oils. They are made up mostly of carbon and hydrogen. They are used to store energy.

- *Proteins* are very complex molecules. Proteins take part in many chemical reactions. They also help build and repair living cells.

- *Nucleic acids* are molecules that contain information about the characteristics of an organism. They also play a role in the production of proteins.

### ✓ Check Your Understanding

**1.** What are the seven characteristics of living things?

**2.** What are organisms called that are made up of just one cell?

**3.** What are four elements found in living things?

**Check Your Understanding**
1. can move, use energy, grow and develop, respond, reproduce, are made of one or more cells, maintain stable internal conditions
2. unicellular
3. carbon, hydrogen, nitrogen, and oxygen

# LAB ACTIVITY
## Investigating Yeast Reproduction

### BACKGROUND
Like all living things, microscopic organisms reproduce. Yeast is a type of microscopic organism. It reproduces by forming offspring called buds. The buds form on the side of the parent yeast cell.

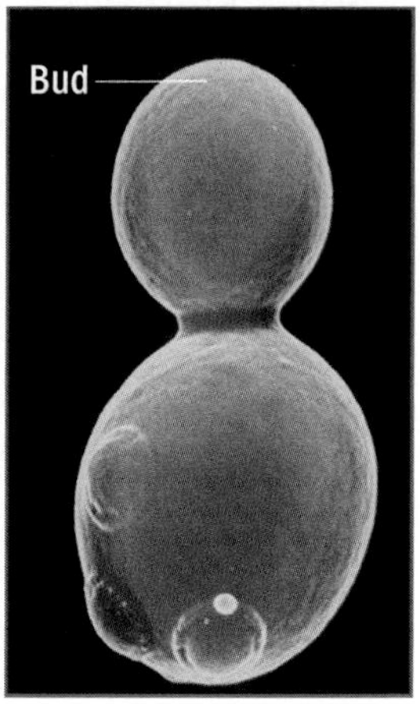

### PURPOSE
In this activity, you will observe reproduction in yeast.

### MATERIALS
safety goggles, apron, warm water, cup or beaker, tablespoon, corn syrup, yeast, methylene blue stain, slide, cover slip, microscope, pencil, paper

Bud

Figure 2-13 *Yeast cells reproduce by budding.*

### WHAT TO DO
1. Fill a cup or beaker with warm water.

2. Add a tablespoon of corn syrup. Mix the two liquids together.

3. Add a tablespoon of dry yeast. Stir the solution. Let the solution sit for 15 to 20 minutes.

4. Put a few drops of the yeast solution onto a slide. Put one drop of stain onto the specimen. Place a cover slip on top.

5. Observe the slide using the high power of a microscope. Try to find a few yeast cells with buds. Use the fine-adjustment knob to focus the image. Draw what you see on a separate sheet of paper.

### DRAW CONCLUSIONS
- What did you see?  Students should be able to see several yeast cells with buds forming.

- How can you show that yeast is a living thing?  Yeasts must be living things because they can reproduce.

# BIOLOGY IN YOUR LIFE
## Choosing a Dog

Having a dog as a pet is a big responsibility. You are taking care of a living thing. As the owner of a pet, you must help it stay healthy. You must provide food and shelter. Also make sure your dog gets enough exercise.

Before you get a dog, you should find out if you can care for it. Too often, people get a dog and find out it is not what they expected. Before getting a dog, you should ask yourself some questions. Can you provide your dog with enough space to exercise? Can you afford the costs of food and medical care? Who will care for your dog when you are away? Different breeds have different needs.

**Figure 2-14** *Golden retrievers like to run and play.*

**Look at the following chart. It provides information about some types of dogs. Use the chart to answer the questions.**

1. Which dogs might you choose if you want a large dog?
   Golden retriever or German shepherd
2. Which dogs are good for apartments? Why?
   Chihuahua or Shih Tzu. They are small and don't require a lot of outdoor exercise.
3. Which dog might a family with young children avoid? Why?
   Chihuahua. They don't like rough play and are not always good with children.

| Characteristics of Dogs | | |
|---|---|---|
| **Breed** | **Weight (pounds)** | **Notes** |
| Golden retriever | 55–80 | Friendly; needs lots of outdoor exercise; good with children |
| German shepherd | 75–90 | Needs lots of outdoor exercise; can be good with children and good protectors |
| Chihuahua | 2–5 | Does not like rough play, so not always good with children; good indoor pet |
| Shih Tzu | 9–16 | Long hair needs daily brushing; good with children; likes being indoors |

**Figure 2-15** *Different dogs have different characteristics.*

## Critical Thinking

Which dog would you choose? Explain your answer.

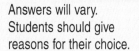

Answers will vary. Students should give reasons for their choice.

# Chapter

## 2 ▶ Review

## Summary

- Living things are called organisms.

- Most organisms have seven characteristics in common: the ability to move, use energy, grow and develop, respond to their environment, reproduce, maintain stable body conditions, and are made of cells.

- Living things are made up of elements. Examples of elements found in living things are hydrogen, oxygen, and nitrogen. A combination of two or more elements is called a compound.

- Elements form compounds. The most common compounds found in living things are carbohydrates, proteins, fats, and nucleic acids.

cell

compound

element

homeostasis

molecule

offspring

organism

reproduction

stimulus

unicellular

## Vocabulary Review

Write the word from the list that matches each definition.

1. any living thing   organism

2. a change in the environment that causes a response   stimulus

3. a new organism produced by a living thing   offspring

4. the way organisms make more of their own kind   reproductio

5. the basic unit of structure and function in all living things   cell

6. a simple substance that cannot be broken down into another substance   element

7. having one cell   unicellular

8. the smallest part of a compound   molecule

9. a combination of two or more elements   compound

10. the process by which organisms keep their internal conditions relatively stable   homeostasis

# Chapter Quiz

Answer the following questions in complete sentences. Write your answers on a separate sheet of paper.

**Test Tip**
Before taking the chapter quiz, review the information in the summary.

1. Name three kinds of organisms that live in your community. Possible answer: tree, squirrel, bird

2. Name three nonliving things that can be found in your community. Possible answer: stone building, concrete sidewalk, water

3. Give an example of a way that a plant moves. Possible answer: leaning toward sunlight and catching insects

4. What are offspring? new organisms produced by a living thing

5. What are all living things made of? cells

6. What is an element? a simple substance that cannot be broken down into other substances

7. What are some elements found in living things? carbon, oxygen, nitrogen, and hydrogen

8. Name the seven characteristics of organisms.

8. movement; reproduction; use energy; grow and develop; respond to the environment; maintain stable body conditions; are made of cells

## CRITICAL THINKING

9. When you are sitting still, what kind of work is your body doing?

9. Possible answer: breathing, respiration, digestion, circulation, thinking

10. Explain at least three ways you respond to your environment.

10. Possible answer: jump when you hear a loud noise, stop when you see a red light, pull hand away when you touch something hot

**Research Project** Students' reports should contain a detailed explanation of why viruses are or are not considered living things. Reports should also contain a visual.

*SCiLINKS*
Go online to www.scilinks.org. Enter the code PMB102 to research **viruses**.

## Research Project

Some scientists debate on whether viruses are living or nonliving. Write a report on viruses. Include an explanation of how a virus does or does not meet each of the characteristics of living things. Also include a drawing or picture of a virus in your report.

See the *Classroom Resource Binder* for a scoring rubric for the Research Project.

# Unit 1 Review

**Choose the letter of the correct answer to each question.**

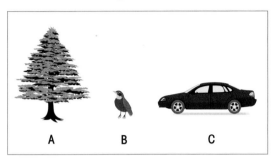

A      B        C

Figure U1-2 *Tree, bird, car*

Use the diagram in Figure U1-2 to answer Questions 1 to 3.

**1.** What makes the tree different from the bird? D (p. 20)
  A. It can move.
  B. It can grow and develop.
  C. It uses energy.
  D. It can make its own food.

**2.** Which of the following is not a characteristic of objects A and B?
  A. They are made of cells. B (p. 20)
  B. They stay the same size their entire lives.
  C. They reproduce.
  D. They react to changes in their environment.

**3.** How is the object labeled C different from the other two objects? A (p. 18)
  A. It is not alive.
  B. It is made of only one cell.
  C. It is warmblooded.
  D. It can reproduce.

**4.** Which of the following is *not* a step in the scientific method? C (pp. 8–10)
  A. gathering information
  B. suggesting a possible answer to a question
  C. making up results so you do not have to do an experiment
  D. asking a question

**5.** What do biologists study? D (pp. 3–4, 10–11)
  A. stars and planets
  B. the age of rocks
  C. the weather
  D. living things

**6.** Which means to change as you grow? A (pp. 20–21)
  A. develop
  B. respond
  C. stimulate
  D. reproduce

**7.** What do you call an idea that has not yet been proven, but has been tested many times by scientists? B (p. 6)
  A. a hypothesis
  B. a theory
  C. a scientific law
  D. a fact

**Critical Thinking**

Design an experiment to answer the following question: Will I have more energy in the morning if I eat breakfast or if I skip breakfast?

Possible experiment: Set up two groups of people; one group eats breakfast, one group does not. Record how they feel each morning for a week when the school day begins. (pp. 8–10)

# Unit 2 ▶ Cells and Simple Organisms

Chapter 3 **Cell Structure and Function**

Chapter 4 **Cells and Energy**

Chapter 5 **Heredity and Genetics**

Chapter 6 **Simple Organisms**

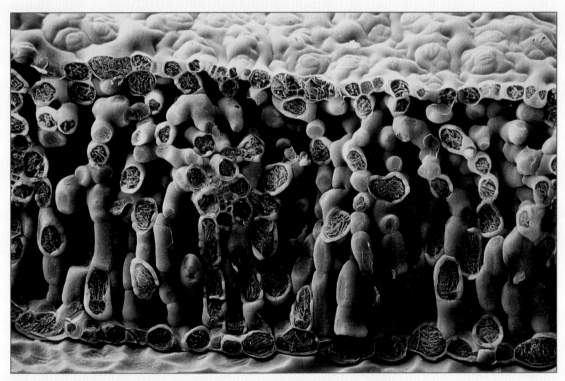

Figure U2-1 *The leaf of a turnip plant is made of layers of cells. The different kinds of cells have different functions. Together they help the turnip to live.*

### Biology Journal

In this unit, you will learn about cells. In your journal, write a list of questions that you have about cells. As you read each chapter, go back and answer your questions. You can also use the Internet or other references to help answer your questions.

**Biology Journal** The journal activity can be an alternative assessment, a portfolio project, or an enrichment exercise. Have students write at least five questions. As they read the chapters, have them answer the questions.

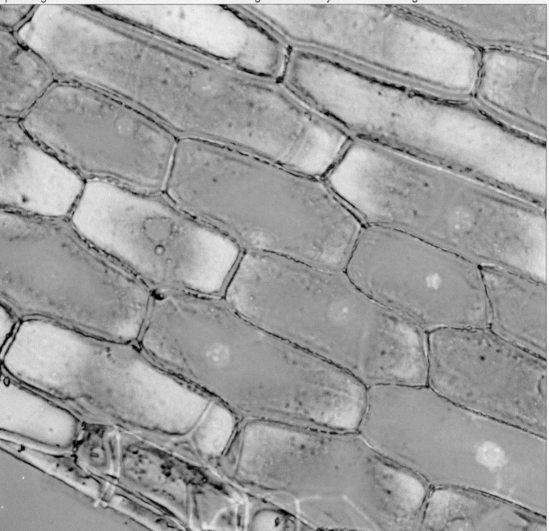

**Figure 3-1** *This photograph may look like a brick wall, but it is not. It shows onion cells. Just like bricks are the basic units of a building, cells are the basic units of living things. What do you think cells do for a living thing?*

Possible answer: Cells carry out life functions.

## Learning Objectives

- List the parts of a cell.
- Explain how plant cells differ from animal cells.
- Describe the structure of organisms.
- Explain the process by which cells reproduce.
- **LAB ACTIVITY:** Identify parts of a plant cell.
- **ON-THE-JOB BIOLOGY:** Describe the job of a cytotechnologist.

## Words to Know

| | |
|---|---|
| **cell membrane** | a thin, protective covering around a cell |
| **cytoplasm** | a jellylike substance that fills a cell |
| **organelle** | a part in a cell that floats in the cytoplasm and helps the cell to carry out life functions |
| **nucleus** | the part of a cell that controls all the other parts —plural *nuclei* |
| **mitochondria** | (myt-oh-KAHN-dree-uh) structures in a cell that release most of a cell's energy—singular *mitochondrion*) |
| **vacuole** | a structure in a cell that stores food, water, or wastes |
| **chloroplast** | a part in a plant cell that uses sunlight to make food |
| **cell wall** | a thick, stiff covering around some cells, such as plant cells |
| **tissue** | a group of cells that work together to do a certain job |
| **organ** | a group of tissues that work together to do a certain job |
| **mitosis** | the division of the nucleus of a cell |
| **chromosome** | a thick strand of nuclear material that passes on traits from the parent cell to the daughter cells during cell reproduction |
| **meiosis** | the type of cellular reproduction that produces sex cells |

## Cells Are the Building Blocks of Life

You began life as a single cell. When you were born, that single cell had become an organism of many cells. Now, your body is made up of trillions of cells.

**Linking Prior Knowledge**
Students should recall that all living things are made of cells (Chapter 2).

Think of your body as a brick building. Just like bricks are the basic building blocks in a building, your cells are the basic building blocks in your body.

## Different Kinds of Cells

Some organisms are made up of only one cell. Within the one cell, these organisms can carry out all the life functions. For example, bacteria are made of only one cell. These single cells can move, reproduce, and carry out the other life functions.

Other organisms are made up of many cells. Most of the organisms that you know about have many cells. But, not all cells are alike. Some cells release energy. Some cells carry messages to your brain. Some cells make new skin when you need it. There are many different types of cells in your body that do many different things. However, all your cells carry out the work that your body needs.

## Discovery of the Cell

In 1665, a scientist named Robert Hooke placed a thin piece of cork under his microscope. Cork comes from the bark of a certain kind of oak tree. Robert Hooke saw many tiny openings in the cork. They looked to him like small rooms that were called cells. So, he named these tiny structures "cells." The name is still used today.

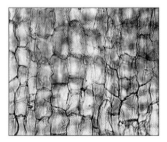

Figure 3-2 *These cork cells are like the cells that Robert Hooke saw through his microscope.*

What Robert Hooke saw were really dead cells. Today, with the help of high-powered microscopes, scientists can study living cells very closely.

Scientists can see how certain types of cells carry out life functions. They can see how certain cells move. They can see how other types of cells reproduce. They can also see the many parts inside a cell. You will learn about many of these parts.

## ✓ Check Your Understanding

Answer the questions in complete sentences.

**1.** What are the building blocks in your body?

**2.** Give an example of how cells are not all alike.

**3.** What did Robert Hooke see under his microscope?

**1.** cells

**2.** Possible answer: Some cells release energy while some cells carry messages to the brain.

**3.** tiny openings in cork

## Great Moments in Biology

### THE CELL THEORY

By the 1800s, microscopes had been invented that were more powerful than the one that Robert Hooke used. These microscopes allowed other scientists to study cells in greater detail.

Scientists discovered many things about cells. They were able to see some structures inside cells. They were also able to see some of the activities going on inside cells. These discoveries about cells led to the cell theory.

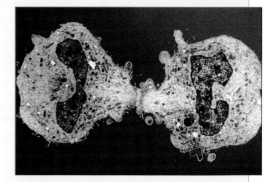

Figure 3-3 *Two new cells form when a living cell splits. New cells come only from other living cells.*

The main points of the cell theory are:

• All living things are made of one or more cells.

• Cells carry out all life functions.

• New cells come only from other living cells.

Although the cell theory was developed in the 1800s, biologists today agree with the main points of the theory. Modern microscopes are even more powerful than they were in the 1800s. Today, the information that biologists discover about cells still supports the cell theory.

**CRITICAL THINKING** Why do you think powerful microscopes were important in the development of the cell theory? Possible answer: The information that biologists discovered using powerful microscopes led to the development of the cell theory.

# Parts of a Cell

Most cells are very small. However, many parts can be found inside a cell. Some cell parts may be found in only certain types of cells. Cell parts work together to carry out life functions for the organism.

### Cell Membrane

Around the outside of each cell is a **cell membrane**. This cell part is a thin, flexible covering. The cell membrane has several important jobs. It holds the cell together. It protects the cell. It keeps harmful substances from entering the cell. At the same time, the cell membrane controls the movement of needed materials, such as food, oxygen, and water into the cell. It also controls the movement of wastes out of the cell.

### Cytoplasm

The **cytoplasm** is a clear, jellylike substance within a cell. This substance is mostly water. Most chemical reactions in a cell take place in the cytoplasm. All of the structures inside a cell float in the cytoplasm. These structures are called **organelles**.

### Nucleus

Inside each cell is an organelle called the **nucleus**. The nucleus is surrounded by another membrane. This membrane is called the *nuclear membrane*. The nucleus is the cell's control center. Molecules in the nucleus control most of the activities that take place in the cell.

These molecules contain genes. Genes store information about an organism's characteristics. For example, your genes store information about the color of your eyes, your height, and many other things.

**Think About It**

Why do you think scientists study living cells?

Possible answer: so that they can understand how living cells carry out life functions

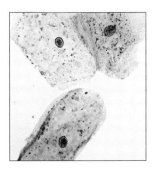

Figure 3-4 *The cell membrane, cytoplasm, and nucleus can be seen inside each of these human cheek cells.*

### Mitochondria

The **mitochondria** are known as the "powerhouses" of cells. These bean-shaped structures act as the cell's power plant. They break down molecules from food to release energy. Cells use the energy to carry out other life processes.

### Vacuoles

**Vacuoles** are the storerooms of cells. Food and water are stored in the vacuoles. Waste materials also can be stored in the vacuoles.

The cell membrane, cytoplasm, nucleus, mitochondria, and vacuoles are only some parts of a cell. Figure 3-5 shows these cell parts in an animal cell.

**ANIMAL CELL**

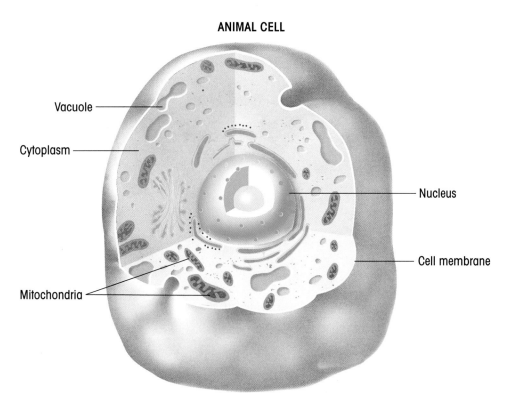

Figure 3-5 *An animal cell has many different organelles.*

# More Cell Parts

**Remember**
Recall from Chapter 2 that proteins are complex molecules that take part in many chemical reactions.

There are many parts to a cell. You have already learned about some cell parts. There are, however, more parts that help a cell to carry out life functions. Some of these cell parts include ribosomes, the endoplasmic reticulum, Golgi bodies, and lysosomes.

### Ribosomes

*Ribosomes* are very small structures that make proteins for the cell. Cells need proteins to grow and to carry out other life functions.

### Endoplasmic Reticulum

The *endoplasmic reticulum*, or ER, is a network of tubes inside a cell. Substances move along these tubes to get from the nucleus to the other organelles.

### Golgi Bodies

*Golgi bodies* are flattened, folded sacs that package and move materials. These cell parts allow materials to travel easily through the cytoplasm and out of the cell.

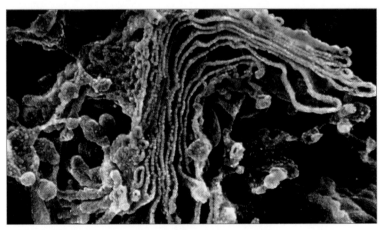

Figure 3-6 *The flattened, folded sacs are a Golgi body.*

### Lysosomes

*Lysosomes* are round packets. They contain chemicals that break down substances and old cell parts that the cell does not need.

Each cell part has a special job. All of the organelles in a cell work together to keep the cell alive. Figure 3-7 shows all of the parts you just read about.

**ANIMAL CELL**

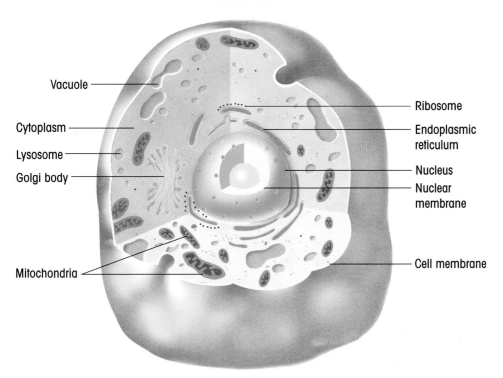

Figure 3-7 *The organelles in a cell work together to carry out life functions.*

### ✓ Check Your Understanding

Answer the questions in complete sentences. Write your answers on a separate sheet of paper.

1. In which part of a cell do most chemical reactions take place?

2. Which part of a cell controls the other parts?

3. Which part of a cell stores wastes?

4. **CRITICAL THINKING** Explain why cells are called building blocks of life.

**Check Your Understanding**
1. in the cytoplasm
2. the nucleus
3. the vacuoles
4. Possible answer: because all living things are made of one or more cells

# Plant Cells Are Different From Animal Cells

All plants and animals are made up of cells. Animal cells and plant cells have some similar cell parts. However, plant cells are different from animal cells in several important ways.

**Science Fact**

Plant cell walls are made mostly of a material called cellulose. The fibers of cellulose are made of carbohydrates. Cellulose gives the cell wall strength and support.

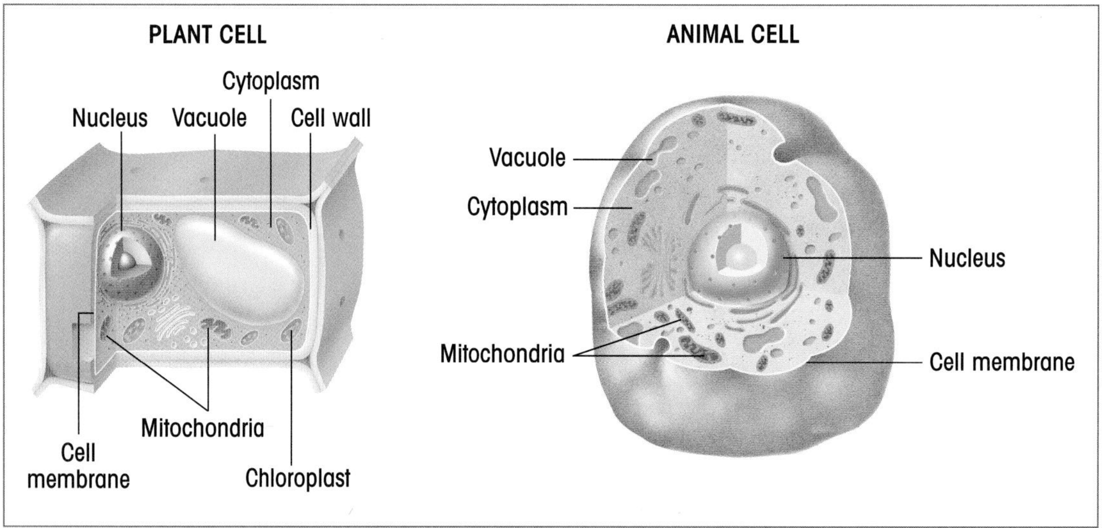

1) Plant cells have special structures called **chloroplasts**. These structures help plants make food using light during photosynthesis. Animals cannot make their own food.

2) Plant cells have a **cell wall** around their cell membranes. This structure is rigid and protects the plant cell. Animal cells do not have a cell wall.

3) Plant cells usually have fewer vacuoles than animal cells. The one or two vacuoles that plant cells do have are very large. Plant cells need these large vacuoles to hold water.

Figure 3-8 shows some parts in a plant cell and in an animal cell. Compare the parts in the two diagrams.

**PLANT CELL**

Cytoplasm
Nucleus   Vacuole   Cell wall

Cell membrane
Mitochondria
Chloroplast

**ANIMAL CELL**

Vacuole
Cytoplasm
Nucleus
Mitochondria
Cell membrane

Figure 3-8 *A plant cell has chloroplasts, a cell wall, and fewer but larger vacuoles than an animal cell does.*

# Cells of Tissues and Organs

In multicellular organisms, groups of cells can work together to do a certain job. These groups of cells are called **tissues**. There are four main kinds of tissues in most animals: muscle tissue, covering tissue, connective tissue, and nerve tissue. Skin is a covering tissue. Bone is a connective tissue.

**Think About It**

What kind of tissue makes up the scales of a snake?

Snake scales are part of the skin of a snake. Snake scales are a covering tissue.

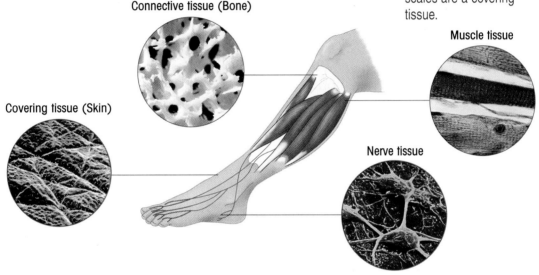

Connective tissue (Bone)

Covering tissue (Skin)

Muscle tissue

Nerve tissue

Figure 3-9 *Different cells make up different tissues in animals.*

Groups of tissues can work together to do a certain job. These groups of tissues make an **organ**. Your heart, brain, and eyes are all organs. Your heart, for example, is made up of muscle tissue, nerve tissue, and connective tissue. Muscle tissue makes your heart pump. Nerve tissue tells it when to pump. Connective tissue holds the parts together.

Plants also have cells, tissues, and organs. Plant cells form different kinds of tissues. For example, one tissue type covers the outside of a leaf. Another tissue transports water and nutrients. Special kinds of tissue form plant organs. Flowers, leaves, stems, and roots are all plant organs.

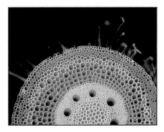

Figure 3-10 *A corn root has different types of plant tissues.*

## Cell Reproduction

When you cut yourself, the cut heals. How does healing occur? The cells around the cut can reproduce. Each cell can produce two new cells. The new cells are exactly like the cells that they came from.

Most of your body cells reproduce. These cells reproduce all of the time. Most body cells die. New cells must replace the old ones.

In cell reproduction, old cells are replaced by new cells. Cell reproduction also increases the number of cells in your body. Your body has grown since you were born. However, your cells have not only increased in size. They have also increased in number. Much of this growth is a result of cell reproduction.

### Dividing in Two

A cell can reproduce by dividing its nucleus and forming two new cells. This process is called **mitosis**. The two new cells are called daughter cells. Daughter cells have the same traits as the parent cell.

Before mitosis takes place, the chemical in the nucleus of the parent cell makes exact copies of itself. There are two copies so that each daughter cell will get the same traits as the parent. The chemical then forms thick strands called **chromosomes**.

Chromosomes play an important role in mitosis. Chromosomes contain a chemical called DNA. DNA on chromosomes passes the characteristics of a living thing on to its offspring.

Figure 3-11
*Chromosomes contain DNA. Strands of human DNA are held tightly together in an X shape.*

## The Four Stages of Mitosis

There are four important stages in mitosis: prophase, metaphase, anaphase, and telophase.

1) During *prophase*, the nuclear membrane in the parent cell disappears. The chromosomes can be seen under a microscope.

2) During *metaphase*, the chromosomes line up in the center of the cell.

3) During *anaphase*, the chromosomes split up and separate. One set of chromosomes is pulled to one end of the cell. The other set of chromosomes is pulled to the other end of the cell.

4) During *telophase*, the middle of an animal cell pinches together. A nuclear membrane forms around each set of chromosomes. In plant cells, a new cell wall and a new cell membrane form down the middle of the cell.

The parent cell has become two new daughter cells. Daughter cells are exactly like the parent cell but smaller. Daughter cells are like the parent cell because they are made with exact copies of the parent cell's chromosomes.

**MITOSIS**

1) PROPHASE

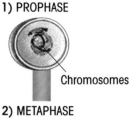

Chromosomes

2) METAPHASE

3) ANAPHASE

4) TELOPHASE

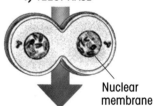

Nuclear membrane

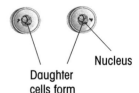

Nucleus

Daughter cells form

Figure 3-12 *The four stages of mitosis in an animal cell.*

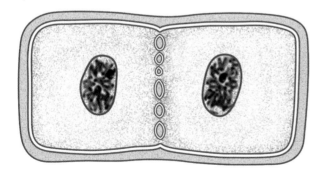

Figure 3-13 *During telophase in a plant cell, a new cell wall and a new cell membrane form down the middle of the cell.*

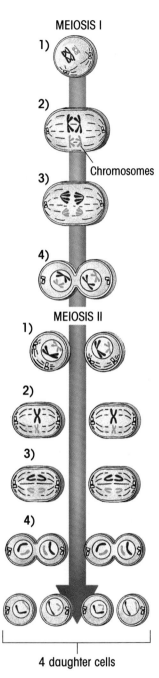

MEIOSIS I

1)

2)

3) Chromosomes

4)

MEIOSIS II

1)

2)

3)

4)

4 daughter cells

Figure 3-14 *Meiosis I and Meiosis II*

# Meiosis

Body cells are reproduced through mitosis. However, there is another type of cellular reproduction. This type of reproduction is called **meiosis**. The process of meiosis produces sex cells. Sex cells can eventually join to form offspring that are different from the parent cells. Meiosis occurs in two parts: meiosis I and meiosis II.

### Meiosis I

1) During prophase I, chromosomes have already duplicated. The nuclear membrane disappears.

2) During metaphase I, chromosomes line up in pairs.

3) During anaphase I, chromosome pairs separate.

4) During telophase I, cells pinch apart.

### Meiosis II

1) During prophase II, chromosomes do not duplicate.

2) During metaphase II, chromosomes line up in the center of the cell.

3) During anaphase II, chromosomes split and separate.

4) During telophase II, cells pinch together. Nuclear membranes form.

Four daughter cells are formed. Each cell has only half the number of chromosomes as the parent cell.

The phases in both parts of meiosis are similar to mitosis. However, in mitosis there is one cell division. In meiosis, there are two cell divisions. So, four daughter cells are produced instead of two. Also, in meiosis, the four daughter cells are not identical to the parent cell. They have half the number of chromosomes as the parent cell. These daughter cells are very important in the formation of offspring.

## ✓ Check Your Understanding

Answer the questions in complete sentences. Write your answers on a separate sheet of paper.

1. Name three ways in which plant cells are different from animal cells. Plant cells have chloroplasts, a cell wall around their cell membranes, and fewer but larger vacuoles.

2. What are two different groups that body cells are organized into? tissues and organs

3. What are the four stages of mitosis? prophase, metaphase, anaphase, and telophase

4. **CRITICAL THINKING** How is meiosis different from mitosis? In mitosis, there is only one cell division. In meiosis, there are two cell divisions. Mitosis produces two daughter cells that are identical to the parent cell. Meiosis produces four daughter cells that have only half the number of chromosomes as the parent cell.

## A Closer Look

### CANCER CELLS

Sometimes cells reproduce when they are not needed. These abnormal cells are called cancer cells. Cancer cells usually go through mitosis much more quickly than healthy cells. After a while, cancer cells can block the healthy functions of the body.

Biologists and researchers are working hard to understand cancer. They do know some of the causes that lead to cancer. For example, smoking can cause lung cancer. Too much sunlight can cause skin cancer.

Figure 3-15 *Cancer cells, shown in pink, can block the healthy functions of the body.*

Scientists have found treatments for some types of cancer. Some of these treatments include surgery and medicines. Early detection and treatment may result in a cure for some cancers. Perhaps in your lifetime, scientists will discover a cure for all types of cancer.

**CRITICAL THINKING** How do you think one cancer cell can produce many cancer cells? Explain your answer. Through mitosis, a cancer cell can reproduce two daughter cells. These daughter cells can produce more cancer cells.

# LAB ACTIVITY
## Identifying Parts of a Plant Cell

**BACKGROUND**
Plants are living things made up of cells.

**PURPOSE**
In this activity, you will identify the different parts of a plant cell.

**MATERIALS**
safety goggles; glass slide; cover slip; lens paper; dropper; iodine; clear, thin onion layer; microscope; pencil; paper

**WHAT TO DO**
1. Copy the chart in Figure 3-16 onto a separate sheet of paper.

2. Clean the slide and cover slip with lens paper.

3. Put on safety goggles. You should take them off when you look into the microscope.

4. Place a drop of iodine on the slide. Carefully place a thin layer of onion on the slide over the iodine. Gently place a cover slip over the onion sample. Make sure students observe the concave, dull side of the onion membrane.

5. Place the slide under the microscope. Focus the microscope under low power. Draw what you see in your chart.

6. Look at the onion cells under high power. Try to find the different parts of the cell. Remind students to use only the fine-adjustment knob when using the high-power lens.

7. Draw what you see under the microscope in your chart. Label the cell parts that you have learned about.

| Observing a Plant Cell | |
| --- | --- |
| Under low power | |
| Under high power | |

Figure 3-16 *Copy this chart onto a separate sheet of paper.*

5. Remind students to only use the coarse-adjustment knob while looking from the side of the microscope. They should focus with the fine-adjustment knob.

**DRAW CONCLUSIONS**
- What was the difference between looking at the cells under low power and then under high power? Possible answers: Under low power, cells are arranged like a brick wall. Under high power, the parts inside of one cell can be seen.
- Which cell parts can you identify? Possible answers: the cell wall, chloroplasts, vacuoles, and the nucleus

# ON-THE-JOB BIOLOGY
## Cytotechnologist

Ron Brown is a cytotechnologist (sy-toh-tehk-NAHL-uh-jihst). He works in a lab. Ron examines human cells with a microscope. He looks for early signs of cancer and other diseases.

The cells are put on a slide and colored with special dyes. Ron studies the cytoplasm and the nucleus inside the cells. If the cells and the parts inside look normal, he writes a final report. If the cells do not look normal, Ron has a special doctor examine them, too.

The doctor can often tell what is wrong. The cells might show signs of cancer or another disease. If found early, some cancers and some diseases can be treated.

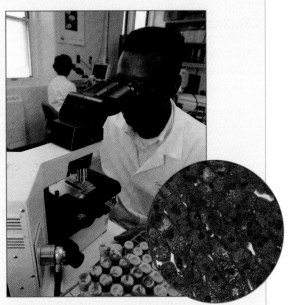

Figure 3-17 *A cytotechnologist looks through a microscope to examine cells like a liver cell.*

**Look at the liver cell in Figure 3-17. The parts of the cell that Ron might study are labeled A and B. Answer the questions that follow.**

1. Which two parts of the cell does Ron Brown examine? the nucleus and the cytoplasm

2. How are each of those parts labeled in the diagram? The nucleus is labeled *A* and the cytoplasm is labeled *B.*

3. What other cell part can you see? Possible answer: the cell membrane

**Critical Thinking**

They look for clues that indicate diseases in cells, and they must pay careful attention to detail.

Critical Thinking

How are cytotechnologists like detectives?

## Summary

- Cells are the basic building blocks of life. Some parts of a cell include the cell membrane, the nucleus, cytoplasm, mitochondria, vacuoles, ribosomes, the endoplasmic reticulum, Golgi bodies, and lysosomes.

- Plant cells differ from animal cells in three important ways. Plant cells have chloroplasts. Plant cells have a cell wall. Plant cells have fewer but larger vacuoles than animal cells do.

- A group of cells that work together to do a certain job is called a tissue. A group of tissues that work together to do a certain job is called an organ.

- Cells can reproduce by dividing in two. This process is called mitosis.

- Meiosis is a type of cellular reproduction that produces sex cells.

cell wall

chloroplast

cytoplasm

mitochondria

mitosis

nucleus

vacuoles

## Vocabulary Review

**Complete each sentence with a term from the list.**

**1.** A jellylike substance that fills a cell is called _____. cytoplasm

**2.** A part in a plant cell that uses sunlight to make food is called a _____. chloroplast

**3.** The _____ is the part of a cell that controls all the other parts. nucleus

**4.** Structures in a cell that release most of a cell's energy are called _____. mitochondria

**5.** The _____ is the thick, stiff covering around plant cells. cell wall

**6.** A structure in a cell that stores food, water, or wastes is called a _____. vacuole

**7.** The division of the nucleus of a cell is called _____. mitosis

# Chapter Quiz

**Answer the questions in complete sentences.**

**Test Tip**
Always reread the questions and your answers at the end of a test if you finish early. Many times you may know the answer, but rushing may cause you to make a mistake.

1. Who first used the word *cell?*  Robert Hooke

2. What is the thin, protective covering around a cell called?  the cell membrane

3. What is the job of the nucleus?  It controls most of the activities that take place in the cell.

4. What is the function of chromosomes?  Possible answer: Chromosomes pass on traits from the parent cell to the daughter cells during cell reproduction.

5. What do chloroplasts in plant cells do?
They help plants make food from light.

6. Why do plants need large vacuoles?
to hold water

7. What is another word for a group of cells that work together to do a certain job?  tissue

8. Why are daughter cells produced by mitosis  Because they both receive exactly like the parent cell?  the same set of chromosomes from the parent cell

9. What cell part makes proteins for the cell?
ribosomes

10. What kind of cells does meiosis produce?
sex cells

**Critical Thinking**

11. because animals do not have chloroplasts, they cannot make their own food

12. Mitosis is the process by which cells reproduce. When cells reproduce, they increase in number. When the cells in an organism increase in number, the organism grows.

## CRITICAL THINKING

11. Why are animals not able to make food the way that plants do?

12. Explain how mitosis helps an organism to grow.

*SCi*LINKS
Go online to www.scilinks.org. Enter the code **PMB104** to research **cell features**.

**Research Project**

Lysosomes are cell parts. Use the Internet and other resources to find out more about lysosomes. Write a paragraph that describes what kinds of substances lysosomes break down and what kinds of substances are contained inside lysosomes.

See the *Classroom Resource Binder* for a scoring rubric for the Research Project.

Figure 4-1 *Humpback whales need a lot of energy to leap into the air. Energy comes from the food that they eat. How much food do you think a humpback whale can eat in one day?*

between 2,000 to 9,000 pounds (900 to 4,050 kilograms) of food

## Learning Objectives

- Explain how organisms get energy from food.
- Explain how materials move in and out of a cell.
- Describe cellular respiration and photosynthesis.
- Compare cellular respiration and photosynthesis.
- LAB ACTIVITY: Get chlorophyll from spinach.
- BIOLOGY IN YOUR LIFE: Relate diet and Calories.

# Chapter 4 ▷ Cells and Energy

## Words to Know

| | |
|---|---|
| **ATP** | a form of chemical energy that can be used by a cell to carry out life processes |
| **diffusion** | the movement of a substance from an area that is crowded to an area that is less crowded |
| **semipermeable** | having tiny pores or holes so that only certain molecules can pass through |
| **osmosis** | (ahs-MOH-sihs) the movement of water through a cell membrane |
| **passive transport** | the movement of materials through a cell membrane without the use of energy |
| **active transport** | the movement of materials through a cell membrane using energy |
| **cellular respiration** | the process by which cells release energy from food |
| **photosynthesis** | the process by which plant cells use carbon dioxide, water, and light energy to make food |
| **chlorophyll** | a substance in chloroplasts that absorbs light energy |

## Amazing Energy

Consider these facts. A peregrine (PER-uh-grihn) falcon can fly up to 200 miles (about 320 kilometers) per hour. Hummingbirds can beat their wings up to 90 times per second.

How do these animals get the energy to do such amazing things? How do you get the energy to do all the things that you do? The answer is that many cells get energy from food. Your cells get energy from the food that you eat. In this chapter, you will learn how animal cells and plant cells get energy.

**Linking Prior Knowledge**
Students should recall the characteristics of life (Chapter 2) and the parts of cells (Chapter 3).

**Remember**
Cells are the basic building blocks of life.

## All Living Things Need Energy

When you kick a ball, you are using energy. When you talk, you are using energy. All the activities you do require energy. Energy is the ability to do work or to cause change.

Your body needs a constant supply of energy. Even when you are asleep, your body uses energy. Your heart beats. Your lungs take in oxygen. Your blood moves throughout your body.

You have learned that all organisms are made up of one or more cells. Cells are the building blocks of life. They carry out life functions for all organisms. Cells can move, reproduce, use energy, grow, and respond to the environment. In order to carry out these life functions, cells must have energy.

## Energy

### Science Fact

ATP stands for adenosine *tri*phosphate. It contains three phosphates. ADP stands for adenosine *di*phosphate. It contains two phosphates.

Energy can be found in many forms. Energy from the Sun is in the form of light energy. Energy in food is in the form of chemical energy.

Think about the foods that you eat. Once you eat something, it is broken down into molecules. Some of these molecules are in the form of sugar. These sugar molecules can be absorbed by your cells.

The sugar molecules contain energy. Your cells transfer the energy in the sugar molecules to a make a special compound. This compound is called **ATP**. ATP is a form of chemical energy. The energy in ATP can be used by a cell to carry out life processes.

Cells can breakdown ATP quickly to release the energy. ATP breaks down to ADP. The formula for this is: ATP → ADP + P + energy.

## How Materials Enter and Leave Cells

Cells need to absorb certain materials to get energy. Cells must also be able to get rid of waste materials. How do these materials get into and out of a cell?

Molecules of a substance move from an area that is crowded to an area that is less crowded. This movement is called **diffusion**.

The cell membrane is **semipermeable.** There are tiny pores or holes in it that allow some molecules to diffuse into the cell. For example, oxygen molecules pass into your blood cells through the membranes of cells in your lungs.

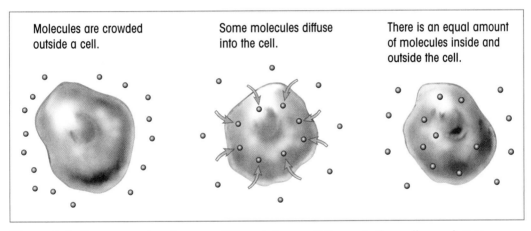

Figure 4-2 *Oxygen molecules can diffuse into a cell through the cell membrane.*

The food that you eat must be broken down into tiny molecules before it can enter your cells. Suppose you eat a plate of spaghetti. The spaghetti is broken down into nutrients in your body. These nutrients are now small enough to pass through the membranes of your body cells. Some of these nutrients are used as an energy source.

# Osmosis

### Science Fact

Osmosis maintains the balance of water inside and outside of a cell.

A cell is made up mostly of water and dissolved substances. Water can move into or out of a cell through the cell membrane. The diffusion of water through a cell membrane is called **osmosis**.

The direction that water moves depends on the amount of dissolved substances inside and outside of a cell. If there are more dissolved substances inside the cell, water will move into the cell.

**Remember**

Plant cells store water in vacuoles.

## Osmosis in Plant Cells

Plants whose cells do not have enough water appear dry and may droop. However, once the plant receives water, its cells may absorb the water through osmosis. Once water is absorbed into its cells, the plant swells. It will probably stand up straight and appear healthy. Figure 4-3 shows the effect of osmosis on a plant.

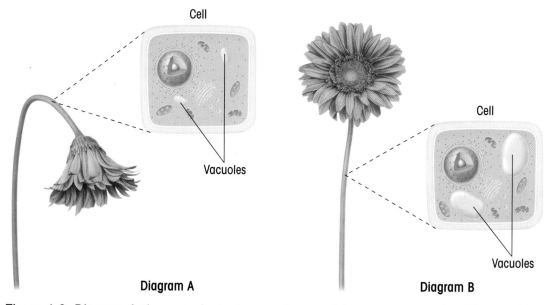

Cell

Vacuoles

Diagram A

Cell

Vacuoles

Diagram B

**Figure 4-3** *Diagram A shows a plant whose cells do not have enough water. Diagram B shows the same plant after its cells have absorbed water through osmosis.*

## Passive Transport and Active Transport

The diffusion of certain molecules through the cell membrane occurs passively, or without help. It does not require the cell to use any energy. This type of movement is called **passive transport**. Osmosis takes place through passive transport.

However, there is movement through the cell membrane that is the opposite of diffusion. The molecules of certain substances move through the cell membrane from an area that is less crowded to an area that is *more* crowded. Energy is needed to move these materials across the cell membrane. This type of movement is called **active transport**.

Active transport allows certain minerals to enter a plant's roots. There are more minerals in the plant's roots than in the soil. Cells in the plant's roots have to use energy in order to move more minerals into its cells.

Molecules of certain materials that help make proteins move into a cell through active transport, too. Active transport is also needed to move certain waste materials out of a cell.

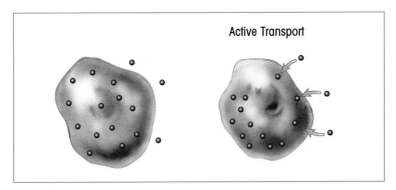

Figure 4-4 *During active transport, energy is needed to move molecules through the cell membrane from an area that is less crowded to an area that is more crowded.*

## Cellular Respiration

In most organisms, two things are needed for cells to get energy: food and oxygen. While you are eating that plate of spaghetti, you are also breathing. Breathing gets oxygen into your body.

The mitochondria are where cellular respiration takes place. Cellular respiration releases energy.

Many foods break down into sugar molecules. The mitochondria in cells use oxygen to release energy stored in sugar molecules. The energy in sugar molecules is converted into usable energy. This energy is ATP. The process by which cells release energy from food is called **cellular respiration**.

Cellular respiration releases energy. Water and carbon dioxide are also produced in the process. Carbon dioxide is a gas. Water and carbon dioxide pass out of your cells as byproducts of cellular respiration.

**Cellular Respiration**

Oxygen + Sugar ⟶ ATP energy + Water + Carbon dioxide

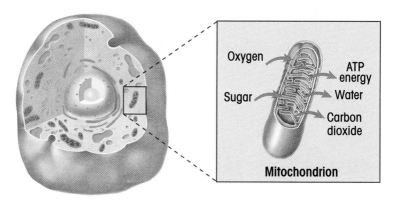

Figure 4-5 *Cellular respiration takes place in the mitochondria of cells.*

Some people confuse breathing and cellular respiration. Breathing is the way you take air into your lungs and let it out. Cellular respiration is the way your cells change food into usable energy called ATP.

Write your answers on a separate sheet of paper. Use complete sentences.

1. What do cells need in order to carry out life functions?

2. What is diffusion?

3. Where in the cell does cellular respiration take place?

4. **CRITICAL THINKING** Why does the energy in sugar need to be converted into ATP?

## Photosynthesis

Cellular respiration releases energy from food molecules forming ATP. Animals have to eat other organisms to get food. However, plants get food in a different way. They make their own food using carbon dioxide, water, and energy from sunlight. This process is called **photosynthesis**.

In Chapter 3, you read that plant cells have special parts that animal cells do not have. One of these parts is chloroplasts. Chloroplasts contain a green substance called **chlorophyll**. Chlorophyll absorbs energy from a light source, such as sunlight.

Carbon dioxide is a gas that passes into plant cells. Water passes through the cell membranes of plant cells in the roots. Energy from sunlight, absorbed by chlorophyll, helps the water molecules and carbon dioxide molecules to make food for the plant. This food is a type of sugar. This sugar is stored until it is needed for cellular respiration.

**Science Fact**

An active cell requires about two million molecules of ATP per second to perform its life functions.

Photosynthesis also produces oxygen as a byproduct. The oxygen passes out of plant cells and into air.

**Photosynthesis**
Light energy + Water + Carbon dioxide $\longrightarrow$ Oxygen + Sugar

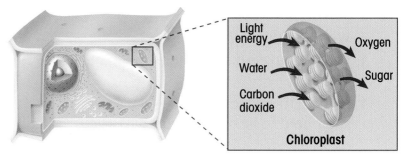

Figure 4-6 *Photosynthesis takes place in the chloroplasts of plant cells.*

Once plant cells have made their own food, they carry out cellular respiration just like animal cells. The mitochondria use oxygen to release the energy in food molecules and form ATP.

## Cellular Respiration and Photosynthesis

Compare cellular respiration and photosynthesis.

Cellular Respiration
Oxygen + Sugar $\rightarrow$ ATP energy + Water + Carbon dioxide

Photosynthesis
Light energy + Water + Carbon dioxide $\rightarrow$ Oxygen + Sugar

Do you notice something about these two processes? They are related to each other. Cellular respiration uses oxygen and sugar to release energy. Water and carbon dioxide are given off as byproducts. Photosynthesis uses energy, carbon dioxide, and water to make oxygen and sugar.

## Check Your Understanding

Write your answers in complete sentences.

1. What three things are needed by plant cells during photosynthesis?

2. Why do plants need to go through the process of photosynthesis?

3. What two things are produced as a result of photosynthesis?

4. **CRITICAL THINKING** What is the relationship between cellular respiration and photosynthesis?

## A Closer Look

### PHOTOSYNTHESIS AND LIFE

Some scientists have a theory about the first cells that appeared on Earth about 3.5 billion years ago. Earth was very different back then. The air was filled with deadly gases. There was no oxygen to breathe. These early cells did not need oxygen. They also did not make their own food. Instead, the materials around them may have provided them with food.

Figure 4-7 *Many different life-forms may have appeared on Earth due to the oxygen in air.*

Over time, some cells were able to start making their own food. These cells used photosynthesis to help make their food. Photosynthesis not only helped these cells make food, it also put oxygen in the air.

Scientists have evidence that because there was oxygen in the air, more life-forms began to appear. Eventually, with the large amount of oxygen in the air, many kinds of life-forms appeared on Earth.

**CRITICAL THINKING** Why do you think photosynthesis might have helped different life-forms to appear on Earth? Photosynthesis put oxygen into the air. Different life-forms need oxygen to breathe.

# LAB ACTIVITY
## Getting Chlorophyll From Spinach

### BACKGROUND
The chloroplasts inside plant cells contain chlorophyll. Chlorophyll is a green coloring.

### PURPOSE
In this activity, you will prove that chlorophyll is found in green plants.

### MATERIALS
Safety goggles, filter paper, scissors, baby-food jar with lid, spinach leaves, plastic spoons, rubbing alcohol, paper, pencil

Figure 4-8 *Carefully place the filter paper in the jar.*

### WHAT TO DO
1. Cut a 1-inch (2.5-cm) wide and 3-inch (7.5-cm) long strip of filter paper.

2. Fold the strip one-half inch (1.25-cm) from one end. This end is the top end. The top part forms a flap. Tape the flap to the bottom surface of the jar's lid.

4. Place four to five spinach leaves inside the jar.

5. Put on safety goggles. You can take them off after Step 7.

6. Add 3 to 4 tablespoons of alcohol to the jar. Use a plastic spoon to gently mash the spinach leaves.

7. Holding the lid of the jar, carefully place the filter paper into the jar. Make sure that the filter paper stands upright in the jar. The bottom end of the filter paper should touch the liquid in the jar. Turn the lid to tighten it.

8. Let the filter paper stand in the jar for at least half an hour.

9. After half an hour, observe the filter paper.

### DRAW CONCLUSIONS
- Describe what appeared on your filter paper. Possible answer: A green line appeared from the bottom of the paper.
- Why did you mash the spinach leaves? Possible answer: to get the chlorophyll out of the plant cells

# BIOLOGY IN YOUR LIFE
## Counting Calories

A Calorie is a measure of food energy. Some foods supply more energy than others do. A glass of orange juice has about 110 Calories. Twenty french fries have about 300 Calories.

Eating well is more than just counting Calories. You need to eat a balanced diet. A balanced diet has lots of grains, fruits, and vegetables. It also includes the right amount of protein and dairy foods. Foods high in protein include meat, eggs, beans, and fish. Dairy foods include milk, cheese, and yogurt. Foods with fat and oils, such as french fries and potato chips, should be eaten in small amounts.

Figure 4-10 *Food gives you energy. The energy in food is measured in Calories.*

Inform students that one Calorie is actually 1000 calories. One calorie is the amount of energy needed to raise the temperature of one gram of water one degree Celsius.

The following table contains the recommended number of Calories that active people of average size should eat in one day. Study the table and then, answer the questions.

| Number of Calories in One Day | |
| --- | --- |
| Men | 2,300–3,000 Calories |
| Women | 1,900–2,200 Calories |
| Teen boys | About 2,800 Calories |
| Teen girls | About 2,200 Calories |

Figure 4-9 *Calorie chart*

**1.** About how many Calories should you have each day? about 2,800 for teen boys and about 2,200 for teen girls
**2.** Do you eat the right amount of Calories each day? How could you find out?
Students could keep a food diary and count the Calories they consume.

**Critical Thinking**
Possible answers: eat low-Calorie foods during the day; exercise more that day; eat small amounts of high-Calorie foods.

**Critical Thinking**
Eating large meals is very common during holidays. How can you make sure that you do not take in more Calories than your body needs?

## Summary

- All living things need energy. All living things get energy from food.

- Many materials enter and leave cells through diffusion. Osmosis is the diffusion of water through a cell membrane.

- Cellular respiration is the process in which cells obtain energy from food. ATP is a compound in a cell that releases energy. Cellular respiration takes place in the mitochondria of cells.

- Plants make their own food through photosynthesis. Plants need carbon dioxide, water, and energy from light to make their own food.

- Cellular respiration and photosynthesis are related processes.

active transport

cellular respiration

diffusion

osmosis

passive transport

photosynthesis

semipermeable

## Vocabulary Review

**Complete each sentence with a term from the list.**

1. The movement of materials through a cell membrane without the use of energy is called _____. passive transport

2. The process by which plants use carbon dioxide, water, and light energy to make food is called _____. photosynthesis

3. The movement of a substance from an an area that is crowded to an area that is less crowded is called _____. diffusion

4. The movement of water through a cell membrane is called _____. osmosis

5. The cell membrane is _____ because it has tiny pores that only certain molecules can pass through. semipermeable

6. The process by which cells obtain energy from food is called _____. cellular respiration

7. The movement of materials through a cell membrane using energy is called _____. active transport

Research Project  Students' reports may include the following information: glucose is the fuel of all living things; the production of glucose begins in plants through photosynthesis; sucrose is commonly called "sugar"; sucrose is the main sugar in most sweeteners.

# Chapter Quiz

**Answer the questions in complete sentences.**

**Test Tip**
Always try to write answers in complete sentences. It will help improve the quality of your thinking as well as that of your writing.

1.  Where does energy for a cell come from? food, sugar, or ATP

2.  What form of energy is found in food? chemical energy

3.  What two things do cells need in order to go through the process of cellular respiration? food and oxygen

4.  How is breathing different from cellular respiration?

4. Breathing is the way that you take in air into your lungs and let it out. Cellular respiration is the way that your cells change food into usable energy.

5.  During photosynthesis, what green substance absorbs energy from the Sun? chlorophyll

6.  What are the products of cellular respiration? What are the products of photosynthesis?

6. ATP energy, carbon dioxide, and water; sugar and oxygen

7.  What type of diffusion takes place when a plant cell absorbs water? osmosis

8.  What structures in the cell release energy from food? the mitochondria

**CRITICAL THINKING**

9.  How do you know that your body uses energy while you are sleeping?

9. Possible answers: because my heart beats while I am sleeping; because my lungs take in oxygen when I am sleeping; because blood moves through my body when I am sleeping

10. How do animals and plants differ in the way that they get energy?
    Animals eat food. Plants make their own food.

## Research Project

Sugars are carbohydrates. Two types of sugars are glucose and sucrose. Use the Internet and other resources to find out more about these types of sugar. Write a report that tells some of the differences between glucose and sucrose.

Go online to www.scilinks.org. Enter the code **PMB106** to research **carbohydrates**.

See the *Classroom Resource Binder* for a scoring rubric for the Research Project.

**Figure 5-1** *Parents pass certain traits to their offspring. Look at the photograph of wolves above. What traits have been passed from the parent wolf to its offspring?*

Possible answer: the dark outline around the wolves' ears

## Learning Objectives

- Describe how the study of genetics began.

- Define *heredity* and explain how genes control inherited traits.

- Explain how gene combinations can be predicted using a Punnett square.

- Explain how a DNA molecule is replicated.

- Give examples of genetic engineering.

- **LAB ACTIVITY:** Calculate how chance plays a role in genetics.

- **ON-THE-JOB BIOLOGY:** Learn about the job of a rose breeder.

# Chapter 5 Heredity and Genetics

## Words to Know

| | |
|---|---|
| **trait** | an inherited characteristic |
| **heredity** | the passing of traits from parents to offspring |
| **genetics** | the study of heredity |
| **gene** | a factor that controls inherited traits |
| **dominant** | describes a gene or trait that will always show itself |
| **recessive** | describes a gene or trait that will be hidden when the dominant gene or trait is present |
| **homozygous** | having two like genes for the same trait |
| **heterozygous** | having two unlike genes for the same trait |
| **genotype** | the gene combination for a given trait in an organism |
| **phenotype** | the appearance of a trait in an organism |
| **Punnett square** | a chart that shows possible gene combinations |
| **DNA** | the large molecules in a cell that store genetic information |
| **RNA** | a molecule in a cell used in the making of proteins |
| **pedigree** | a chart that shows how a certain trait is passed from generation to generation |
| **mutation** | a change in a gene |
| **genetic engineering** | the methods used to change the genetic information in the DNA of cells or organisms |

## Linking Prior Knowledge

Students should recall that genes contain hereditary information (Ch. 3). They should also recall the role of chromosomes during meiosis (Ch. 3).

## Heredity

**Everyday Science**

Look in a mirror. On a sheet of paper, list all of the traits that you see. Can you think of some traits that cannot be seen?

Possible answer: blood type

You are a special organism. No other organism has all of your characteristics or **traits**. However, where did you get your traits from? Your parents, of course! Parents pass many of their own traits to their children, or offspring. The color of your hair, the size of your feet, and the shape of your nose are some of your traits.

The passing of traits from parents to offspring is called **heredity**. The study of heredity is called **genetics**. Biologists who study heredity are called geneticists.

## The History of Genetics

One of the first people to study genetics was an Austrian monk named Gregor Johann Mendel. Mendel studied traits in pea plants. He grew pea plants that had different traits from one another. Some of his pea plants were short. Some were tall. Some had white flowers, and some had purple flowers. Some of his pea plants produced round seeds, and others produced wrinkled seeds.

Mendel crossbred the pea plants. In crossbreeding the plants, Mendel had different kinds of pea plants reproduce together. He crossbred tall plants with short plants. He crossbred plants that had white flowers with plants that had purple flowers. He crossbred plants that had round seeds with plants that had wrinkled seeds.

In his first crosses, Mendel found that only one of the two traits appeared in the offspring plants. For example, when he crossbred tall pea plants with short pea plants, he always got tall pea plants. After his first crosses, Mendel took the offspring plants and crossed them. In these second crosses, he found that both traits appeared again.

Figure 5-2 *Gregor Johann Mendel studied traits in pea plants.*

## Dominant and Recessive Traits

From his experiments, Mendel discovered that traits from one parent may hide traits from the other parent. He found that certain traits would always show themselves over other traits. These traits he called **dominant**. The hidden or masked traits he called **recessive**. Some of the results from Mendel's experiments can be seen in Figure 5-3.

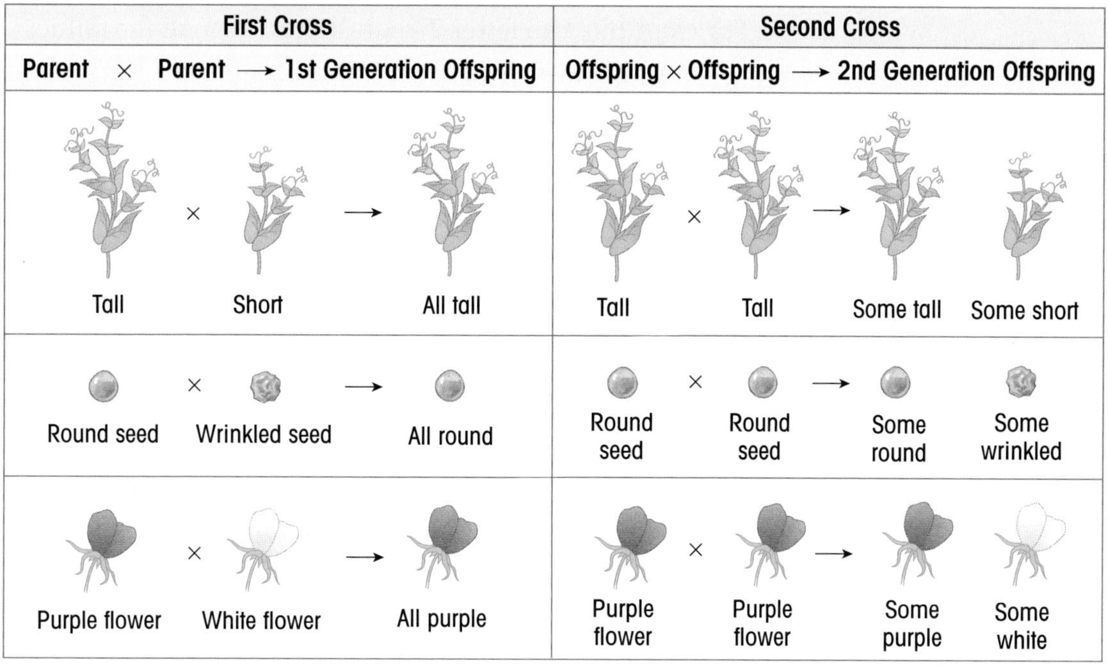

| First Cross | | Second Cross | | | |
|---|---|---|---|---|---|
| **Parent × Parent ⟶ 1st Generation Offspring** | | **Offspring × Offspring ⟶ 2nd Generation Offspring** | | | |
| Tall | Short | All tall | Tall | Tall | Some tall  Some short |
| Round seed | Wrinkled seed | All round | Round seed | Round seed | Some round  Some wrinkled |
| Purple flower | White flower | All purple | Purple flower | Purple flower | Some purple  Some white |

Figure 5-3 *Mendel crossbred different types of pea plants. He found dominant traits and recessive traits.*

## Genes

Mendel used the word *traits* when he described his pea plants. He did not know that traits were controlled by genes. **Genes** contain information about inherited traits. Dominant traits are controlled by dominant genes. Recessive traits are controlled by recessive genes. Genes come in pairs. Half of each pair comes from the mother and the other half from the father. It is the combination of genes that determines a specific trait.

## Representing Genes

### Science Fact

A tall pea plant is about 7 to 8 feet tall (2 to 2.5 meters). A short pea plant is about 1 foot tall (0.3 meter).

In pea plants, the trait for tallness is dominant over the trait for shortness. Purple flowers are dominant over white flowers. Round seeds are dominant over wrinkled ones.

Letters are used to represent the traits carried on genes. A letter represents a trait. If the gene that controls the trait is dominant, the letter is written in uppercase. For example, the letter *T* represents the gene for tallness. If the gene is recessive, the letter is written in lowercase. The letter *t* represents the gene for shortness.

### Two Like Genes

Organisms that have two like genes for a given trait are called **homozygous**. When Mendel crossed certain tall pea plants, the offspring were always tall. The offspring received two dominant genes for tallness from the parent plants. Mendel called these offspring homozygous tall. The symbol for these offspring is *TT*.

Mendel also found that when he crossed two short plants, the offspring were always short. The short offspring received two recessive genes from the parent plants. Mendel called these offspring plants homozygous short. The symbol for these offspring is *tt*. Figures 5-4 and 5-5 show the homozygous plants that Mendel crossed.

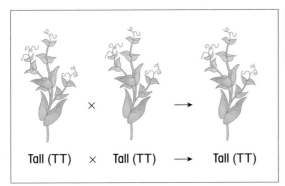

Tall (TT)  ×  Tall (TT)  ⟶  Tall (TT)

**Figure 5-4** *This cross resulted in homozygous tall offspring (TT).*

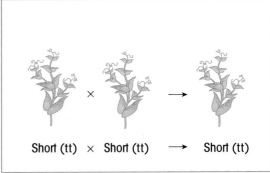

Short (tt)  ×  Short (tt)  ⟶  Short (tt)

**Figure 5-5** *This cross resulted in homozygous short offspring (tt).*

## Two Unlike Genes

Mendel wanted to know what the offspring of two different homozygous plants would be like. He crossbred homozygous tall plants *(TT)* with homozygous short plants *(tt)*. All of the offspring were tall!

All of the offspring had received a tall gene *(T)* from one parent and a short gene *(t)* from the other parent. Organisms that have two unlike genes for a given trait are called **heterozygous**. Because the tall gene was dominant, all of the offspring were tall. The symbol for these offspring is *Tt*. These offspring are called heterozygous tall.

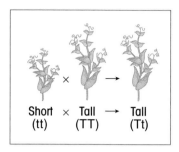

Short  ×  Tall  →  Tall
(tt)      (TT)      (Tt)

**Figure 5-6** *This cross resulted in a heterozygous tall offspring (Tt).*

## Genotype and Phenotype

The genetic combination, or gene pair, for a certain trait in an organism is called the **genotype**. The appearance of a trait in an organism is called the **phenotype**. A homozygous tall plant's genotype is *TT*. Its phenotype is tall.

Organisms with different genotypes may have the same phenotype. For example, a homozygous tall plant *(TT)* and a heterozygous tall plant *(Tt)* have different genotypes. However, they have the same phenotype, which is tall.

✓ **Check Your Understanding**

Write your answers on a separate sheet of paper. Use complete sentences.

**1.** What is heredity?

**2.** Describe one way in which Mendel crossbred pea plants.

**3.** What is a dominant trait?

**4.** CRITICAL THINKING What is the genotype of a heterozygous tall plant?

**Check Your Understanding**
**1.** the passing of traits from parents to offspring
**2.** Possible answers: Mendel crossbred tall pea plants with short pea plants. Mendel crossbred purple-flowered pea plants with white-flowered pea plants. Mendel crossbred round-seeded pea plants with wrinkled-seeded pea plants.
**3.** a trait that always shows itself over another trait
**4.** *Tt*

# Punnett Squares

A **Punnett square** is a chart that shows possible gene combinations. The possible gene combinations in the offspring of two organisms can be predicted using a Punnett square.

### How to Use a Punnett Square

The results of Mendel's experiments with two homozygous plants can be shown using a Punnett square. Look at Figure 5-7 and follow the steps listed below.

**STEP 1** Draw a chart with four boxes.

**STEP 2** Above the two boxes going across, write the symbol for a homozygous tall plant *(TT)* with one letter over each box.

**STEP 3** Write the symbol for a homozygous short plant *(tt)* going down the left side with one letter next to each box.

**STEP 4** Fill in the left side of each box with the letter at the top of the column *(T)*.

**STEP 5** Fill in the right side of each box with the letter at the left of the row *(t)*.

You should see that all of the boxes are heterozygous tall *(Tt)*.

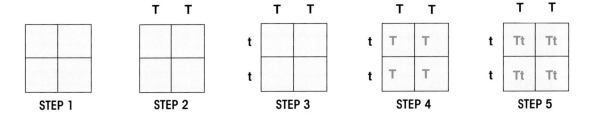

Figure 5-7 *Follow these steps to make a Punnett square.*

## Predicting Gene Combinations

The combination of genes from organisms with heterozygous genes can also be shown in a Punnett square. Figure 5-8 shows the possible gene combinations when crossing two heterozygous tall pea plants *(Tt)*.

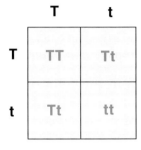

|   | T | t |
|---|---|---|
| **T** | TT | Tt |
| **t** | Tt | tt |

*Figure 5-8 The genotypes for tall offspring and short offspring can be seen in this Punnett square.*

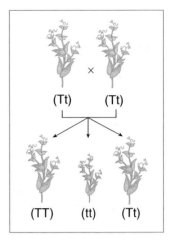

(Tt)    (Tt)

(TT)    (tt)    (Tt)

*Figure 5-9 Two heterozygous tall pea plants can produce tall offspring and short offspring.*

A Punnett square can help you predict the chances that a certain combination may occur. The chance of a certain combination occurring is called *probability*. The Punnett square in Figure 5-8 shows that there is a 1 out of 4 chance that the offspring will be homozygous tall *(TT)*. It also shows a 1 out of 4 chance that the offspring will be homozygous short *(tt)*. There is also a 2 out of 4 chance that each offspring will be heterozygous tall *(Tt)*.

You can show probability as a percent by using this formula:

$$\frac{\text{Number of times an event occurs}}{\text{Number of total possible events}} \times 100 = \text{Probability \%}$$

For example the probability for heterozygous tall *(Tt)* is:

$$\frac{2}{4} \times 100 = \frac{2}{4} \times \frac{100}{1} = \frac{200}{4} = 50\%$$

**Remember**

$\frac{200}{4}$ is the same as $200 \div 4$.

$1 \div 4$ is the same as $\frac{1}{4}$.

The Punnett square also shows that there is a 25% chance $(1 \div 4 \times 100)$ that the offspring will be homozygous tall *(TT)* or homozygous short *(tt)*.

# More Complex Patterns of Heredity

The way genes control traits can show complex patterns of heredity. Some genes do not follow the patterns that Mendel found. These genes interact in different ways.

## Incomplete Dominance

When one gene for a certain trait is not completely dominant over the other gene, a blending effect occurs. This pattern of heredity is called *incomplete dominance*.

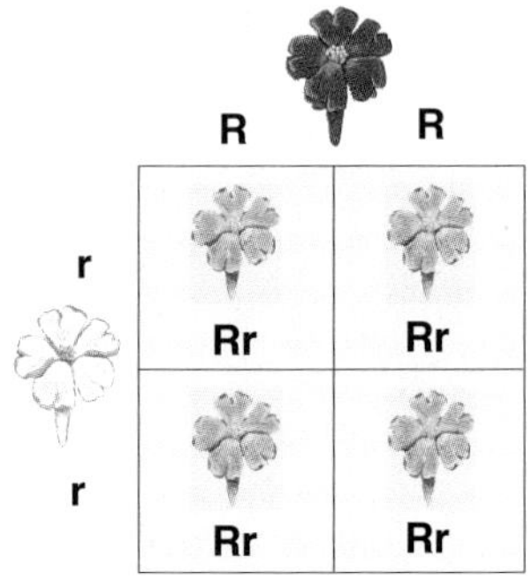

Figure 5-10 *Incomplete dominance in four o'clock plants produces pink flowers.*

Incomplete dominance can be seen in the color of the flowers in four o'clock plants. Four o'clock plants can have red flowers, white flowers, or pink flowers. Red flowers are homozygous *(RR)*. White flowers are homozygous *(rr)*. Look at Figure 5-10. When four o'clocks that have red flowers are crossed with four o'clocks that have white flowers, a blending effect occurs. Heterozygous pink flowers *(Rr)* are produced.

## Codominance

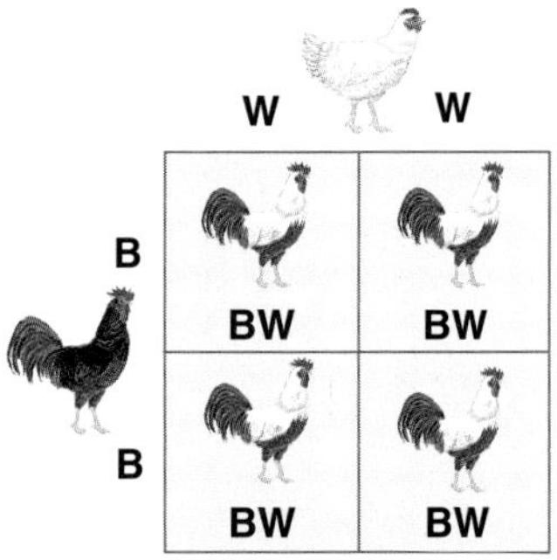

Another pattern of heredity can occur when two dominant genes are present for a certain trait. For example, when a homozygous chicken with black feathers *(BB)* and a homozygous chicken with white feathers *(WW)* reproduce, the offspring is heterozygous *(BW)*. Look at Figure 5-11. It shows that offspring will have some black feathers and some white feathers. This pattern of heredity is called *codominance*.

Figure 5-11 *Codominance in Erminette chickens produces black feathers and white feathers.*

## Multiple Alleles and Polygenic Traits

Many traits are controlled by one gene that has more than two possible variations. These traits are controlled by *multiple alleles*. Human blood groups are controlled by multiple alleles.

Some traits are controlled by many genes interacting with one another. This type of trait is called a *polygenic trait*. Many human traits, such as height and skin color, are caused by many genes working together.

Write your answers in complete sentences.

**1.** How does a Punnett square show gene combinations?

**2.** How are pink four o'clock flowers examples of incomplete dominance?

# DNA

An organism's genetic information is stored in large molecules. These large molecules are called **DNA.** DNA stands for *deoxyribonucleic acid.*

A DNA molecule looks like a twisted ladder. The "steps" of the ladder are made up of smaller molecules called bases. There are four kinds of bases. These bases are adenine, thymine, guanine, and cytosine. They are often labeled *A, T, G,* and *C.* Each base always pairs with another base to form the steps of the ladder. Base A always pairs with base T. Base C always pairs with base G.

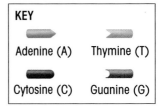

**KEY**

Adenine (A)    Thymine (T)

Cytosine (C)    Guanine (G)

A DNA molecule can have millions of steps. These steps are made up of different combinations of the base pairs. The number and pattern of steps form a genetic "code" that makes up genes. This genetic code controls the activities in the cell. It directs how the cell functions and how it divides.

## Replication

During mitosis, a cell produces new daughter cells. However, before a cell goes through mitosis, a DNA molecule must make a copy of itself. This process is called *replication.* The base pairs that make up the DNA ladder break apart. Each of these bases will pair up with another base. Two new molecules of DNA identical to the first DNA molecule will be formed.

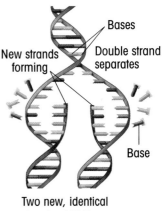

Bases

New strands forming

Double strand separates

Base

Two new, identical strands of DNA

**Figure 5-12** *During replication, two new molecules of DNA are formed.*

## Making Proteins

**Remember**
Proteins help build and repair living cells.

DNA molecules direct the formation of proteins. DNA molecules are found in the nucleus of a cell. Proteins are made by the ribosomes of cells. A special molecule is needed to carry the information from the nucleus to ribosomes. That molecule is called **RNA**.

RNA differs from DNA in several ways. An RNA molecule has only one side of the ladder shape. Also, RNA has the bases A, G, and C. However, it has a base U, or uracil, instead of base T. When forming base pairs, base A pairs with base U.

Proteins are made up of special molecules called amino acids. RNA carries the information from DNA to a ribosome. The ribosome reads the information and puts together the correct amino acids to make proteins.

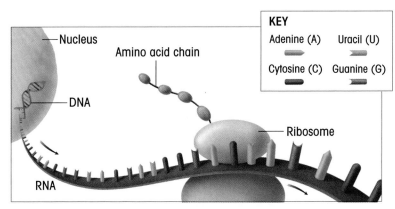

Figure 5-13 *RNA carries the information from DNA to a ribosome. The ribosome puts amino acids together to make proteins.*

## Science Fact

Different species have a different number of chromosomes in their body cells. Dogs have 78 chromosomes. Cats have 38 chromosomes. Carrots have 18 chromosomes.

## Reproductive Cells and Chromosomes

During meiosis, daughter cells, called sex cells, are produced from a parent cell. Sex cells have only half the number of chromosomes that other body cells have. For example, in human sex cells, there are only 23 chromosomes. Human body cells normally have 46 chromosomes.

Sex cells need to have half the normal number of chromosomes. During sexual reproduction, a male sex cell and a female sex cell join. Each parent gives only half of its chromosomes to the offspring. This way the offspring has the exact number of chromosomes it should have.

The 46 chromosomes in humans are arranged in 23 pairs. Scientists can show pairs of chromosomes in an organized display called a *karyotype*. Figure 5-14 shows a karyotype for a human. Notice that there are 23 pairs of chromosomes. Each pair contains one chromosome from the male parent and one chromosome from the female parent.

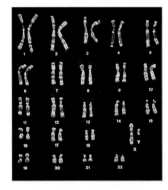

Figure 5-14 *A karyotype for a human*

### Sex Chromosomes

The gender, or sex, of an organism is determined by one pair of chromosomes. This pair is made up of sex chromosomes. There are two kinds of sex chromosomes. The larger sex chromosome is called the *X chromosome*. The smaller chromosome is called the *Y chromosome*.

During meiosis, each sex cell receives only one chromosome for gender from each parent cell. A human female sex cell always contains an *X* chromosome. A human male sex cell may contain an *X* chromosome or a *Y* chromosome. During fertilization, when the male sex cell and female sex cell join, the gender of the offspring is determined. Two *X* chromosomes *(XX)*, one from each parent, produce a female offspring. An *X* chromosome from the mother and a *Y* chromosome from the father *(XY)* produce a male offspring.

Figure 5-15 *Females have two X chromosomes.*

### Sex-Linked Disorders

The X chromosome and Y chromosome carry genes for other traits, not just gender. The *X* chromosome carries many genes. The *Y* chromosome carries few genes. The traits controlled by genes in the *X* and *Y* chromosomes are called *sex-linked traits*.

Figure 5-16 *Males have an X chromosome and a Y chromosome.*

|  | **X** | **X**$^c$ |
|---|---|---|
| **X** | Normal<br>**XX** | Carrier<br>**XX**$^c$ |
| **Y** | Normal<br>**XY** | Colorblind<br>**X**$^c$**Y** |

Figure 5-17
*Colorblindness, found on the X chromosome, can be passed to offspring.*

There are certain hereditary disorders that are sex linked. The genes that control them are on one of the sex chromosomes. Some of these disorders include colorblindness and night blindness. Most of the genes for sex-linked disorders are recessive. Most of them are found on the *X* chromosome.

Sex-linked disorders are more often found in men than in women. Women have two *X* chromosomes. A woman who has one recessive gene for a sex-linked disorder will not have the disorder. Her genotype is $XX^C$. The dominant gene on her other *X* chromosome masks the disorder. However, she can pass the recessive gene ($X^C$) for the disorder to her offspring. A woman who has one normal gene and one gene for a sex-linked disorder is called a carrier. Figure 5-17 shows the possible gene combinations that can occur when the mother is a carrier of colorblindness.

Men, however, have only one X chromosome. If a man has a gene that causes a disorder on his X chromosome, there is no dominant gene on his Y chromosome to mask the trait. He will have the disorder ($X^C Y$). This man will also pass the gene to all of his daughters. However, he will not pass the gene to any of his sons.

**Check Your Understanding**

**1.** by directing the production of certain proteins at certain times

**2.** During sexual reproduction, each parent cell gives half of its chromosomes to the offspring. This way the offspring has the exact number of chromosomes it should have.

**3.** a hereditary disorder whose gene is found on one of the sex chromosomes

**4.** because sex-linked disorders are usually found on the *X* chromosome; a man passes an *X* chromosome to his daughters but not to his sons.

✓ **Check Your Understanding**

Write your answers on a separate sheet of paper. Use complete sentences.

**1.** How does a DNA molecule control the activities of a cell?

**2.** Why do reproductive cells need to have half the normal number of chromosomes?

**3.** What is a sex-linked disorder?

**4.** **CRITICAL THINKING** Why can a man who has a sex-linked disorder pass the disorder to his daughters but not to his sons?

# Pedigrees and Human Genetic Disorders

Genetic traits can be traced in a family using a pedigree. A **pedigree** is a chart that shows how a certain trait is passed from generation to generation. Carriers of the trait can be identified as well as those family members who have the disorder.

Hemophilia is a sex-linked blood disorder. A person with hemophilia lacks a protein that helps blood to clot. A simple cut or bruise can be life threatening to a person with hemophilia.

Queen Victoria of England was known to be a carrier of the disorder. Figure 5-18 shows a pedigree that traces how Queen Victoria passed the disorder to her offspring. As you can see, two of her daughters were carriers and one of her sons had the disorder. Four of Queen Victoria's granddaughters were carriers and three of her grandsons had the disorder.

**PEDIGREE**

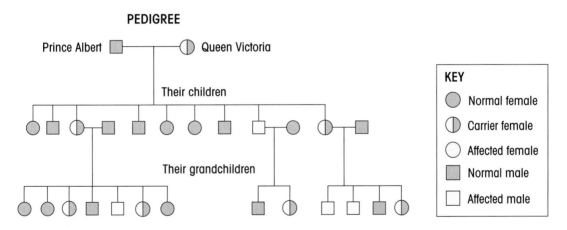

Figure 5-18 *This pedigree shows how hemophilia was passed from Queen Victoria to her family.*

## Mutations: Changes in Genes

Genes can change. A change in a gene is called a **mutation**. A mutation occurs if the order of the bases in a DNA molecule changes. If this change occurs, the genetic code of the cell changes. The activities in the cell also change.

Mutations can occur in a single gene or in an entire chromosome. They can be caused by conditions in the environment. Some examples include radiation from X-rays, ultraviolet radiation from the Sun, and radiation from nuclear energy. Mutations can also be caused by inhaling chemicals. Some of these chemicals can be found in cigarettes.

### Effects of Mutations

The chance that a mutation helps an organism is unlikely. Very often, a mutation does not have much of an effect on the organism at all. A mutation is harmful if it prevents an organism from carrying out its life functions.

*Figure 5-19 The extra foot of the frog is a mutation.*

## The Environment Can Affect Traits

The genetic code of an organism does not control everything about the organism. An organism's environment can also affect the organism's phenotype. For example, an apple tree may have genes to grow a lot of apples. However, a cold, dry spring may not allow the tree to grow any apples that year.

The same is true for people. Most human phenotypes are the result of genetic combinations, as well as environmental factors. Perhaps, a person has genes for being very tall. However, that person may eat poorly or may not get enough sleep. Then, that person may grow only to average height.

*Figure 5-20 The windy environment has caused the divi-divi tree to grow sideways.*

# Using Genetics

People have been experimenting with genetics for many years. Over time, farmers discovered that plants with certain traits could be used to produce more plants with those traits. Animals with traits such as speed or strength could be bred to produce offspring that had the same traits. The breeding of parent organisms to produce certain traits in the next generation is called *selective breeding.* Plants and animals have been bred to get the best traits for thousands of years.

### Inbreeding

Another way of producing organisms with certain traits is through inbreeding. *Inbreeding* is the crossing of closely related organisms. These organisms have certain similar traits. They are crossed so that their offspring will have these traits. Purebred dogs and purebred racehorses are produced through inbreeding.

There is a danger to inbreeding. Recessive genes may appear as more inbreeding occurs. Disorders may appear more often.

Figure 5-21 *A purebred bulldog is an example of inbreeding.*

### Hybridization

The breeding of two organisms that have different kinds of genes is called *hybridization.* Both parent organisms have useful traits. These traits are passed to the offspring. A mule is an example of a hybrid. It is the cross between a female horse and a male donkey.

Sometimes the offspring can show traits that are more useful than those traits of either parent. A certain type of plant is produced by crossing wheat and rye plants. The hybrid plant is more nutritious than either parent plant.

Figure 5-22 *A mule is an example of a hybrid.*

# Genetic Engineering

Selective breeding has been practiced to produce organisms that have certain characteristics for thousands of years. Today, however, scientists have a better understanding of DNA. New organisms can now be produced by directly changing genetic material. The methods used to produce new forms of DNA are called **genetic engineering**.

### Plasmids and Gene Splicing

In one type of genetic engineering, pieces of DNA that contain certain genes are removed from an organism. The genes are then inserted into the DNA of another organism. Chemical substances that cut a strand of DNA at a certain location are used to remove the genes. These substances are called *restriction enzymes*.

In addition to chromosomes, bacterial cells often contain small circle-shaped DNA molecules. These molecules are called *plasmids*. Plasmids can be removed from bacterial cells. They can be cut apart using restriction enzymes. The genes from another organism can be joined with the cut pieces of a plasmid. This process is called *gene splicing*.

A new plasmid forms. It contains a new combination of DNA. DNA formed from the DNA of different species is called *recombinant DNA*.

### Insertion of the New Plasmid

The new plasmid with the recombinant DNA is then inserted into a bacterial cell. Inside the cell, the plasmid makes hundreds of copies of itself. Each new plasmid contains the new form of DNA. The cell reproduces quickly, too. Therefore, thousands of copies of the plasmid can be made very quickly. Figure 5-23 shows the splicing of a gene and the insertion of a new plasmid into a bacterial cell.

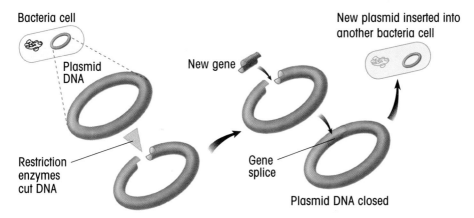

Bacteria cell

New plasmid inserted into another bacteria cell

Plasmid DNA

New gene

Restriction enzymes cut DNA

Gene splice

Plasmid DNA closed

Figure 5-23 *In gene splicing, new DNA that contains a certain gene can be spliced into the plasmid of a bacterial cell. The new plasmid is then inserted into another bacterial cell.*

### Producing Insulin

One way this technique in genetic engineering is used is to produce insulin. Insulin is a chemical that controls the level of glucose in the body. It is usually made by a body organ called the pancreas. A gene that produces insulin in human cells can be inserted into the plasmid of a bacterial cell. The recombinant DNA in the new plasmid can be inserted into another bacterial cell. A large amount of human insulin can be made when the cell divides.

The insulin produced by these cells can help people who have a certain form of the disease diabetes. These people are not able to produce their own insulin.

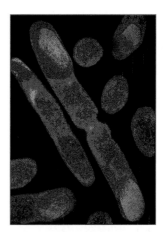

Figure 5-24 *These bacteria have been inserted with genes that produce insulin. Now the bacteria make insulin.*

# Cloning

The new bacterial cells that contain the recombinant DNA are clones. *Clones* are organisms that have the same genetic information. They are all produced from the same single cell.

Figure 5-25 *Being a cloned organism has not prevented Dolly from having offspring—a lamb named Bonnie.*

Today, entire organisms can can be cloned. One way of cloning animals is to take a body cell from an adult animal and an egg cell of another animal. The nucleus of the egg cell is removed. The nucleus from the body cell is inserted into the empty cell. The new clone cell is then placed into the reproductive part of a female animal. This animal carries the cloned animal until birth. The clone cell divides and grows normally. It becomes a clone of the animal whose body cell was used.

In 1997, the first mammal was cloned using this method. This animal is a sheep named Dolly. The success of this cloned organism has led to the cloning of other animals. Cloned cows, mice, and pigs have been produced in similar ways to Dolly.

## Concerns About Genetic Engineering

A better understanding of genetics has allowed scientists to create organisms that have certain traits. The traits of some of these organisms can have many benefits. For example, the production of insulin and the greater nutritional value of some plants can help many people.

However, the creation of new organisms is also something that concerns many people. These new organisms show traits that they have been bred for. They may also have traits that scientists are not aware of. These traits may cause harm.

Substances can be produced to improve the lives of some people. The cures for diseases may be found. The genes that cause inherited disorders may be changed, and people may be healed. Some people believe that the cloning of animals may lead to the cloning of people. All of these possibilities are issues that people will have to think about. We will also have to determine how genetic engineering is to be used. The ability to cause change in organisms must be applied wisely.

Students should list ethical behaviors and attitudes.

**Think About It**

Some concerns about genetic engineering are based on a code of values, or ethics, that people believe in. What values do you apply in your life? Write a short list of these values.

## Great Moments in Biology

### THE HUMAN GENOME PROJECT

Genes control inherited traits and the activities of cells. The human body holds many genes. Identifying the genes in the 23 pairs of human chromosomes could provide useful information.

The Human Genome Project was started in 1990. The word *genome* refers to all of the genes in an organism. One of the goals of this project is to identify all human genes.

You have read that genes contain the bases A, T, G, and C. One gene can contain about 3,000 bases. A chromosome can have 100 million bases. Special technologies are used to find the order of bases that make up genes. Researchers around the world are working to identify the order of the bases in all the genes in human chromosomes. By the year 2000, about 30,000 genes had been identified.

Figure 5-26 *This scientist is making a DNA sequencing gel. The patterns formed show the order of the bases on pieces of chromosomes.*

Information gathered by this project is helping create the human genome map. This map provides information about the location of genes on chromosomes. Some of these genes may cause certain diseases. The Human Genome map may help doctors and researchers to find cures for diseases. Prevention of diseases may also be found.

CRITICAL THINKING  How might the Human Genome map change health care?  Possible answers: it may help doctors prevent a disease before it occurs; health insurance may be denied to a person with a gene for a disease

---

### ✓ Check Your Understanding

Write your answers on a separate sheet of paper. Use complete sentences.

1. When does a mutation occur?

2. What is inbreeding?

3. What is hybridization?

4. CRITICAL THINKING  Why is recombinant DNA a new form of DNA?

### Check Your Understanding

1. when the order of the bases in a DNA molecule changes

2. the crossing of closely related organisms

3. the crossing of two organisms that have different kinds of genes

4. because DNA from different species combined to form recombinant DNA

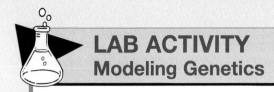

# LAB ACTIVITY
## Modeling Genetics

**BACKGROUND**

The chance of a certain combination of genes occurring is called probability.

**PURPOSE**

In this activity, you will record how chance can determine the genotype and phenotype of the offspring of two heterozygous tall *(Tt)* parents.

| Modeling Genetics | | | | |
|---|---|---|---|---|
| Flip | Parent 1 Disk | Parent 2 Disk | Genotype | Phenotype |
| 1 | | | | |
| 2 | | | | |
| 3 | | | | |
| 4 | | | | |
| 5 | | | | |

**Figure 5-27** *Copy this chart onto a separate sheet of paper.*

**MATERIALS**

two different-colored disks, permanent marker, pencil, paper

**WHAT TO DO**

1.  Copy the chart in Figure 5-27 onto a separate sheet of paper.

2.  Pick one disk to be the "Parent 1" disk. Use a permanent marker to label one side of the disk *T*. Label the other side *t*.

3.  The other disk will be the "Parent 2" disk. Label one side *T* and the other side *t*.

4.  You will flip both disks at the same time, ten times. Each flip represents a gene for height given from each parent organism to an offspring. Flip both disks once. Record in your chart the genes, *T* or *t*, that are face up on each disk.

5.  Flip the disks nine more times. Record the results for each flip.

6.  Fill in the genotype and phenotype that the offspring received from each flip.

**DRAW CONCLUSIONS**

*   What are the possible genotypes in the offspring plants?  *TT, Tt, tt*

*   As a percent, how often do the offspring appear tall? How often do they appear short? (To find probability as a percent, look back on page 71.)  tall, 75%; short, 25%

# ON-THE-JOB BIOLOGY
## Rose Breeder

David Walker is a rose breeder. He creates new kinds of roses by crossbreeding two different types of roses. These new roses are called hybrids.

There are thousands of kinds of hybrid roses. Some roses are bred for their scent, their color, or the size of their flowers. Some hybrids are valued for their ability to flower all season long.

**Figure 5-28** *A rose breeder crossbreeds different roses to produce hybrids.*

To create a hybrid, David crossbreeds two different roses. Then, he takes the seeds that these two different roses produce and plants them. The offspring may show traits of one parent, both parents, or neither parent. Sometimes traits not shown in the offspring will show up in future generations.

Figure 5-29 shows the traits of some roses. Use the chart to answer the questions.

| The Traits of Some Roses | | | | |
|---|---|---|---|---|
| **Rose** | **Blooms All Season** | **Hardy** | **Thorny** | **Fragrance** |
| Rose A | Yes | Yes | No | Strong |
| Rose B | No | Yes | No | None |
| Rose C | Yes | Yes | Yes | Moderate |

**Figure 5-29** *Roses to be crossbred*

1. Suppose you want to produce a hardy rose that blooms all season. Which roses might make good parents?  Rose A and Rose C

2. Suppose you want to breed a rose without many thorns. Which roses might you breed? Rose A and Rose B

**Critical Thinking**

Possible answer: a rose that smells good and is not prone to disease; the parent flowers would need to smell good and not be prone to disease.

**Critical Thinking**

Suppose you want to produce a new breed of rose. What kind of rose would you try to produce? What traits would you look for in the parents?

# Chapter

## 5 ▷ Review

## Summary

- Heredity is the passing of traits from parents to offspring. Genetics is the study of heredity. Gregor Johann Mendel discovered that when he crossbred pea plants, certain dominant traits masked recessive traits.

- Genes contain information about inherited traits. Genes come in pairs. The combination of genes can be predicted using a Punnett square.

- An organism's genetic information is stored in large molecules called DNA. A DNA molecule controls the activities of a cell by directing when and how certain proteins will be made.

- During sexual reproduction, each parent cell gives half of its chromosomes to the offspring. This way the offspring has the exact number of chromosomes it should have.

- Females have two *X* chromosomes. Males have one *X* chromosome and one *Y* chromosome. The genes for sex-linked disorders are found on the sex chromosomes.

- The methods used to produce new forms of DNA are called genetic engineering. The ability to cause change in organisms through genetic engineering must be applied wisely.

dominant

genetics

mutation

Punnett square

recessive

trait

## Vocabulary Review

**Complete each sentence with a term from the list.**

1. The study of heredity is called _____. genetics

2. A gene or trait that will always show itself is _____. dominant

3. A gene or trait that will be hidden when the dominant trait is present is _____. recessive

4. A change in a gene is called a _____. mutation

5. A _____ is a chart that shows possible gene combinations. Punnett square

6. An inherited characteristic is a _____. trait

# Chapter Quiz

Write your answers in complete sentences.

1. What is heredity?
   the passing of traits from parents to offspring
2. Where do genes come from? Genes come from your
   parents. Half come from your mother and the other half come from your father.
3. What is a polygenic trait?
   a trait that is controlled by more than one pair of genes.
4. How many chromosomes are there in a human
   cell? 23 pairs or 46 chromosomes

5. What is a sex-linked trait?
   a trait controlled by genes found on the *X* or *Y* chromosome.
6. If a plant has homozygous tall genes, what kind of
   genes does this plant have? two tall genes

7. Using letters, how would you represent a plant
   that is heterozygous tall? *Tt*

8. What happens during DNA replication?
   Two new identical molecules of DNA are formed.

## CRITICAL THINKING

9. A Punnett square shows that there is a 1 out of 4
   chance for tall offspring to be produced by a
   certain gene combination. What is the probability
   for tallness in this gene combination? 25%

10. Why are plasmids important in genetic
    engineering? Possible answer: because they can be
    cut apart and inserted with new genes

**Test Tip**
To prepare for a test, be sure
you can discuss all the
Learning Objectives listed at
the beginning of the chapter.

**Research Project**
Students should look up the
four major blood groups: A,
B, AB, and O. They should
find that the gene that
controls Group A is
dominant. The gene that
controls Group B is also
dominant. However, when
both genes are present, they
are both expressed as AB.
Group O is controlled by a
recessive gene. Students
should write the following
combinations for blood
types: Group A blood type:
AA or AO; Group B blood
type: BB or BO; Group O
blood type: OO; Group AB
blood type: AB.

*SCLINKS*
Go online to www.scilinks.org.
Enter the code PMB108 to
research **blood types**.

## Research Project

Blood type is a trait controlled by different genes.
Research the four major blood groups. Find out about
the genes that control blood type. Then, write down
the possible gene combinations for all blood types using symbols.

See the *Classroom Resource Binder* for a scoring rubric for the Research Project.

**Figure 6-1** *There is a world of life all around us that we never see. These microscopic organisms have been magnified 65 times. Where do you think these organisms live?*

Possible answers: Diatoms are found in the ocean, lakes, and streams.

## Learning Objectives

- List and describe the six kingdoms of living things.
- List several examples of each kingdom.
- Describe viruses.
- Explain how bacteria can be helpful and harmful.
- LAB ACTIVITY: Identify some protists that can be found in pond water.
- BIOLOGY IN YOUR LIFE: List some ways simple organisms affect your daily life.

# 6 Simple Organisms

## Words to Know

| | |
|---|---|
| **kingdom** | the most general of the seven levels of classification |
| **classification** | a system of grouping things according to similarities |
| **bacteria** | simple, unicellular organisms that do not have nuclei —singular *bacterium* |
| **flagella** | long, thin, threadlike structures that are used for movement —singular *flagellum* |
| **archaebacteria** | (ahr-kee-bak-TEER-ee-uh) bacteria that live in extreme environments |
| **protist** | a simple organism that is neither plant nor animal but that often has characteristics of both |
| **protozoan** | an animal-like protist |
| **cilia** | tiny hairs used for movement—singular *cilium* |
| **pseudopod** | (SOO-duh-pahd) an extension of cytoplasm used for movement |
| **algae** | plantlike protists |
| **fungi** | organisms that may be unicellular or multicellular, have cells with a nucleus, and do not have chlorophyll—singular *fungus* |
| **hyphae** | (HY-fee) chains of cells that form fibers in fungi |
| **spore** | the reproductive cell of organisms such as ferns, fungi, and algae |

## Classifying Living Things

You have already read about the things large animals such as bears and microscopic organisms, have in common. They share the seven characteristics of living things. However, they still have many differences.

**Linking Prior Knowledge**
Students should recall the characteristics of plant cells and animal cells (Chapter 3). Have students relate these characteristics as they learn about bacteria, protists, and fungi.

**Everyday Science**

Classification is a part of everyday life. Make a list of the things you classify into groups at home and at school.

Possible answers: clothing, shoes, food, textbooks, notebooks

Biologists group organisms into **kingdoms**. This grouping is called **classification**. Classification allows biologists to organize living things. This makes it easier to study living things. It also shows how living things are related. Living things are grouped by the structures of their bodies, the types of cells they have, and how they get energy.

## The Six Kingdoms

The way in which biologists classify organisms changes all the time. These changes are based on new discoveries. Many years ago, biologists classified living things into only two kingdoms: plant and animal. However, scientists soon discovered organisms that did not quite fit into either of these kingdoms. Other kingdoms were created. Many scientists today classify organisms into six kingdoms. Look at Figure 6-2 to learn more about these kingdoms.

| The Six Kingdoms | | |
|---|---|---|
| **Kingdom: Archaebacteria**<br>• single-celled, have no cell nucleus 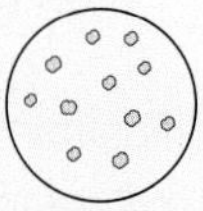<br>**Examples:**<br>• bacteria that live in extreme environments, such as salt marshes and hot springs | **Kingdom: Bacteria**<br>• single-celled, have no cell nucleus <br>**Examples:**<br>• all other bacteria, such as bacteria that cause strep throat or Lyme disease | **Kingdom: Protists**<br>• many are single-celled, have cell walls 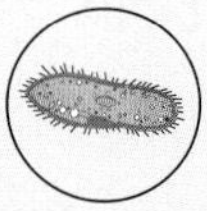<br>**Examples:**<br>• protozoans, algae, slime molds |
| **Kingdom: Fungi**<br>• cannot move, have no chlorophyll, have cell walls <br>**Examples:**<br>• yeasts, molds, mushrooms, mildew | **Kingdom: Plants**<br>• have many specialized cells, use chlorophyll and sunlight to make food, have cell walls <br>**Examples:**<br>• ferns, mosses, seed plants | **Kingdom: Animals** <br>• have many specialized cells, do not have chlorophyll, eat other organisms, do not have cell walls<br>**Examples:**<br>• insects, mammals, birds, fish |

Figure 6-2 *Living things can be classified into six kingdoms.*

# Levels of Classification

A kingdom is the largest and most general level of classification. Since there are so many different kinds of living things, biologists classify organisms into groups that are smaller than kingdoms. Figure 6-3 shows the different levels of classification used by many biologists.

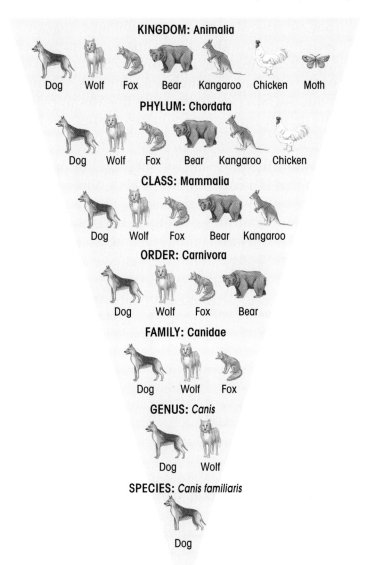

KINGDOM: Animalia

Dog   Wolf   Fox   Bear   Kangaroo   Chicken   Moth

PHYLUM: Chordata

Dog   Wolf   Fox   Bear   Kangaroo   Chicken

CLASS: Mammalia

Dog   Wolf   Fox   Bear   Kangaroo

ORDER: Carnivora

Dog   Wolf   Fox   Bear

FAMILY: Canidae

Dog   Wolf   Fox

GENUS: *Canis*

Dog   Wolf

SPECIES: *Canis familiaris*

Dog

**Think About It**

Which two animals would be more closely related, two from the same phylum or two from the same genus? Use Figure 6-3 to help you.

Animals from the same genus would be more closely related.

Figure 6-3 *The seven levels of classification*

All animals belong to the same kingdom. Biologists break down each kingdom into smaller groups called *phyla*. Each phylum is then broken down into smaller groups. The more levels that two organisms have in common, the more closely related they are.

## Classifying Viruses

Viruses are structures that many biologists do not know how to classify. Therefore, viruses are not grouped in any kingdom. Viruses cause many diseases in humans. Some of these are the common cold, measles, mumps, the flu, and AIDS. Viruses also cause diseases in plants. Biologists do not agree about whether viruses are living things or not.

Viruses are structures with an outer coat of protein. Inside this coat is a core of DNA or RNA. Viruses are not made of cells. Viruses can only reproduce when inside another organism's cells. Look at Figure 6-4 to learn more about reproduction in viruses.

1. A virus attaches to the cell.  2. A virus injects genetic material into the cell.  3. The genetic material directs the cell to make new virus parts.  4. The parts form new viruses.  5. The cell membrane bursts, releasing new viruses.

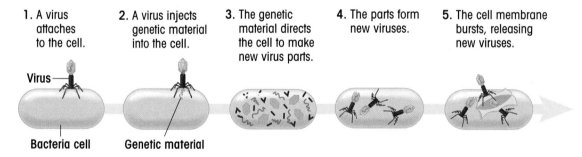

Figure 6-4 *Viruses reproduce by invading, and killing, another organism's cells.*

## Characteristics of Bacteria

**Bacteria** are microscopic, unicellular organisms. Bacteria are the most numerous organisms on Earth. They live everywhere—in the sea, on mountaintops, in the desert, in the Arctic, and in hot springs.

Bacteria come in many different shapes, but they all have a few common characteristics. Most bacteria have a cell wall, but they lack other cell parts. Bacteria cells do not have a nucleus. They also do not have mitochondria, lysosomes, or Golgi bodies.

Bacteria have DNA. The DNA of bacteria is contained in circular strands that lie inside the cytoplasm. When bacteria reproduce, they split in two. This is a form of asexual reproduction called *fission*. Some bacteria have **flagella**. Flagella are long, threadlike structures that help an organism to move.

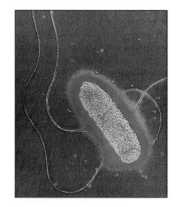

Figure 6-5 *This salmonella bacterium has been magnified 12,600 times. Salmonella have flagella that help them move.*

## Two Kingdoms of Bacteria

There are two different types of bacteria. The two kingdoms of bacteria are the archaebacteria kingdom and the bacteria kingdom.

### Archaebacteria

The **archaebacteria** include types of organisms that have lived on Earth for millions of years. These organisms are able to live in very extreme environments. These environments are places with high levels of harsh chemicals, or very hot or very cold temperatures. They live in hot springs, deep-sea vents in the ocean, and salt marshes.

## Science Fact

Archaebacteria that live in deep-sea vents carry out a process similar to photosynthesis. This process is called *chemosynthesis*. Instead of sunlight energy, they use chemical energy to make their own food.

Figure 6-6 *Archaebacteria live in harsh environments, such as this hot spring.*

## Bacteria

All other types of bacteria belong in the second bacteria kingdom. When people talk about bacteria, they are usually talking about this group. There are three basic shapes of bacteria. They are round, rod-like, or spiral shaped.

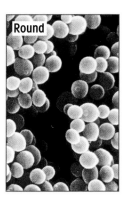 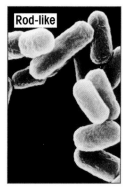 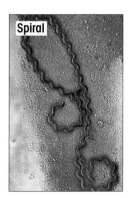

Figure 6-7 *Bacteria come in three basic shapes.*

One type of bacteria, the blue-green bacteria, contains chlorophyll. Blue-green bacteria are unusual because they can carry out photosynthesis. The process of photosynthesis carried out by bacteria is similar to the photosynthesis in plants.

## The Role of Bacteria

There are thousands of different types of bacteria. Some of these bacteria are helpful to us. Other types of bacteria are harmful to us.

### Helpful Bacteria

Most bacteria are harmless. There are many living on your skin and inside your body right now. Some types of bacteria live in your digestive tract. These bacteria help to digest food and produce vitamins.

Bacteria are used to make many dairy foods such as yogurt and cheese. Bacteria are also used to make sauerkraut and pickles.

Some bacteria feed on dead matter. This includes leaves, dead animals, and the food you eat. Bacteria break down dead matter into smaller molecules. This process is called *decomposition*. Decomposition is one way in which the nutrients in dead plants and animals are returned to the soil. This helps plants get the nutrients they need.

Bacteria are helpful to plants in another way. All plants need nitrogen. However, they are not able to use the nitrogen in the air. They need the nitrogen to be changed into a form they can use. Bacteria do this job. They change nitrogen from the air into usable compounds in the soil. This process is called *nitrogen-fixation*.

Figure 6-8 *The bumps on the roots of this plant contain helpful nitrogen-fixing bacteria.*

### Harmful Bacteria

Not all bacteria are helpful. Some bacteria cause infection and disease. Strep throat and pneumonia are diseases caused by bacteria. Bacteria can cause infections in cuts and scrapes. Bacteria can also cause tooth decay.

Some bacteria can cause food poisoning if food is not handled properly. One of the most common kinds of food poisoning is caused by salmonella bacteria. It is very important to cook foods thoroughly. This kills harmful bacteria that could cause disease.

botulinum

Figure 6-9 *A type of bacteria called botulinum can be found in poorly canned foods. These bacteria cause the deadly disease botulism. Cans that contain the bacteria often bulge.*

**Safety Alert**

It is important to wash all foods thoroughly before eating. This will reduce the amount of harmful microscopic organisms on your food. Botulinum is a very dangerous bacteria. Never open a bulging can. Dispose of it properly.

## On the Cutting Edge

### MODERN USES OF BACTERIA

In recent years, bacteria have been used for things that some people are not even aware of. Oil spills can be devastating to plants and animals. Bacteria have been used to break down oil in water. When there is an oil leak from a ship, these bacteria are released to feed on the oil. The oil then turns into carbon dioxide and other byproducts that do not harm the environment.

Research on bacteria continues. A bacterial rinse to prevent tooth decay has been tested by researchers. A bacterial spray to use in the treatment of children's ear infections has also been tested. This spray could reduce the amount of medicine needed for treatment.

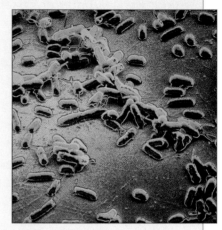

Figure 6-10 *These bacteria are used to help clean up oil spills.*

CRITICAL THINKING Why do you think oil spills in the ocean are harmful? Possible answer: If there was an oil leak and the oil was to remain in the water, it would be harmful to fish, sea animals, and plant life. It might also harm drinking water.

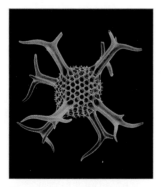

Figure 6-11 *Protists, such as this radiolarian, are microscopic organisms.*

## The Protist Kingdom

When people use makeup, they may be smearing the remains of microscopic organisms on their face. That is because microscopic organisms are sometimes used in producing makeup. Other microscopic organisms are used to make toothpaste, cleaning products, and even ice cream.

These organisms belong to the protist kingdom. **Protists** are another group of simple organisms. They are different from bacteria because they have nuclei in their cells.

Protists come in all shapes and sizes. Some have only one cell. Others have more than one cell. Some protists carry out photosynthesis. Others must take in food to get energy. Most protists live in water or in a moist environment. Protists can reproduce both sexually and asexually. The three main groups of protists are the protozoans, the algae, and the slime molds.

| Types of Protists | |
| --- | --- |
| Protozoans | • animal-like<br>• move around and eat other organisms |
| Algae | • plantlike<br>• contain chlorophyll for photosynthesis |
| Slime molds | • funguslike<br>• made of threadlike fibers |

Figure 6-12 *There are three types of protists.*

## Protozoans: Animal-like Protists

**Protozoans** are often called animal-like protists. They are more like animals than plants. Protozoans eat bacteria and other microscopic organisms.

There are different ways protozoans move. Some protozoans, such as the paramecium, move by means of **cilia.** These are thousands of tiny hairs on the organism. As these hairs beat like tiny oars, the organism moves forward. Other protozoans move using flagella.

**Remember**
Flagella are long, threadlike structures used for movement. Both bacteria and protists can have flagella.

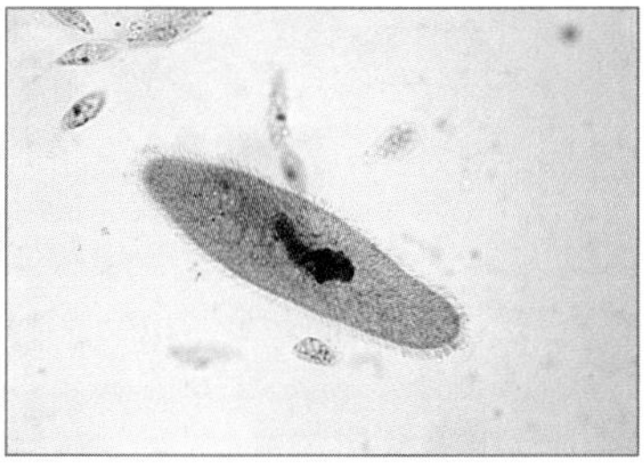

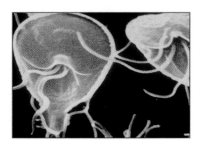

Figure 6-13 *The paramecium moves using cilia. (100×)*

Figure 6-14 *The giardia uses flagella to move. (9,800×)*

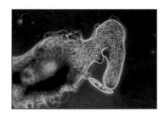

Figure 6-15 *This amoeba is using pseudopods to capture a paramecium.*

One type of protozoan, an amoeba, moves by using its pseudopods. A **pseudopod** is an extension of an organism's cytoplasm. An amoeba moves its pseudopods forward and slides along. An amoeba also catches its food by surrounding it with two pseudopods. An amoeba then pulls the food into its body. Then, the amoeba digests the food.

### Algae: Plantlike Protists

**Algae** are plantlike protists. There are several thousand species of algae. Almost all algae contain chlorophyll. This means they can carry out photosynthesis. Algae live in water. They all have nuclei in their cells.

Some algae are unicellular, but others are made of many cells. Most of the cells of multicellular algae all do the same things. They are not specialized. Different kinds of algae include red algae, green algae, and brown algae. Algae are very important to life on Earth. Many different kinds of organisms feed on algae. They take part in many food chains. Algae also produce much of the oxygen on Earth.

Figure 6-16 *The kelp shown here are a type of brown algae. Kelp are part of a food chain that includes fish and other animals.*

Most algae are helpful. However, sometimes there can be too much algae in a given area. From time to time, algae blooms grow in a body of water. An algae bloom is a large increase in the size of a population of algae. This increase can occur when there is too much waste material in the water. Algae blooms can be harmful to fish, birds, and other organisms. This is because the algae take up too much of the water's oxygen.

Figure 6-17 *This lake has an algae bloom.*

### Algae: Euglena

One kind of algae, the euglena, seems to have the characteristics of several different organisms. The euglena moves around using a flagellum. Like bacteria, the euglena can feed off dead matter. Yet, the euglena also has chloroplasts in its cell. It carries out photosynthesis to make some of its own food. The euglena has an eye spot. This spot is sensitive to light. The eye spot helps lead the euglena to the light it needs for photosynthesis. The euglena also has a contractile vacuole. This structure helps control the amount of water in the cell.

**Remember**
Homeostasis is the ability of an organism to maintain its body conditions. The action of the contractile vacuole is part of the euglena's homeostasis.

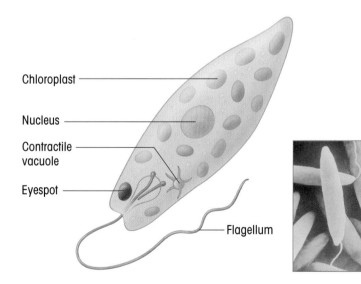

Chloroplast

Nucleus

Contractile vacuole

Eyespot

Flagellum

Figure 6-18 *The euglena is a protist that has both plantlike and animal-like characteristics.*

**Remember**

Specialized cells are those that do different jobs.

Slime molds are a type of protist that are similar to fungi. They have threadlike fibers that make up their bodies, which fungi also have. However, they are not considered fungi because they do not have specialized cells. You may have seen a slime mold growing on the side of a dead tree or on the ground near decaying leaves.

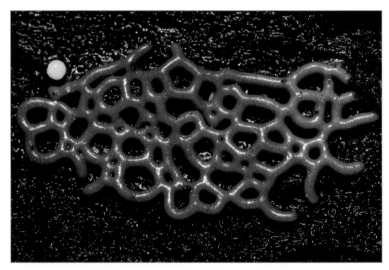

Figure 6-19 *This pretzel slime mold is a type of funguslike protist.*

**Check Your Understanding**

1. They classify them into groups based on similarities.
2. They help make foods and recycle nutrients.
3. protozoans, algae, and slime molds
4. algae
5. a funguslike protist

✓ **Check Your Understanding**

Answer the questions in complete sentences. Write your answers on a separate sheet of paper.

**1.** How do scientists group living things?

**2.** Why are bacteria important?

**3.** What are the main groups of protists?

**4.** What types of protists are most like plants?

**5.** What is a slime mold?

# The Fungus Kingdom

You have probably eaten mushrooms in salad or on pizza. Mushrooms are part of the fungus kingdom. **Fungi** are not plants. They do not have chlorophyll in their cells. However, each of their cells does have a nucleus and other cell parts. They also have specialized cells.

Most fungi live in moist, dark places. Most fungi have hairy structures called **hyphae**. Hyphae are chains of cells that form fibers in fungi. These fibers are easy to see on bread or fruit mold.

Like bacteria, fungi use dead matter as their food. Dead matter is the source of energy for fungi. Fungi feed on dead organisms, wood, and foods. Some fungi, such as corn smut, destroy crops.

Figure 6-20 *The corn smut growing on this ear of corn is part of the fungus kingdom.*

## Reproduction in Fungi

Fungi reproduce both asexually and sexually. Many fungi reproduce by means of spores. This is a form of asexual reproduction. **Spores** are special reproductive cells in round cases called spore cases. When the spore cases break open, the spores are released. If the spores land somewhere that is wet and warm enough, new fungi can grow. In sexual reproduction, the adult fungi produce different forms of sex cells. If the two sex cells join, a new fungus forms.

Yeast reproduce asexually. The form of asexual reproduction they use is called *budding*. A small offspring organism, or bud, forms as part of the yeast cell. Eventually the offspring pinches off and becomes a separate organism.

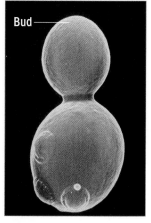

Figure 6-21 *Yeast reproduce by budding.*

## Everyday Science

Yeast helps bread rise by creating bubbles of carbon dioxide gas. These bubbles push the bread up. Try making bread at home and watch the action of yeast.

Kinds of Fungi

There are several kinds of fungi. Mildew is a fungus that grows in damp places. It often grows on bathroom tiles around the bathtub or shower. Molds are a type of fungus that often grow on food products. Yeast is another kind of fungus. It is used in making bread. Mushrooms are probably the most familiar type of fungus.

Figure 6-22 *Mushrooms, like these toadstool mushrooms, are probably the most well-known type of fungus.*

## Safety Alert

Never eat any mushrooms you find outside. Many poisonous fungi look just like the kind you find in grocery stores. One bite of some kinds of mushrooms can be deadly.

Mushroom Structure

Mushrooms belong to a group of fungi called club fungi. They grow in dark, moist areas, such as a forest floor. There are many different kinds of mushrooms. Some mushrooms can be eaten. Other mushrooms are poisonous. Mushrooms come in many different sizes, shapes, and colors.

Most mushrooms have a round cap on top of a stalk. Under the cap there are fibers called gills. The stalk, cap, and gills are all made out of hyphae. Hyphae also grow underground and act as "roots."

Mushrooms reproduce using spores. The spores are released from the gills. The spores can later become new mushrooms if the conditions are right. Look at Figure 6-23 to learn more about the parts of a mushroom.

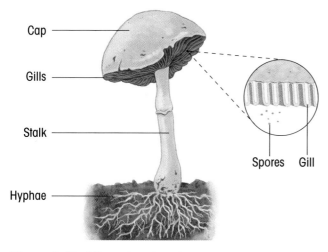

Cap

Gills

Stalk

Spores   Gill

Hyphae

**Figure 6-23** *Like other fungi, mushrooms can reproduce using spores.*

## Lichens: A Fungi-Algae Combination

Some fungi grow in close company with other organisms. One example of this combination is the lichen. Lichens are not single organisms. Instead, they are a structure formed by the relationship between a fungus and algae. The fungus and algae live attached to one another. Lichens can be found living on rocks, on trees, and on the ground.

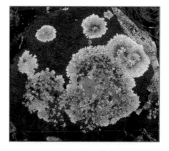

**Figure 6-24** *Lichens, like the ones on this rock, are important in many food chains. In Alaska, herds of caribou survive on lichens for food.*

✓ **Check Your Understanding**

Answer the questions on a separate sheet of paper.

**1.** What are three examples of fungi?

**2.** Where do most fungi live?

**3.** CRITICAL THINKING How are fungi different from protists?

**Check Your Understanding**
**1.** mushrooms, yeast, mold
**2.** dark, moist places
**3.** Possible answer: They have specialized cells.

# LAB ACTIVITY
## Identifying Protists

### BACKGROUND
Many different kinds of protists live in fresh water. They are microscopic. Protists can be found in lakes, ponds, and streams.

### PURPOSE
In this activity, you will observe and identify protists in their natural habitat.

### MATERIALS
safety goggles, glass jar or beaker, leaves and stones, water collected from a lake or pond, dropper, slide, cover slip, microscope, cotton ball

Figure 6-25 *Use a dropper to take a sample of pond water.*

### WHAT TO DO
1. Place leaves and stones in the bottom of the glass jar. Add several cups of pond water. Leave the jar in a warm place for 2 to 3 days.

2. Use a dropper to place one to two drops of pond water onto a clean slide.

3. Pull apart a cotton ball so that you have only a few fibers. Place a few fibers onto the pond water on the slide.

4. Gently place a cover slip onto the slide.

5. Observe your slide under a microscope. Draw what you see on a separate sheet of paper. Then, try to identify the protozoans you observed. Use reference materials if you need help.

### DRAW CONCLUSIONS
- What protists did you identify?
  Answers may include paramecium, amoeba, stentor, spirogyra, or other protists.
- Many protists sometimes attach to rocks or leaves of plants. What do you think is the purpose of using the cotton fibers?
  The cotton fibers give the protists a place to attach to, which makes it easier to observe them.

# BIOLOGY IN YOUR LIFE
## Living with Simple Organisms

When you eat ice cream, you probably do not think of seaweed. However, the two items are related. Substances from seaweed help make ice cream smooth and creamy. Seaweed is a form of algae. Algae extracts are used in many other foods and household products. In Japan, people eat many types of seaweed. One type, called nori, is a type of red algae. It is used to season and wrap sushi.

Figure 6-26 *Red algae is used to make agar.*

Agar is an extract from red algae. It keeps breads from drying out. Agar is common in puddings and jellies. Agar is also used in skin lotions and in certain medicines.

Other simple organisms are important in foods and other products. Use Figure 6-27 to answer the questions.

| Uses of Simple Organisms | | |
|---|---|---|
| **Organism** | **Kingdom** | **Some Uses** |
| Red algae | Protist | Sushi, frozen foods, candies, milk, cheeses, ice cream, medicine capsules, shampoos, lotions |
| Brown algae | Protist | Frozen foods, pudding, salad dressing, milk shakes, lotions |
| Mold | Fungus | Causes blue veins in some cheeses |
| Yeast | Fungus | Makes bread rise, ferments grapes into wine and apples into cider |

Figure 6-27 *Simple organisms are used in many products.*

**Critical Thinking**

Mold can be used to make cheese, but spoiled food on which mold has grown can make you sick.

**Critical Thinking**

Explain how mold can be useful in food and harmful in food.

1. Which type of simple organism is used to make blue cheeses?  mold

2. Which type of simple organism is used to make bread?  yeast

3. In which kingdom do red and brown algae belong?  protist

## Summary

- Viruses are not classified into any kingdom. They are made up of DNA or RNA with an outer coating of protein. Viruses cause many diseases, including AIDS, mumps, the common cold, polio, and the flu.

- Organisms can be classified into six kingdoms: archaebacteria, bacteria, protist, fungi, plant, and animal.

- Bacteria are unicellular organisms without a nucleus in their cell. They can live just about everywhere on Earth.

- Protists are simple organisms that live in water. They have nuclei in their cells. Protozoans are animal-like protists. Algae are plantlike protists. Slime molds are funguslike protists.

- The fungus kingdom is made up of mushrooms, molds, yeasts, and mildew. These organisms do not carry out photosynthesis. They feed on dead organic matter. Most fungi live in dark, warm places.

| |
| --- |
| algae |
| bacteria |
| cilia |
| classification |
| flagella |
| protozoan |
| pseudopod |

## Vocabulary Review

**Complete each sentence with a term from the box.**

1. A _____ is an animal-like protist.  protozoan

2. Simple unicellular organisms that do not have nuclei are _____.  bacteria

3. Tiny hairs used for movement are called _____.  cilia

4. A system of grouping things according to similarities is called _____.  classification

5. Plantlike protists are called _____.  algae

6. Long, thin, threadlike structures that some unicellular organisms use to move are _____.  flagella

7. An extension of the cytoplasm that is used for movement is called _____.  pseudopod

# Chapter Quiz

**Answer the questions in complete sentences. Write your answers on a separate sheet of paper.**

1. What are the six kingdoms many biologists use?
   archaebacteria, bacteria, protist, fungi, plant, and animal
2. What are viruses?
   structures that are made of an outer protein coat with DNA or RNA inside
3. How does a virus reproduce?
   by taking control of another cell
4. How are bacteria cells different from the cells of other organisms?  They do not have a nucleus.

5. Name two ways bacteria are helpful.  recycle nutrients, help make food

6. Name two ways bacteria can be harmful.  cause disease, spoil food

7. In what ways is the euglena similar to both a plant and an animal?  It carries out photosynthesis but also moves around and takes in food.

8. What are three structures protozoans use to move?  flagella, cilia, and pseudopods

9. How do yeast reproduce?  by budding

10. How do mushrooms reproduce?  by producing spores

## CRITICAL THINKING

11. Why is classification useful to biologists?
    It organizes living things in a logical way, which makes it easier to study them.
12. How are fungi different from plants?
    They do not have chlorophyll and cannot make their own food.

## Research Project

Archaebacteria that live near deep-sea vents carry out chemosynthesis. Use the Internet and other references to find out more about chemosynthesis. Find out how these bacteria survive. How do they relate to other organisms that live near the deep-sea vent as well? Write a report about your findings.

*SCiLINKS*
Go online to www.scilinks.org.
Enter the code **PMB110** to research **chemosynthesis**.

See the *Classroom Resource Binder* for a scoring rubric for the Research Project.

# Unit 2 Review

**Choose the letter of the correct answer to each question.**

Use Figure U2-2 to answer Questions 1 to 3.

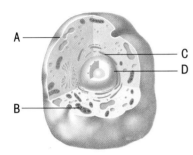

Figure U2-2 *An animal cell*

**1.** Which cell part is sometimes called the powerhouse of the cell? B (p. 37)

A. part labeled A

B. part labeled B

C. part labeled C

D. part labeled D

**2.** Which cell part controls the activities of the cell? D (p. 36)

A. part labeled A

B. part labeled B

C. part labeled C

D. part labeled D

**3.** What is the name of the cell part labeled A? A (p. 37)

A. cell membrane

B. nucleus

C. mitochondrion

D. vacuole

**4.** Gregor Mendel studied which branch of biology? A (p. 66)

A. genetics

B. zoology

C. ecology

D. microbiology

**5.** What is the diffusion of water molecules called? A (p. 54)

A. osmosis

B. diffusion

C. passive transport

D. active transport

**6.** What are plantlike protists called? C (pp. 98–9

A. slime molds

B. protozoans

C. algae

D. archaebacteria

**7.** What is the most general level of the seven levels of classification? B (pp. 90–91

A. phylum

B. kingdom

C. species

D. genus

**Critical Thinking** (pp. 70–71)

Students answers should match the following Punnett square.

TT – 25%; Tt – 50%; tt – 2

**Critical Thinking**

Draw a Punnett square to show the cross of two pea plants that are heterozygous tall. Write the percents of each possible gene combination in the offspring.

Chapter 7  **The Study of Plants**

Chapter 8  **Types of Plants**

Chapter 9  **Plants at Work**

Chapter 10  **Reproduction in Seed Plants**

Figure U3-1 *Redwood trees, like the ones shown here, are among the largest and oldest organisms on Earth.*

## Biology Journal

Redwood trees are plants. In this unit, you will learn many important features of plants, the different parts of plants and how some plants reproduce. In your journal, write a list of questions that you have about plants. As you read each chapter, go back and answer your questions. You also can use the Internet or other references to help answer your questions.

**Biology Journal** The journal activity can be an alternative assessment, a portfolio project, or an enrichment exercise. Have students write at least five questions. As they read the chapters, have them answer the questions.

**Figure 7-1** *Plants provide many important food crops. These apples will provide people with important vitamins and minerals. They are also a source of income for farmers. What other foods come from plants?*

Possible answers: oranges, peaches, tomatoes, peppers, corn, wheat, rice

## Learning Objectives

- List three reasons why plants are important to humans.

- Name the six important characteristics of plants.

- Describe what plants need in order to grow.

- **LAB ACTIVITY:** Compare the ability of different materials to hold water.

- **ON-THE-JOB BIOLOGY:** Analyze the farming and harvesting times of different crops.

## Words to Know

| | |
|---|---|
| **photosynthesis** | the process by which light energy is used to make food |
| **chlorophyll** | (KLAWR-uh-fihl) substance in chloroplasts that absorbs light energy |
| **soil** | a mixture of sand, silt, and clay |
| **loam** | soil that has a well-balanced mixture of sand, silt, and clay |
| **humus** | the organic material in soil |

## Careers Using Plants

Have you ever considered a job working with plants? If you are artistic, you might get a job arranging flowers for flower shops. Farmers work with plants every day. Other plant lovers work as gardeners in parks. Designing landscapes is another good job.

Each of these jobs is related to plants in some way. In this chapter, you will learn about the importance of plants. You will also learn about some of the characteristics of plants.

## The Importance of Plants

Chocolate comes from the seeds of a plant. Your favorite pair of jeans was made from a cotton plant. Plants may even keep your house warm. If you use a fireplace, you probably burn wood. Coal and oil are fuels made from plant products. Many important drugs for treating diseases come from plants. In fact, the pain reliever aspirin was originally made using the bark of a willow tree.

**Linking Prior Knowledge**
Students should recall the six characteristics of living things (Chapter 2). They should also recall the kingdoms of living things and give an example of each (Chapter 6).

Figure 7-2 *Chocolate comes from the seeds of a cocoa plant.*

Almost everything you eat is a product of plants, either directly or indirectly. Fruits, vegetables, and grains come directly from plants. Meats, such as steak or bacon, come from animals that eat plants. After all, if plants did not exist, cows and pigs would have nothing to eat. Without cows and pigs, there would be no steak or bacon.

Finally, plants release the oxygen you need to breathe and carry out cellular respiration. Remember that oxygen is a byproduct of photosynthesis. The oxygen released by plants becomes part of the air. When you breathe, you take in this oxygen. Without oxygen, you could not survive.

## The Characteristics of Plants

You know that biologists group all organisms into kingdoms. You have already studied four kingdoms of life: the two kingdoms of bacteria, the protist kingdom, and the fungus kingdom.

So what is a plant? We know that trees, bushes, grasses, and flowers are examples of plants. However, plants have a few special characteristics that you should know about. These six characteristics put an organism into the plant kingdom.

1) Plants are able to make their own food by using light energy. This process is known as **photosynthesis.**

2) Plants are multicellular. Some unicellular organisms carry out photosynthesis. For example, blue-green bacteria and euglena can carry out photosynthesis. However, they are not plants because they are unicellular.

**Figure 7-3** *This barrel cactus is a plant.*

3) Plants have specialized cells. This means that each plant organism has several different kinds of cells. The different cells perform different jobs. Like animals, plants have tissues and organs.

4) The cells of plants have cell walls. The cell walls give a plant strength and support.

5) Plants cannot move from place to place. That is to say, they cannot walk, swim, or fly. However, plants do move. Plants often turn to face sunlight. Some plants that live in water float from place to place. Most plants stay rooted in the ground.

6) Plants have three basic parts: roots, stems, and leaves. However, each of these parts may have a different size, shape, or color in different plants. For example, the roots of a very large elm tree will be much larger than the roots of a pea plant.

Leaves make food.

The **stem** transports water and nutrients and supports leaves.

**Roots** take in water and minerals.

Figure 7-4 *All plants have a few basic parts, such as leaves, stems, and roots. Each part has basic functions.*

# The Needs of Plants

## Everyday Science

Like all plants, houseplants need certain things to survive. Look at the tag on a houseplant from a garden store. Write down how much sunlight, water, and minerals the plant needs to be healthy.

Ask students to bring in their list of plant requirements or the actual labels.

**Remember**
Chlorophyll is located in the chloroplasts of a plant cell. Chloroplasts are small, green cell parts that can be seen using a microscope.

Like all living things, plants need food, oxygen, and water in order to grow. They also need light, carbon dioxide, and minerals.

Plants need light for photosynthesis. They usually get light from the Sun. Plants also need water, which is required for photosynthesis. Most plants get water from the ground. Plants need carbon dioxide for photosynthesis. They need oxygen for respiration. They get both of these gases from the air or water that they live in.

You have already read how plants get food. They make their own. Plant cells contain a substance called chlorophyll. The **chlorophyll** absorbs light energy, which usually comes from the Sun. Carbon dioxide from the air enters plant cells. The light energy allows plants to change carbon dioxide and the hydrogen from water into other products. These are sugar and oxygen.

Chloroplasts

Figure 7-5 *Most of the photosynthesis a plant carries out takes place inside the leaves.*

On a separate sheet of paper, make a chart like the one shown below. List the things a plant needs to grow. Then, write the source of each need.

| Needs of Plants | |
|---|---|
| Plant Need | Source |
| 1. | |
| 2. | |
| 3. | |
| 4. | |

**Check Your Understanding**
1. light; from the Sun or artificial light
2. carbon dioxide and oxygen; from the air
3. water; from the soil
4. minerals; from the soil

## Modern Leaders in Biology

**MICHAEL KASPERBAUER,**
**Plant Researcher**

Growing up on a farm in Iowa, Michael Kasperbauer was interested in plants. He noticed that plants that grew alone were shorter than plants that grew crowded together. He came up with the hypothesis that plants crowded together grew taller to stay in the sunlight.

When Kasperbauer became a researcher for the U.S. Department of Agriculture, he designed some experiments to test this hypothesis. He found out that the leaves could absorb light reflected from nearby plants. This signal made the plant grow larger shoots.

Figure 7-6 *Michael Kasperbauer performs research on the growth of plants.*

Kasperbauer used this information to design a plastic mulch that reflects light to the underside of leaves. This mulch helps strawberry and tomato plants produce more fruit. It also improves the flavor and nutrient content of these crops.

**CRITICAL THINKING** How do you think the research of Kasperbauer could help gardeners? It could help gardeners to grow larger and more nutritious crops.

## Plants Grow in Soil

Clay

Silt

Sand

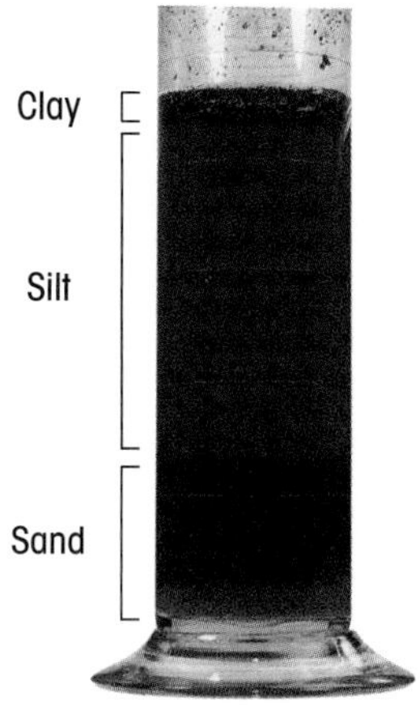

**Figure 7-7** *You can see the different particles of soil by mixing it with water. Sand will settle to the bottom because it is the heaviest.*

**Soil** is a mixture of sand, silt, and clay. Soil also contains decayed organisms, living organisms, minerals, and water. Almost all plants must grow in soil. The soil gives the plants important minerals, and other nutrients.

Soil is made of a mixture of different-sized particles. The largest particles are called sand. Medium particles are called silt. The smallest particles are called clay. If soil is too sandy, water will run through it very quickly. If soil has too much clay, water will not drain fast enough.

Some plants will rot if they stay rooted in wet ground for too long. The best soil for most plants has a good mixture of sand, silt, and clay. This kind of soil is called **loam**.

**Humus** is the organic material in soil. The term *organic* means from living things or things that were once living. Humus contains dead plant and animal materials that have been broken down by bacteria and other organisms. Humus provides very important nutrients for plants.

## Science Fact

Different plants need different amounts of nutrients, such as nitrogen or phosphorus. The nutrients in the soil in your area may determine what plants will grow best there.

**Figure 7-8** *Most farm crops grow very well in loam.*

Although most plants grow well in loam, not all do. Desert plants, for example, grow very well in dry, sandy soil. Some plants, such as moss, can only grow where the soil stays very moist and it is shady.

## ✓ Check Your Understanding

Answer the questions in complete sentences. Write your answers on a separate sheet of paper.

1. What are the components of soil?

2. What is loam?

3. What is humus?

4. **CRITICAL THINKING** Many gardeners use decaying leaves and kitchen wastes to improve their soil. Why do you think this is helpful?

**Check Your Understanding**

1. sand, silt, and clay

2. a soil with a good mixture of sand, silt, and clay

3. the organic matter found in soil

4. It adds nutrients to the soil.

## On the Cutting Edge

**PLANTS AS HEALERS**

Plants provide food, oxygen, shelter, and fabric. Plants do a number of other amazing things as well.

One of the most used cancer drugs is taxol. Taxol is made from the bark of the Pacific yew tree. About six 100-year-old Pacific yew trees are needed to produce enough taxol to treat one patient for a year. Taxol fights cancer by preventing mitosis in cancer cells. It has been used to treat breast cancer, ovarian cancer, and brain cancer. In the past, the yews were protected by the U.S. Forest Service for their taxol. Today, some taxol is made in the laboratory.

Figure 7-9 *The bark of a yew tree is used to make taxol.*

**CRITICAL THINKING** Why do you think it is important to study plants? It is important to study plants because they can be a source of new medicines.

# LAB ACTIVITY
## Investigating Properties of Soil

**BACKGROUND**

The size of soil particles determines the amount of air pockets in the soil. These air pockets control how water passes through the soil. Different soils hold different amounts of water.

**PURPOSE**

In this activity, you will investigate the ability of different growth materials to hold water.

**MATERIALS**

vermiculite, potting soil, sand, clay soil, funnels, cotton, graduated cylinders, beakers, water, stopwatch or timer

| Properties of Soil | |
|---|---|
| **Material** | **Amount of Water (after 1 minute)** |
| Vermiculite | |
| Potting soil | |
| Sand | |
| Clay soil | |

Figure 7-10 *Data chart*

**WHAT TO DO**

1. Make a chart like the one shown in Figure 7-10 on a separate sheet of paper.

2. Plug the neck of a funnel with a small amount of cotton. Fill the funnel half way with vermiculite. Place the funnel in a beaker.

3. Measure 10 ml of water in a graduated cylinder. Add the water to the surface of the vermiculite. Let stand for 1 minute.

4. After 1 minute, measure the amount of water that passed into the beaker. Record this information in your chart.

5. Follow Steps 2–4 for each of the other plant growth materials.

**DRAW CONCLUSIONS**

- Which material held the most water? Which material held the least amount of water? Answers will vary, but students should expect to see that the vermiculite holds the most water and the sand held the least amount of water.
- Plants do not grow well in a clay soil because there are not enough air pockets to allow the roots to get oxygen. What could a gardener who has clay soil do to the soil to help plants grow there? Possible answer: The gardener could mix sand or vermiculite into the regular soil.

# ON-THE-JOB BIOLOGY
## Farmer

When you eat a salad, do you think about where it came from? Most of our fruits and vegetables come from farms. Farmers provide us with nutritious foods that keep us healthy.

Martha Kelly is a vegetable farmer. She also knows something about biology. Martha knows that plants need sunlight to carry out photosynthesis. She also knows that plants need water. Martha often uses watering tools called irrigation systems. They help keep her plants watered when there is not enough rain. Martha also knows that different plants need different types of soil. She plants some crops in sandy soil, and other crops in loam.

Farmers must time the picking, or harvesting, of their crops. Some crops grow faster than others. Look at the table. It shows how long it takes for seeds to grow to full-grown plants. Use the table to answer the questions.

1. Which plant matures the fastest?
   summer squash
2. Which takes longest? celery

3. If you had planted cucumbers in mid-May, about when would they be ready to pick? mid-July

### Critical Thinking

Why do you think it is important for farmers to know about biology?

Figure 7-11 *Farmers produce many different kinds of healthy foods.*

| Growth Times for Plants | |
|---|---|
| **Plant** | **Time to Full-grown** |
| Beets | 55 days |
| Carrots | 70 days |
| Celery | 105 days |
| Corn | 80 days |
| Cucumbers | 60 days |
| Green beans | 54 days |
| Green peppers | 65 days |
| Potatoes | 90 days |
| Summer squash | 53 days |
| Tomatoes | 75 days |

Figure 7-12 *Different plants grow at different rates.*

**Critical Thinking** because it will help them to produce a better crop

## Summary

- Plants are very important to our lives because they provide food, clothing, shelter, and fuel. Not only do we eat plants directly, but we eat animals that eat plants. Also, plants release oxygen for people and other animals to use for cellular respiration.

- There are six important characteristics of plants: 1) Plants carry out photosynthesis. 2) Plants are multicellular. 3) Plants have specialized cells. 4) The cells of plants have cell walls. 5) Plants usually stay rooted in one place. 6) Plants have three basic parts—roots, stems, and leaves.

- To grow, plants need food, oxygen, and water. They also need light, carbon dioxide, and minerals. The light usually comes from the Sun. The food comes from photosynthesis. The oxygen and carbon dioxide come from the air. The water and minerals come from the soil.

- Almost all plants must grow in soil. Soil is made up of rocks, minerals, water, air, and decayed plant and animal material called humus. For most plants, the best soil is loam, which is a good mixture of sand, silt, and clay.

| |
|---|
| chlorophyll |
| humus |
| loam |
| photosynthesis |
| soil |

## Vocabulary Review

**Complete each sentence with a term from the box.**

1. The organic material in soil is called _____. humus

2. Soil that has a well-balanced mixture of sand, silt, and clay is called _____. loam

3. A mixture of sand, silt and clay is called _____. soil

4. The process by which light energy is used to make food is called _____. photosynthesis

5. A substance in chloroplasts that absorbs light energy is called _____. chlorophyll

# Chapter Quiz

**Answer the questions in complete sentences. Write your answers on a separate sheet of paper.**

1. What are three different kinds of jobs that involve working with plants?
landscaper, florist, gardener

2. What are three reasons why plants are important?
Plants provide food, oxygen, medicine, wood, paper, and fabrics.

3. What fuels come from plants?
oil and coal

4. What are six characteristics of plants?

5. Why is plant photosynthesis important to you?
because it adds oxygen to the air for us to breathe

6. What do plants need in order to grow?
oxygen, water, food, light, carbon dioxide, and minerals

7. What do plants get from soil?
minerals and water

8. What things are contained in soil?
rocks, minerals, air, water, decaying plant and animal remains, sand, silt, and clay

9. What is the organic material in soil called?
humus

10. What type of soil is good for many plants? loam

**Test Tip**
It is a good idea to make an outline of the chapter to help you study for a test.

4. Plants carry out photosynthesis. Plants are multicellular. Plants have specialized cells. The cells of plants have cell walls. Plants cannot move from place to place. Plants have three basic parts: roots, stems, leaves.

## CRITICAL THINKING

11. Why do you think it might be important to save plants that are in danger of becoming extinct?

12. How are plants different from plantlike protists, known as algae?

**Critical Thinking**
11. because they are important to other living things and they may provide medicines for people to use

12. Plants are multicellular and have specialized cells.

*SciLINKS*
Go online to www.scilinks.org. Enter the code **PMB112** to research **classifying plants**.

## Research Project

Use the Internet or other references to find out about plants that grow in your area. Pick one of these plants to research. Write a report that includes a description of the plant, how the plants are classified, a list of its requirements, and how it is important. Also include a photo or drawing of the plant.

See the *Classroom Resource Binder* for a scoring rubric for the Research Project.

**Figure 8-1** *There is an enormous variety of plant life on Earth. In what ways do the plants shown here differ from each other?*

Possible answers: They vary in size, shape, color, number of leaves, types of flowers, and width of stem.

## Learning Objectives

- Describe the differences between vascular and nonvascular plants.
- List examples of vascular and nonvascular plants.
- Explain why seed plants are divided into two groups: flowering plants and cone-bearing plants.
- Describe how seed plants and nonseed plants reproduce.
- **LAB ACTIVITY:** Classify plants by their characteristics.
- **BIOLOGY IN YOUR LIFE:** Relate the importance of cone-bearing plants to everyday life.

# Chapter 8 ▶ Types of Plants

## Words to Know

| | |
|---|---|
| **vascular plants** | plants that have structures for transporting water |
| **nonvascular plants** | plants that do not have well-developed structures for transporting water |
| **rhizome** | (RY-zohm) the underground stem of a fern |
| **frond** | the leaf of a fern |
| **seed** | a protective covering that surrounds a young plant and its stored food |
| **plant embryo** | the early, undeveloped stage of a new plant |
| **germination** | the process by which a plant embryo develops and breaks out of the seed |
| **gymnosperm** | a plant that produces uncovered seeds |
| **angiosperm** | a plant that produces covered seeds and flowers |
| **monocot** | a flowering plant with one seed leaf in its seeds |
| **dicot** | a flowering plant with two seed leaves in its seeds |

## Classifying Plants

The plant kingdom includes a great deal of variety, from tiny mosses to huge redwoods. There are more than 270,000 species of plants known today. Biologists group plants in order to study them. In this chapter, you will learn about these different groups.

The plant kingdom has two main groups. One group of plants has special cells for transporting water, food, and minerals. These cells connect to form strands of transport tissue. Roots absorb water. Then, the water is moved up these strands to other parts of the plant.

The water molecules will be absorbed by the string, eventually wetting the string throughout.

### Think About It
The strands of transport tissue are like a piece of string. If you put one end of a string in water, what do you think will happen to the other end over time?

At the same time, plant leaves make food during photosynthesis. Some of this food can be moved down other strands of transport tissue to feed the rest of the plant. Plants that have these structures for transporting water and food are called **vascular plants**. Vascular plants are divided into three groups. Plants that do not have these special transport cells are called **nonvascular plants**. Water cannot be stored or transported in these plants. Look at Figure 8-2 to learn more about these plant groups.

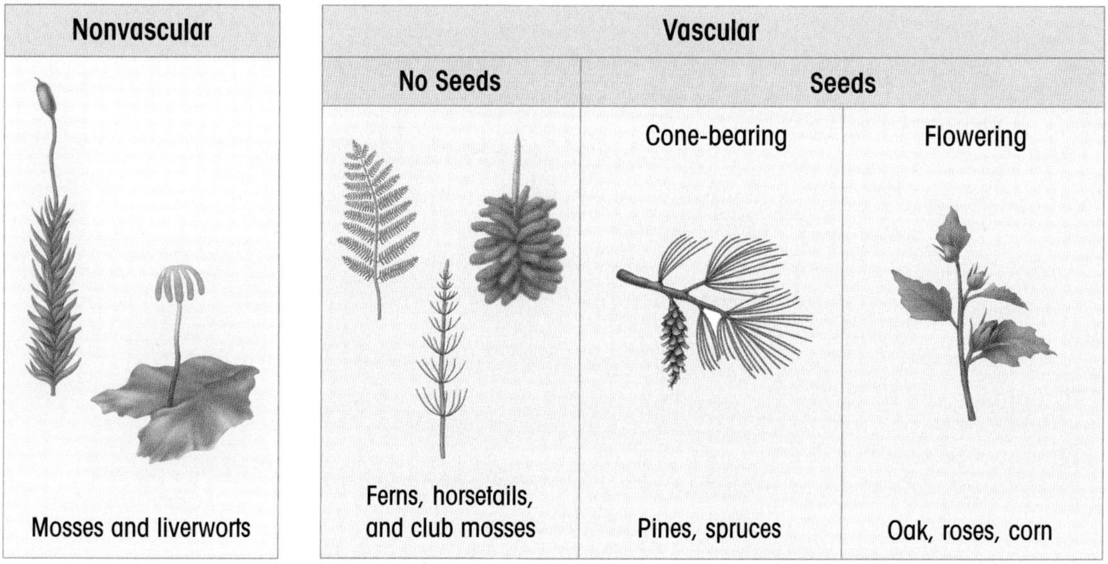

Nonvascular
Mosses and liverworts
Vascular
No Seeds
Ferns, horsetails, and club mosses
Seeds
Cone-bearing
Pines, spruces
Flowering
Oak, roses, corn


**Figure 8-2** *Plants can be divided into two main groups, nonvascular and vascular.*

## Nonvascular Plants: Mosses and Liverworts

Nonvascular plants do not have true roots, stems, or leaves. They do not have a way of storing or transporting water. Therefore, they must live in places that have a constant supply of water. These plants are usually very small. *Mosses* and *liverworts* are two kinds of nonvascular plants. These are some of the simplest plants on land. Biologists believe that mosses were the first plants to grow on land.

### Science Fact

Nonvascular plants are often the first plants to grow in a place. They are often called *pioneer plants.*

## Mosses

You may be familiar with mosses. These green plants often form soft carpets on forest floors. Sometimes they grow on logs or on wet rocks. Mosses grow well in moist, shady places.

Figure 8-3 *Mosses grow in moist, shady places.*

Mosses are different from algae because their cells are specialized. All cells in algae do the same work. In mosses, however, the cells of the rootlike parts are better at soaking up water than the cells of the leaflike parts are. Also, these rootlike parts are underground. The leaflike structures are found above ground. They carry out photosynthesis.

## Liverworts

Liverworts also grow in moist places. They can be found along stream banks, near springs, or in rain forests. A liverwort looks like a leathery leaf lying flat on the ground. It has hairlike structures that anchor the plant to the ground. These hairlike structures also absorb water from the soil.

**Everyday Science**

You often find moss at the base of a shady tree. The moss helps to keep the soil moist. Look around your neighborhood for moss. Keep a record of where you find it.

**Remember**
Algae are simple organisms that belong in the protist kingdom.

Figure 8-4 *Liverworts are small nonvascular plants.*

## The Life Cycle of Nonvascular Plants

The life cycle of nonvascular plants has two stages. One stage is a spore-producing stage. The other stage produces *gametes,* which are sex cells. This means that nonvascular plants reproduce using both asexual and sexual reproduction. Look at Figure 8-5 as you read about reproduction in nonvascular plants.

During the asexual stage, nonvascular plants produce spores. The spores are released and land on the ground. If the spores land in a moist place, they can grow into an adult. Then, the adult plants produce egg and sperm cells, or gametes. If the egg is fertilized, it can grow into a spore-producing plant. The spore-producing plant grows out of the gamete-producing plant. Then, the cycle starts all over again.

**Science Fact**

Reproduction that occurs in two stages is called *alternation of generations*. Many plants carry out this type of reproduction.

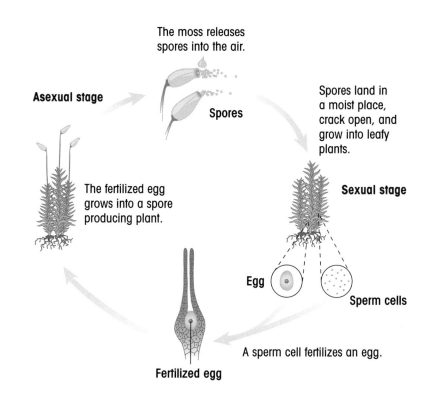

Figure 8-5 *The life cycle of a moss has two stages.*

## Simple Vascular Plants: Ferns and Horsetails

Ferns and horsetails are simple vascular plants. They have specialized cells for transporting food and water. They also have true roots, stems, and leaves. These plants grow in woods, swamps, and gardens where there is a lot of water. Some ferns grow several feet tall. There are tree ferns in tropical areas that grow up to 30 feet (9 meters) tall!

### Fern Structure and Function

The stems of ferns are underground. These grow horizontally just beneath the soil surface. The underground stems of ferns are called **rhizomes**. The leaves of ferns are called **fronds**. They grow up from the stem. The roots grow down from the stem.

Figure 8-6 *Ferns are simple vascular plants.*

## A Closer Look

**PLANTS AS FUEL**

According to the geologic record, 300 million years ago, giant fern forests covered Earth. The land in those days was wet and marshy. The climate was warmer than it is now. Tree ferns, 30 feet high, were common. The greenery then was very lush and thick.

These plants of millions of years ago are the source of coal today. Here is how the coal formed. As the giant ferns died, they fell to the ground. Eventually, they became covered with soil. The soil pressed down on the plant matter. More ferns died and more soil was packed down. Over millions of years, the compressed plants became coal. This coal is still found underground. Coal miners dig up the coal. Then, the coal is burned for energy.

Figure 8-7 *Plants that grew millions of years ago included ferns, horsetails, and other simple plants.*

**CRITICAL THINKING** Oil and coal are sometimes called fossil fuels. Why is this an appropriate name?
The coal comes from plant remains that are millions of years old and found underground, like fossils.

Figure 8-8 *Spore cases form on the underside of a fern frond.*

## The Life Cycle of Ferns

Ferns reproduce in a way that is similar to the way mosses and liverworts reproduce. Reproduction happens in two stages. Look at Figure 8-9 while you read about reproduction in ferns.

The first stage in fern reproduction is asexual. There is only one parent cell, the spore. The second stage is sexual. There are two parent cells, the sperm and the egg.

The brown spots on the underside of a frond are spore cases. When these cases open, spores are released. If a spore lands in a good place for growing, a small plant grows. This small plant does not look like the parent fern, however. It is a small, heart-shaped plant.

Then, the second stage in fern reproduction begins. The heart-shaped plant forms sperm and egg cells. The sperm cells swim to the egg cells through dew or rainwater on the plant. A new fern grows from a fertilized egg. Then, the cycle starts all over again.

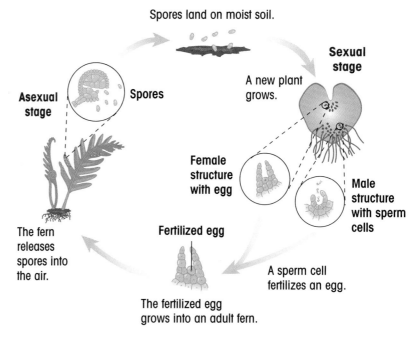

Spores land on moist soil.

**Sexual stage**

A new plant grows.

**Asexual stage**

**Spores**

**Female structure with egg**

**Male structure with sperm cells**

The fern releases spores into the air.

**Fertilized egg**

A sperm cell fertilizes an egg.

The fertilized egg grows into an adult fern.

Figure 8-9 *The life cycle of a fern has two stages.*

✓ **Check Your Understanding**

Answer the questions on a separate sheet of paper.

1. What are transport cells?
   cells that carry water and nutrients
2. Where do mosses grow?
   in moist, shady places
3. How are mosses different from algae?
   They have specialized cells.
4. How do ferns reproduce?
   Ferns reproduce asexually by using spores but also reproduce in a sexual stage with egg and sperm cells.

## Complex Vascular Plants: The Seed Plants

Ferns are a simple kind of vascular plant. Seed plants are more complex vascular plants. Most of the plants you know are seed plants. Wheat, peas, broccoli, apples, and oranges all come from seed plants. Seed plants produce seeds instead of spores.

### Seed Structure and Function

A **seed** is a reproductive structure that contains the first stage of a tiny, new plant. A seed coat surrounds and protects the seed. The tiny, new plant inside the seed is called a **plant embryo**. Most of the seed consists of a food supply for the embryo. The food gives the embryo a quick start when conditions are right for it to grow into a mature plant.

To grow, seeds must have water. Most plants produce many seeds. This is because many seeds do not land in a place that has enough water. Therefore, the seeds dry out and die.

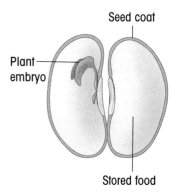

Figure 8-10 *A seed contains a plant embryo. It also contains a source of stored food.*

### Germination

A seed becomes a plant by a process called **germination**. When a seed germinates, the embryo inside it starts developing and growing.

In order to germinate, the seed needs the right temperature and moisture. It also needs enough oxygen. Special chemicals in the seed tell it when it is time to germinate.

**Figure 8-11** *After a seed germinates, it grows into an adult plant.*

## Grouping Seed Plants: Cones and Flowers

Seed plants are divided into two groups: cone-bearing plants and flowering plants. Cone-bearing plants belong to a group of plants called **gymnosperms**. They produce seeds that are uncovered. Many cone-bearing plants have needles for leaves. The seeds are found in the cones.

### Think About It

Some cone-bearing plants do not lose their leaves, or needles, in the winter. Why do you think the term *evergreen* is used to describe these plants?

Possible answer: because they "stay green," or keep their needles, all year

**Figure 8-12** *The seeds of a cone-bearing plant are found inside its cones. Pairs of seeds have wings to help them scatter.*

Flowering plants are also called **angiosperms.** The seeds of flowering plants are considered covered seeds. This is because the seed is surrounded by some kind of fruit.

### Monocots and Dicots

There are two main types of seeds in flowering plants. All seeds from flowering plants contain at least one seed leaf, or *cotyledon.* This tiny leaf is located inside the plant embryo. Plants that contain one seed leaf are called **monocots.** Plants that contain two seed leaves are called **dicots.**

Figure 8-13 *This flower comes from a dogwood tree. Dogwoods are angiosperms, or flowering plants.*

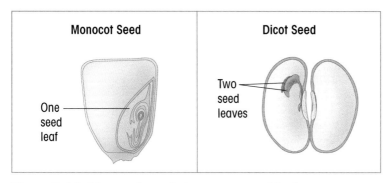

| Monocot Seed | Dicot Seed |
|---|---|
| One seed leaf | Two seed leaves |

Figure 8-14 *Monocot seeds have one seed leaf. Dicot seeds have two.*

There are many more kinds of flowering plants than other kinds of plants. Daisies, grass, corn, roses, lemon trees, and maple trees are all flowering plants. You will read more about cone-bearing and flowering plants in the next two chapters.

## Science Fact

There are over 250,000 different species of flowering plants on Earth! Flowering plants produce much of our food, including corn, wheat, apples, and tomatoes.

### ✓ Check Your Understanding

Answer the questions on a separate sheet of paper.

1. What are three examples of vascular plants?

2. What is the difference between monocot and dicot seeds?

3. What do seeds need in order to germinate?

4. **CRITICAL THINKING** Why do seed plants produce so many seeds?

### Check Your Understanding

1. ferns, wheat, pine tree

2. monocot seeds have one seed leaf, dicot seeds have two seed leaves

3. warmth and water

4. because many of the seeds will not live to become full-grown plants

# LAB ACTIVITY
## Classifying Plants by Their Characteristics

### BACKGROUND
Plants are grouped by how they transport water and how they reproduce.

### PURPOSE
In this activity, you will classify plants by their characteristics.

### MATERIALS
examples of mosses, ferns, cone-bearing plants, and flowering plants

### WHAT TO DO
1. Examine the plants your teacher gives you. Use the questions in the chart below to classify the plants.

2. Draw and label the plants you classified on a separate sheet of paper.

Figure 8-15 *How would you classify this plant?*

| Plant Classification Key | | |
|---|---|---|
| **Question** | **Answer** | |
| | Yes | No |
| A. Is the plant small, with no real roots? | It is a moss. | Go to B. |
| B. Does the plant have fronds? | It is a fern. | Go to C. |
| C. Does the plant have needles and cones? | It is a cone-bearing plant. | Go to D. |
| D. Does it have flowers or fruit? | It is a flowering plant. | Review A, B, and C again. |

Figure 8-16 *Use this key to classify plants.*

### DRAW CONCLUSIONS
- How many of each type of plant did you identify?
  Students should identify four to five different types of plants
- The plant in Figure 8-15 is called *Equesetum*. This plant has transport cells. It does not make cones or seeds. In which group of plants do you think this plant belongs?
  Possible answer: The plant is a simple vascular plant. It is in the same category that ferns are in. Note: The plant shown is a horsetail.

# BIOLOGY IN YOUR LIFE
## Studying Cone-Bearing Plants

The next time you look through a book or magazine, think of a spruce tree. The wood pulp from spruce trees is used to make paper. Spruce trees are cone-bearing plants. Pines, cedars, and hemlocks are examples of cone-bearing plants.

Figure 8-17 *The wood from cone-bearing trees is used to make paper.*

We use products from cone-bearing plants every day. Most lumber comes from cone-bearing trees. In many houses, the structure, inside and outside walls, doors, and furniture are all made from lumber. Fence posts, broomsticks, toys, and pencils are also made from lumber.

Cone-bearing trees also are used to produce many types of paper. The fibers in wood pulp are used to make rayon, film, and solid rocket fuel. Pine saps are used in turpentine, asphalt, paint, chewing gum, rubber, and soaps. Pine oils give some cleaning products a fresh scent.

The use of cone-bearing trees for products must be managed wisely. Many forests made up of these trees are being destroyed. We do not want to lose this important resource.

**Answer the questions in complete sentences. Write your answers on a separate sheet of paper.**

1. What are three products made using lumber? houses, furniture, fences

2. What are three products made using fibers from cone-bearing trees? tissue, wrapping paper, writing paper

**Critical Thinking**
to conserve forest resources

**Critical Thinking**
Why is it important to recycle paper and cardboard?

## Summary

- The plant kingdom is divided into two big groups: nonvascular plants and vascular plants. Nonvascular plants cannot move water and food between different parts of the plant. Vascular plants have special cell tissues for transporting food and water.

- Mosses and liverworts are nonvascular plants. Mosses reproduce by spores. Mosses and liverworts are usually found in wet, shady places.

- Ferns are simple vascular plants. Ferns reproduce by spores. They are usually found in wet, shady places.

- Seed plants are complex vascular plants. They reproduce by seeds. A seed contains a plant embryo and a food supply. It is covered by a protective coat. A seed can live a long time before germinating.

- Seed plants are divided into two groups: cone-bearing plants and flowering plants. Flowering plants have two types of seeds: monocot seeds and dicot seeds.

---

angiosperm

frond

germination

gymnosperm

nonvascular plants

rhizome

vascular plants

---

# Vocabulary Review

Write the term from the list that matches each of the following definitions.

1. plants that have structures for transporting water vascular plants

2. the process by which a plant embryo develops and breaks out of the seed germination

3. plants that do not have structures for transporting water nonvascular plants

4. the underground stem of a fern rhizome

5. the leaf of a fern frond

6. a plant that produces uncovered seeds gymnosperm

7. a plant that produces covered seeds and flowers angiosperm

# Chapter Quiz

**Answer the questions in complete sentences. Write your answers on a separate sheet of paper.**

1. What are the differences between vascular plants and nonvascular plants?
   Vascular plants have transport tubes, nonvascular plants do not.
2. Why do you think nonvascular plants are small and close to the ground?
   because they cannot transport water
3. Describe the characteristics of mosses and liverworts.
   Mosses and liverworts are small, nonvascular plants that live in moist shady places.
4. How does reproduction happen in ferns?
   Ferns reproduce using spores.
5. How does reproduction happen in seed plants?
   Seed plants reproduce using seeds.
6. List two examples each of vascular and nonvascular plants.
   vascular: apple tree, ferns; nonvascular: mosses, liverworts
7. Name five foods that come from seed plants.
   peas, corn, wheat, apples, pears
8. Describe a seed.   A seed contains a plant embryo and a food source, surrounded by a seed coat.

**CRITICAL THINKING**

9. Make a branching diagram showing the different plant groups. Use the section titles of the chapter to help you.

10. Simple organisms usually appear in geologic time earlier than more complex organisms do. Based on this information, which do you think existed first, ferns and horsetails or seed plants?

**Test Tip**
Always try to write answers in complete sentences. This helps organize your thoughts and improves your writing.

**Critical Thinking**
9. Diagrams should contain the titles *plant kingdom, vascular plants, nonvascular plants, ferns and horsetails, seed plants, flowering plants,* and *cone-bearing plants.*

10. Ferns and horsetails probably existed before seed plants.

*SC*<sub></sub>*LINKS*
Go online to www.scilinks.org. Enter the code **PMB114** to research **vascular and nonvascular plants**.

---

## Research Project

Research vascular and nonvascular plants. Make a poster comparing the two groups of plants that includes a diagram and eight to ten facts about each type of plant. Use the Internet or other sources to find information.

---

See the *Classroom Resource Binder* for a scoring rubric for the Research Project.

Figure 9-1 *Mangrove trees grow in salty water. Special cells help the plants remove salt from their tissues. The roots of these plants are also different from other plants. They support the plants and take in air. What active process do you think the special cells use to remove salt from their tissues?*

active transport

## Learning Objectives

- Explain the structure and functions of roots, stems, and leaves.

- Compare two kinds of roots.

- Compare two kinds of stems.

- Describe the two types of transport tissue.

- Give examples of plant responses.

- **LAB ACTIVITY:** Compare a monocot and a dicot.

- **ON-THE-JOB BIOLOGY:** Explain what a horticulturist does.

## Words to Know

| | |
|---|---|
| **taproot** | one thick root that grows larger than the other roots |
| **fibrous roots** | many small, thin roots |
| **epidermis** | a thin outer layer of cells |
| **root hair** | a tiny hairlike structure on the outer layer of a root |
| **cortex** | the second layer of a root made of soft, loose tissue |
| **root cap** | the mass of cells that covers and protects a root tip |
| **phloem** | (FLOH-em) a type of tissue in roots, stems, and leaves that carries food from the leaves to other plant parts |
| **xylem** | (ZY-luhm) a type of tissue in roots, stems, and leaves that carries water up a plant to the leaves |
| **herbaceous stem** | a smooth, soft, and green stem |
| **woody stem** | a hard stem that is not green |
| **blade** | the broad, flat part of a leaf |
| **petiole** | a stalk that connects the leaf to the plant |
| **vein** | a part of a leaf that contains transport tissues |
| **stoma** | a tiny opening in a leaf that allows gases into and out of the leaf —plural *stomata* |
| **tropism** | the growth of a plant in a certain direction in response to a stimulus |

### Linking Prior Knowledge

Students should recall how plants make food through photosynthesis (Chapter 4), plant organs (Chapter 7), and vascular and nonvascular plants (Chapter 8).

# Roots

Roots are one kind of plant organ. Roots are the parts of a plant that grow underground. Although most roots grow in soil, some roots can grow in water, in rocks, and even in air.

Roots have many functions. They hold a plant in place. They absorb water and nutrients for the plant. Some roots also store food for the plant. Sweet potatoes, carrots, and beets are the roots of certain types of plants.

## Two Kinds of Roots

There are two main kinds of roots: taproots and fibrous roots. A **taproot** is one thick root that grows larger than the other roots. Some taproots store food produced by photosynthesis. Carrots, beets, and radishes are examples of taproots. Taproots often grow deep into the soil.

**Fibrous roots** are made up of many small, thin roots. Fibrous roots spread out. They do not grow as deep into the soil as taproots. Grass, wheat, and barley come from plants that have fibrous roots.

Figure 9-2 *Taproots can grow deep into the ground.*

Figure 9-3 *Fibrous roots spread out in the ground.*

# The Structure of Roots

A root is made up of three layers. The outer part is called the **epidermis**. It is made up of one layer of cells. Many tiny, hairlike structures, called **root hairs** grow out of this layer. Root hairs increase the surface area of a root. A greater surface area helps a plant to absorb more water and nutrients. Each root hair is a single cell. Each cell has a very thin cell wall. Thin cell walls help root hairs absorb water very easily.

The second layer of a root is called the **cortex**. This layer has soft, loose tissue. Food made during photosynthesis can be stored there.

The third layer is the inner part of a root. It is made up of transport tissue. These tissues transport water and food produced by photosynthesis throughout a plant.

New cells form in the tip of a root. It is covered by a mass of cells called the **root cap.** The root cap protects the tip of the growing root. It also helps the root to grow deeper into the soil.

**Think About It**

What do you think
would happen
to a plant if its
root hairs were
damaged?

The plant may not be able
to absorb enough water
and nutrients, and may die.

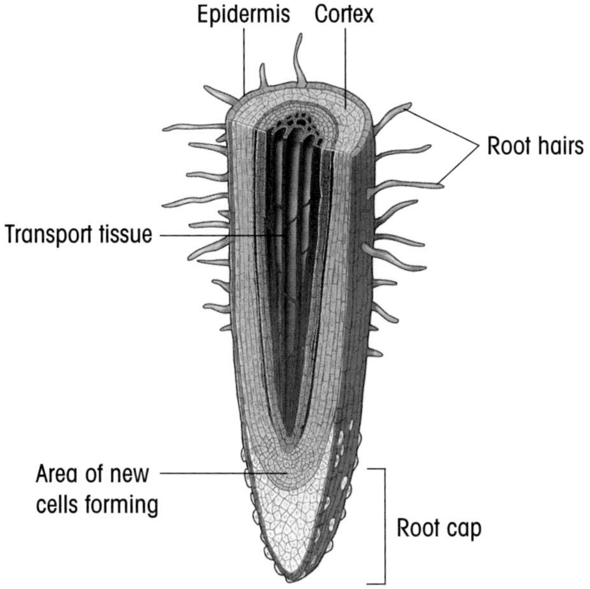

**Figure 9-4** *The parts of a root*

## Everyday Science

Do you eat celery? If you do, have you ever peeled away the "strands" found inside a stalk of celery? The strands are actually xylem and phloem tissues. What do you think the strands do for the plant?

They carry water, nutrients, and food to different parts of the plant.

### Two Types of Transport Tissue

One type of transport tissue carries water and minerals from the roots of a plant up to the leaves. This type of tissue is called **xylem**. The other type of transport tissue carries dissolved food made in the leaves of a plant to other plant parts. This type of tissue is called **phloem**.

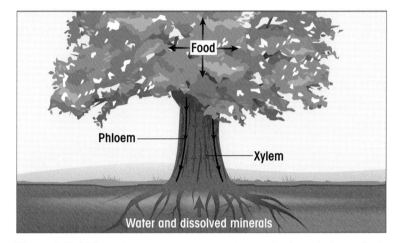

Figure 9-5 *Xylem tissue transports water and minerals in a plant. Phloem tissue transports food made in the leaves of the plant.*

## Stems

Stems are the parts of a plant that connect the roots and the leaves. Stems hold up the leaves so that the leaves can receive sunlight. Photosynthesis takes place in the leaves.

Stems carry water, nutrients, and minerals to different parts of a plant. Stems contain transport tissues. These tissues transport water and nutrients between the roots and the leaves. Stems can also store food.

Transport tissue is arranged differently in the stems of monocots and dicots. In monocots, the tissues are scattered throughout the stems. In dicots, the tissues form a ring. You will see this arrangement in the Lab Activity on page 146.

Figure 9-6 *The stems of a sugar cane plant can store large amounts of sugar.*

## Two Types of Stems

There are two types of stems: herbaceous stems and woody stems. **Herbaceous stems** are smooth, soft, and green. Plants with these kinds of stems do not usually grow very tall. Often, these plants do not usually live for more than one growing season. Tomato plants, daisies, and corn plants have herbaceous stems.

**Woody stems** are hard and usually not green. They can grow taller and wider for many seasons. The trunks of trees are woody stems. The rough outer part of a woody stem is called the bark. Rose bushes and pine trees have woody stems.

**Think About It**

Celery and asparagus are stems that many people eat. Do you think that they are herbaceous stems or woody stems?

herbaceous stems

Figure 9-7 *A tomato plant has an herbaceous stem.*

Figure 9-8 *A rose has a woody stem.*

You read that stems contain xylem and phloem tissues. In woody stems, there is also a special layer of growth tissue. This tissue is called the *cambium*. The cambium produces new layers of xylem and phloem cells. As a plant grows from season to season, new layers of these cells are added to the outermost layer of its woody stem. The stems become wider and taller.

The layers of xylem and phloem cells form ringlike patterns in a stem. These rings are called annual rings. You can count the rings in the trunk of a tree to determine the tree's age. Look at the rings in the tree trunk in Figure 9-9.

Annual rings          Cambium

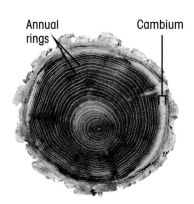

Figure 9-9 *The age of a tree can be determined by counting the tree's annual rings.*

**Check Your Understanding**

**1.** taproots and fibrous roots
**2.** A root cap protects the tip of the growing root.
**3.** herbaceous stems and woody stems **4.** Xylem tissue carries water from the roots up to the leaves. Phloem tissue carries dissolved food made in the leaves to other plant parts.

✓ **Check Your Understanding**

Write your answers in complete sentences.

**1.** What are the two main kinds of roots?

**2.** What does a root cap do for a plant's roots?

**3.** What are the two types of stems?

**4.** CRITICAL THINKING  How is xylem tissue different from phloem tissue?

## On the Cutting Edge

### HYDROPONICS

Soil gives plants minerals and other nutrients. Plants can, however, grow without soil. Growing plants without soil is called hydroponics. In using hydroponics, other types of materials support plants. These materials include water, sand, or gravel. These materials do not give plants the minerals and nutrients that they need. Instead, a chemical mixture of the needed minerals and nutrients is dissolved in water. This chemical mixture is usually pumped into the material that the plants are growing in.

There are many advantages to growing plants using hydroponics. Because soil is not needed, this method can be used to grow plants almost anywhere. Plants that cannot be grown in certain areas can be grown using hydroponics. Also, many plants grown using hydroponics do not have problems with insects and weeds. A disadvantage to this method is that the equipment needed can be very costly.

Figure 9-10 *Plants grown using hydroponics do not need soil.*

**Critical Thinking**  Possible answers: Some countries may have poor soil. Hydroponics does not use soil, so this method can be used in these countries. However, poor countries would not be able to afford the expensive costs.

CRITICAL THINKING  Do you think hydroponics might be a solution to the food production problems in some countries?

# Leaves

Leaves are the green structures that grow from stems. People eat many kinds of leaves, such as lettuce, spinach, and cabbage.

Most leaves are broad and flat. The large surface area helps leaves to absorb as much light as possible. However, a few kinds of plants have needlelike leaves, such as spruce trees and pine trees. These types of leaves help prevent the plant from losing too much water.

Leaves are the food makers for most plants. Some stems can carry out photosynthesis. Very few roots can. Most of a plant's chlorophyll is in its leaves. Chlorophyll absorbs energy from light. The leaf's most important job is to get light energy for photosynthesis.

### Main Parts of a Leaf

There are three main parts to a leaf: the blade, the petiole, and the veins. The broad, flat part of a leaf is called the **blade**. The blade is where most of the food making takes place.

A leaf is connected to the plant by a stalk called a **petiole**. In some plants, the leaves do not have petioles. They grow attached to the stem of the plant.

A leaf contains a system of **veins**. The veins in a leaf are made of xylem and phloem tissues. The veins in a leaf connect to the xylem tissue and phloem tissue in the stem of the plant. The xylem tissue carries water to the leaf. The phloem tissue carries the food made from photosynthesis out of the leaf.

The veins in the leaves of monocots are arranged side by side. The veins in a dicot are branched. Look at the leaf in Figure 9-11. Is it from a monocot or a dicot?

dicot

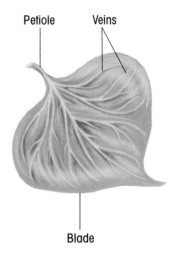

Figure 9-11 *The main parts of a leaf*

# The Structure of a Leaf

The outermost layer of a leaf is called the epidermis. This layer prevents a leaf from losing too much water. The layer under the epidermis is called the *palisade layer*. Most photosynthesis takes place here. The next layer is called the *spongy layer*. This layer contains many air sacs. It is involved in the exchange of carbon dioxide and oxygen in a leaf.

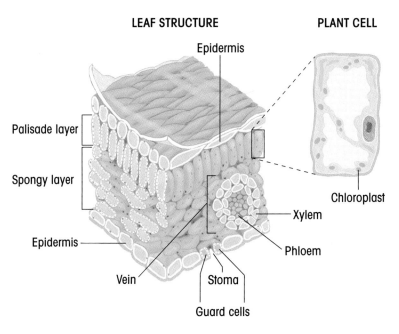

Figure 9-12 *A leaf is made up of three layers of cells.*

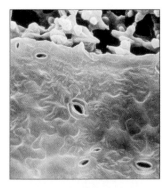

Figure 9-13 *There are usually more stomata found in the underside of a leaf than on the top of a leaf.*

Scattered throughout the epidermis are tiny openings called **stomata**. These openings allow water and oxygen to pass into and out of a leaf. They also absorb oxygen and carbon dioxide from air.

The size of the stomata are controlled by *guard cells*. When the plant is dry, the guard cells around the stomata shrink. The stomata close to prevent the plant from losing water. When the plant is full of water, the guard cells swell. The stomata open to release water. By opening and closing the stomata, guard cells control water loss.

# Plant Responses

Plants respond to changes in their environments. Each response is caused by a stimulus. Plants respond to light, water, and gravity. Some plants can respond to touch. The growth of a plant in a certain direction in response to a stimulus is called a **tropism**.

**Reminder**
A stimulus is a change that causes a response.

### Tropisms

Different plant parts may respond to different kinds of stimuli. The roots of a plant may grow toward water. This type of tropism is called *hydrotropism*.

Gravity is a stimulus that can cause a response in roots and stems. Roots grow down into the ground in response to the pull of gravity. Stems grow above ground away from the pull of gravity. This type of tropism is called *gravitropism*.

Stems and leaves respond to light by growing toward the light. This type of tropism is called *phototropism*. The stems of some plants can also respond to touch. When the stem of a vine touches an object, it grows around the object. This type of tropism is called *thigmotropism*.

### Another Type of Response

Another type of plant response does not involve growth in the direction of a stimulus. The response is due to changes in cell pressure. It is called a *nastic response*. For example, a Venus' flytrap plant closes its leaf when the plant senses an insect crawling on the leaf. The insect causes changes in cell pressure in the leaf.

Figure 9-14 *The Venus' flytrap plant shows a nastic response.*

## ✓ Check Your Understanding

Answer the questions in complete sentences.

**1.** What are the three main parts of a leaf?

**2.** CRITICAL THINKING Describe how phototropism can help a plant to survive.

**Check Your Understanding**
1. the blade, the petiole, and veins  2. By growing toward light, the plant can get more energy to perform photosynthesis.

# LAB ACTIVITY
## Identifying a Monocot and a Dicot

### BACKGROUND
Transport tissue is arranged differently in the stems of monocots and dicots.

### PURPOSE
In this activity, you will identify a monocot and a dicot by examining the arrangement of transport tissue in the stems of some plants.

### MATERIALS
two prepared slides, one labeled *A* and the other labeled *B*; microscope; pencil; paper   Refer to Teacher's Note below

### WHAT TO DO
1. Look at Figure 9-15. It shows the stem tissue of a monocot.

2. Look at Figure 9-16. It shows the stem tissue of a dicot. Notice the difference in the arrangement of the transport tissue in both photos.

3. Look at the prepared slide labeled *A* under a microscope. Observe the slide under low power and then under high power.

4. On a separate sheet of paper, draw what you see. Label the transport tissue. Make sure you label your drawing "Slide A."

5. Repeat Step 3 using the slide labeled *B*.

6. Repeat Step 4 but label your drawing "Slide B."

### DRAW CONCLUSIONS
- Which slide contained stem tissue from a dicot? How do you know?  Slide A; the transport tissues form a ring.

- Which slide contained stem tissue from a monocot? How do you know?  Slide B; the transport tissues are scattered throughout.

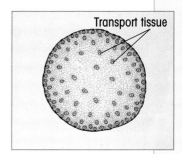

**Figure 9-15**
*A monocot stem*

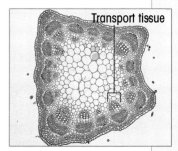

**Figure 9-16**
*A dicot stem*

**Teacher's Note**
Be sure to place masking tape over the information on each slide. Every dicot slide should be labeled *A*. Every monocot slide should be labeled *B*.

# ON-THE-JOB BIOLOGY
## Horticulturist

A horticulturist works with plants. Tina Johnson is a horticulturist. A part of her job is to help many different types of plants grow better. Tina has helped vegetables grow more roots, stems, or leaves. More plant parts can lead to healthier plants. If these parts can be eaten, then more food can be harvested from the plants.

As a horticulturist, Tina tries to find ways of growing more plants in less space. She also looks for ways to keep fruits and vegetables fresh longer. Her work helps both farmers and consumers.

Figure 9-17 *A horticulturist may work to improve the health of plants.*

Horticulturists may have different specialties. This means that they only study certain things about plants. The table below shows some specialties.

**Use the table to answer the questions. Write your answers on a separate sheet of paper.**

| Different Specialties of Horticulturalists | |
|---|---|
| **Horticulture Specialty** | **What Is Studied** |
| Landscape architecture | Design |
| Ornamental horticulture | Landscape plants |
| Post-harvest physiology | Plant storage and processing |
| Viticulture | Grapes |
| Olericulture | Vegetables |

**Critical Thinking**
What kind of specialty or specialties do you think Tina Johnson has?

**Critical Thinking**
Possible answer: olericulture and post-harvest physiology.

1. What is the specialty of a person who creates a new variety of grapes?
   viticulture
2. Combine two horticulture specialties. What does a person with those two specialties do with plants? Possible answer: ornamental horticulture and olericulture; A person with these specialties might use vegetable plants as landscape plants.

# Chapter

# 9 ▶ Review

## Summary

- Roots, stems, and leaves are all plant organs. Roots are the underground part of plants. There are two kinds of roots. A taproot is one thick root that grows larger than the other roots. Fibrous roots are made of many small, thin roots.

- Stems are the parts of plants that connect the roots and leaves. Stems hold up the leaves of a plant. Xylem tissue and phloem tissue found in roots, stems, and leaves transport water, minerals, and food in the plant.

- Photosynthesis takes place in the leaves of most plants. The broad flat part of a leaf is the blade. A leaf contains a system of veins. The stalk that attaches a leaf to a plant is called the petiole.

- Leaves have three layers: the epidermis, the palisade layer, and a spongy layer. Stomata are tiny openings in leaves.

- The growth of a plant in a certain direction in response to a stimulus is called a tropism.

| |
|---|
| blade |
| epidermis |
| herbaceous stems |
| phloem |
| root cap |
| root hair |
| taproot |

## Vocabulary Review

**Complete each sentence with a term from the list.**

1. One thick root that grows larger than the other roots is called the _____. taproot

2. Stems that are smooth, soft, and green are _____. herbaceous stems

3. The thin outer layer of cells is called the _____. epidermis

4. The mass of cells that covers and protects a root tip is called the _____. root cap

5. The broad flat part of a leaf is called the _____. blade

6. A tiny hairlike structure on the outer layer of a root is called a _____. root hair

7. The type of tissue in roots, stems, and leaves that carries food from the leaves to other plant parts is called _____. phloem

# Chapter Quiz

**Write your answers in complete sentences.**

1. What vegetables are roots?
   Accept any of the following: carrots, beets, turnips, and radishes.
2. What plant organ holds up the leaves of a plant?
   the stem
3. What kind of tissue carries water and nutrients from the roots up to the leaves? xylem

4. What type of root is made up of many small, thin roots? fibrous roots

5. How do oxygen, carbon dioxide, and water vapor enter and leave plants? through the stomata

6. Which part of a leaf connects the leaf to the plant? the petiole

7. What kinds of stimuli can cause tropisms in plants? water, light, gravity, and touch

8. What type of tropism causes a plant's roots to grow toward water? hydrotropism

**CRITICAL THINKING**

9. Which type of root do you think absorbs more water, a tap root or fibrous roots? Explain your answer.

10. During a drought, do you think guard cells would swell or shrink? Explain your answer.

**Test tip**
Answer the questions that you are sure of first. Then, go back and answer those questions that you need to think more about. Be sure that the number of each answer matches the number of its question.

**Critical Thinking**
9. Possible answer: fibrous roots because they have more root hairs than taproots
10. Guard cells would shrink so that the plant would not lose water.

**SCLINKS**
Go online to www.scilinks.org. Enter the code **PMB116** to research **autumn leaves**.

## Research Project

During autumn, the leaves of many trees change color and fall from the trees. Find out where the different colors come from. Use the Internet, encyclopedias, and other resources. Write a short report that explains why leaves go through these color changes.

See the *Classroom Resource Binder* for a scoring rubric for the Research Project.

**Figure 10-1** *Flowers are more than pretty parts of a plant. Flowers contain structures that help a plant to reproduce. What is needed to produce a new plant?*

Possible answer: seeds

# Learning Objectives

- Explain how flowers help a plant to reproduce.

- List the parts of a flower.

- Describe pollination and fertilization in plants.

- Give examples of asexual reproduction in plants.

- **LAB ACTIVITY:** Measure root growth in leaf cuttings.

- **BIOLOGY IN YOUR LIFE:** Find out how to attract pollinators to a garden.

# Chapter 10 ▷ Reproduction in Seed Plants

## Words to Know

| | |
|---|---|
| **petal** | a part of a flower that is often brightly colored |
| **sepal** | a special kind of leaf that protects a flower bud |
| **pistil** | the female part of a flower |
| **ovary** | the bottom part of a pistil where female reproductive cells are formed |
| **ovule** | the part of the ovary that contains the female reproductive cells |
| **stamen** | the male part of a flower |
| **pollen** | a light and powdery substance that contains the male reproductive cells of a plant |
| **pollination** | the movement of pollen from a stamen to a pistil |
| **fertilization** | the joining of a male reproductive cell and a female reproductive cell |
| **vegetative propagation** | a kind of asexual reproduction that uses parts of plants to grow new plants |

## Flowers Help a Plant to Reproduce

Flowers are the brightly colored parts of many plants. Some flowers have a pleasant smell to them. The bright colors and the pleasant smell are attractive to certain animals. Attracting animals is very important to some plants. These plants need animals to help them reproduce.

In this chapter, you will learn how flowers help plants to reproduce. You will also learn how flowers, fruits, and cones help in the production of seeds.

**Linking Prior Knowledge**
Students should recall that seed plants are plants that have flowers or cones, and that seeds contain plant embryos (Chapter 8).

### The Parts of Flowers

The parts of flowers that are often brightly colored are called **petals**. Petals protect the reproductive parts inside a flower. Monocots have flowers with petals arranged in groups of three. Dicots have flowers with petals arranged in groups of four or five.

The special leaves found below the petals are called **sepals**. Sepals protect a flower bud.

The parts that help the plant reproduce are in the center of a flower. The two reproductive parts are the stamens and the pistil.

The **pistil** is the female part of a flower. The top part of a pistil is called the *stigma*. Below the stigma is a tube called the *style*. The bottom of a pistil is called the **ovary**. Inside the ovary are **ovules**. The ovules contain the female reproductive cells.

The **stamen** is the male part of a flower. The tip of the stamen is called the *anther*. The anther produces pollen. **Pollen** is a light and powdery substance that contains the male reproductive cells in a plant. The reproductive cells are contained in pollen grains. The *filament* is a thin stalk that holds up the anther.

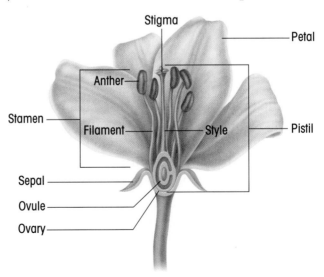

**Figure 10-2** *The parts of a flower*

# Pollination

For reproduction to take place, a male reproductive cell must join with a female reproductive cell. In flowering plants, pollen must get to a pistil. The movement of pollen from a stamen to a pistil is called **pollination**. The tops of pistils are usually sticky to capture pollen.

Pollination can occur in many ways. When an insect brushes against a flower, pollen can stick to the insect. The insect can then move the pollen to the pistil of another flower. Other animals, wind, and water can also pollinate plants.

### Self-Pollination

In some flowers, the stamens are taller than the pistils. Pollen from these stamens can easily fall onto the pistils. The movement of pollen from the stamen of one flower to the pistil of the same flower is called *self-pollination*. Self-pollination can also occur between the stamen of one flower and the pistil of another flower on the same plant.

Figure 10-3 *You can see the pollen on the bee as it pollinates the flower.*

**Safety Alert**

Pollen can cause an allergic reaction in many people. If you are allergic to pollen, try to stay away from flowering plants. If you have serious allergy symptoms, you should see a doctor.

**Science Fact**

Chance plays a big part in pollination. A plant produces many grains of pollen. However, only a very few will find their way to a pistil. Much of the pollen carried by the wind does not land on any plants at all.

Cross-Pollination

Sometimes pollen from the stamen of a flower can land on the pistil of a flower on another similar plant. This type of pollination is called *cross-pollination*. Cross-pollination also can occur between two different plant species.

## Fertilization

After a pollen grain lands on the top of a pistil, the pollen grain begins to grow a long tube. This tube grows down into the ovary. The tube carries the male reproductive cells contained in the pollen grain. Once in the ovary, a male reproductive cell joins with a female reproductive cell in the ovule. The joining of the male reproductive cell and the female reproductive cell is called **fertilization**. The ovule begins to develop into a seed. A new plant can grow from the seed. Figure 10-4 shows fertilization in a flowering plant.

**Remember**
Every seed has an undeveloped plant inside of it called a plant embryo.

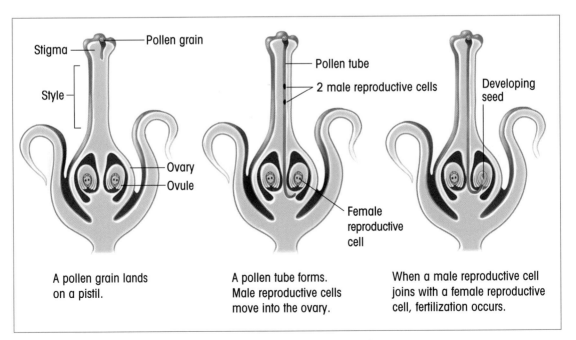

Stigma — Pollen grain
Style
Ovary
Ovule

A pollen grain lands on a pistil.

Pollen tube
2 male reproductive cells

Female reproductive cell

A pollen tube forms. Male reproductive cells move into the ovary.

Developing seed

When a male reproductive cell joins with a female reproductive cell, fertilization occurs.

Figure 10-4 *Fertilization occurs in the pistil of a flowering plant.*

## Fruits Protect Seeds

An ovary can contain one ovule or many ovules. After fertilization, each ovule can become a seed. As seeds develop, the flower falls off the plant. The ovary around the seeds grows larger to protect the seeds inside of it. In time, the ovary grows into a fruit.

Fruits are plant ovaries that contain and protect the seeds inside of them. Apples, oranges, lemons, tomatoes, and pea pods are all fruits.

## Scattering Seeds

If all seeds fell right under the parent plant, they would all have to compete for space, water, sunlight, and nutrients.

However, many seeds are carried away from the parent plant. Wind, water, and animals help carry seeds to other areas. Some seeds are light and can be blown by the wind. Water can carry fruits such as coconuts miles away from the parent plant. Animals can eat fruits and release the seeds away from the parent plant. Some seeds have hooklike parts. These seeds can become stuck to the fur of an animal. The animal can carry the seeds away from the parent plant.

# Reproduction in Cone-bearing Plants

Cone-bearing plants do not have flowers. Instead, the cones contain the reproductive parts. There are two kinds of cones: female cones and male cones.

Female cones are called seed cones. Seed cones contain female reproductive cells. Male cones are called pollen cones. Pollen cones contain male reproductive cells.

Wind carries pollen to seed cones. Fertilization takes place inside the seed cone and seeds develop.

### Everyday Science

Have you ever seen a dandelion get blown by the wind? Have you ever seen winglike pods fall from a maple tree? If you have, you have seen seeds get scattered. Make a list of other ways that you think seeds get scattered. Share your list with your class.

Figure 10-5 *Pollen cones contain male reproductive cells.*

# Asexual Reproduction in Seed Plants

Many plants can be reproduced asexually. When this happens, the new plants have identical genes to the parent plant. The production of organisms with identical genes is called cloning. Cloning in seed plants can occur through **vegetative propagation**. In vegetative propagation, plant parts are used to grow new plants. Roots, stems, and leaves can be used to grow plants that are identical to the parent plant.

### Cuttings

*Cuttings* are pieces of a plant that can be used to grow new plants. Roots, stems, and leaves can all be used as cuttings. For example, the roots of a dandelion plant can be cut into smaller pieces. The smaller pieces of roots can grow to become plants identical to the parent plant.

### Tubers and Bulbs

Plants can also be grown from underground stems called *tubers*. White potatoes are tubers. Small buds called "eyes" may grow on white potatoes. If the eyes are cut from a white potato and planted in soil, new identical white potato plants can grow.

Another type of underground stem is a *bulb*. Bulbs are covered with thick leaves. An onion is an example of a bulb. Bulbs produced by an onion plant can be planted to produce more onion plants identical to the parent plant. Flowers such as daffodils and tulips have bulbs.

### Grafting

Another way of producing a new plant is by joining the stems of two different plants. This method is called *grafting*. In grafting, the stem of one plant grows using the root system of another plant. Certain rose plants and apple plants are produced by grafting.

Figure 10-6 *These stems are cuttings that have been placed in water to grow roots. They can now be placed in soil to grow new plants.*

**✓ Check Your Understanding**

Write your answers in complete sentences.

**1.** What are the male part and the female part of a flower called?

**2.** What is pollination?

**3.** How does fertilization in a flowering plant occur?

**4.** CRITICAL THINKING  Explain why a pea pod is considered a fruit.

**Check Your Understanding**

**1.** The male part is the stamen. The female part is the pistil.

**2.** the movement of pollen from a stamen to a pistil

**3.** A pollen grain lands on the top of a pistil. The grain forms a pollen tube that grows down into the ovary. The tube carries the male reproductive cell to the ovule. Fertilization occurs when the male reproductive cell joins with the female reproductive cell.

**4.** Because a pea pod is the ovary of a flower that protects the seeds inside of it.

## On the Cutting Edge

### T-BUD GRAFTING

Over 200 years ago, George Washington organized the planting of hundreds of trees at his Mount Vernon estate in Virginia. Since then, many of those trees have died. In fact, in the year 2001, only 13 of the original trees were still alive.

In August 2001, tree experts began work to rebuild the Mount Vernon grounds. In order to reproduce the trees that George Washington planted, they used a grafting method to copy the remaining 13 trees.

The method that the tree experts used is called T-bud grafting. In this type of grafting, a T-shape cut is made in the bark of a young tree. The bud of the tree to be copied is put into the flaps of bark in the "T" slice. The bud is tied tightly in place. A new plant grows from the bud. This young tree should be strong enough to be planted in about two years.

CRITICAL THINKING  Why do you think grafting is useful? Possible answer: because it can be used to produce plants that cannot be reproduced in any other way

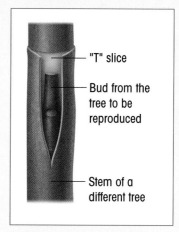

"T" slice

Bud from the tree to be reproduced

Stem of a different tree

**Figure 10-7** *In T-bud grafting, a bud from the tree to be reproduced is placed inside the "T" slice in the stem of a different tree.*

# LAB ACTIVITY
## Growing Plants From Cuttings

**BACKGROUND**
Some plants can be reproduced asexually using cuttings.

**PURPOSE**
In this activity, you will compare how well leaf cuttings from different plants grow.

**MATERIALS**
safety goggles, scissors, philodendron plant, plastic cups, water, jade plant, potting soil, paper plate, ruler, pencil, paper, graph paper

| Length of Roots | | | |
|---|---|---|---|
| Type of Leaf | Soil | Water | Air |
| Philodendron | | | |
| Jade | | | |

Figure 10-8 *Copy this chart onto a separate sheet of paper.*

**WHAT TO DO**

1. Copy the chart in Figure 10-8 onto a separate sheet of paper.

2. Put on safety goggles. Cut three leaves with their petioles from the philodendron plant. Place one petiole in a cup of water. Place one in soil. Place the third petiole on a paper plate.

3. Repeat Step 2 using leaves from a jade plant.

4. Carefully water the soil in both cups. For the next few weeks, water the soils whenever they appear dry. After a few weeks, examine all of the cuttings. You should see roots growing from some of the cuttings.

5. Measure the lengths of the roots for all six leaves. Record your measurements in your chart.

6. Use the data in your chart to make a bar graph. The vertical axis should be the lengths of the roots and each bar on the horizontal axis should be each root.

**DRAW CONCLUSIONS**

- In general, which type of plant grew longer roots from the leaf cuttings? Answers will vary.

- If you were to grow plants from the leaf cuttings, what do you think the plants would look like? Explain your answer. Each new plant would be like the plant the leaf came from because using a leaf cutting to produce a new plant is a form of asexual reproduction.

# BIOLOGY IN YOUR LIFE
## Attracting Pollinators to a Garden

Many kinds of birds and insects are pollinators. Many plants produce nectar, a sweet substance. Pollinators visit the flowers of these plants to get the nectar. While getting the nectar, they move pollen from the stamens to the pistils of flowers. You can design a garden to attract them. Your garden can be in a field, in containers, or on a rooftop. But first, you need to find out what attracts these pollinators.

Figure 10-9
*Hummingbirds are attracted to orange flowers.*

Red or orange flowers shaped like bells often lure hummingbirds. These tiny birds like to drink the sweet nectar from these types of flowers.

Butterflies like flowers that are yellow, orange, pink, or purple. Some butterflies like flowers with a red and yellow bull's-eye pattern. Many butterflies also look for a large landing area, a place to lay eggs, or food for their offspring.

**Look at the chart below. Use it to answer the questions. Write your answers on a separate sheet of paper.**

| Attracting Pollinators | |
|---|---|
| **Butterflies** | **Hummingbirds** |
| • Milkweed | • Clematis |
| • Zinnia | • Butterfly bush |
| • Hollyhock | • Honeysuckle |
| • Black-eyed Susan | • Trumpet creeper |
| • Butterfly bush | • Azalea |

Figure 10-10 *Flowers that attract some pollinators*

Critical Thinking
They provide a large landing area and a good place to lay eggs.

**Critical Thinking**
Why do you think butterflies like flowers with wide petals?

**1.** Choose three plants for a butterfly garden. Accept any three from the butterfly list

**2.** What flower attracts both hummingbirds and butterflies? Butterfly bush

**3.** Which pollinators are attracted to clematis? hummingbirds

## Summary

- Flowers contain the structures that help a plant reproduce. The main parts of a flower are the petals, the sepals, the pistil, and the stamens.

- The movement of pollen from a stamen to a pistil is called pollination. Wind, water, and animals can all pollinate flowers.

- Once pollen has landed on a pistil, it begins to grow a pollen tube. This tube grows down into the ovary. The male reproductive cells travel to the ovary along the tube. When a male reproductive cell joins with a female reproductive cell, fertilization occurs. The ovary grows into a fruit to protect the seeds inside it.

- Seeds need to travel away from the parent plant. Seeds can be scattered by wind, water, or animals.

- Instead of flowers, cone-bearing plants reproduce using cones. The seed cones produce female reproductive cells. The pollen cones produce male reproductive cells. Fertilization takes place in the female seed cones.

- Many seed plants can be reproduced asexually through vegetative propagation.

| |
|---|
| fertilization |
| ovary |
| petal |
| pistil |
| pollen |
| stamen |

## Vocabulary Review

**Complete each sentence with a term from the box.**

1. A light and powdery substance that contains the male reproductive cells of a plant is called _____. pollen

2. A _____ is a part of a flower that is often brightly colored. petal

3. The joining of a male reproductive cell and a female reproductive cell is called _____. fertilization

4. The male part of the flower is the _____. stamen

5. The female part of the flower is the _____. pistil

6. The bottom part of the pistil where female reproductive cells are formed is called the _____. ovary

# Chapter Quiz

**Write your answers in complete sentences.**

1. What are the three parts of a pistil?
the stigma, the style, and the ovary
2. What are the two parts of a stamen?
the anther and the filament
3. What is self-pollination?

4. What is cross-pollination?

5. What plant part grows into a fruit?
the ovary
6. What part of a pistil develops into a seed after fertilization occurs? the ovule

7. Why is it important for seeds to be scattered?

8. What two structures help conifers to reproduce?
seed cones and pollen cones

## CRITICAL THINKING

9. How does a flower help a plant to reproduce?
Flowers contain the reproductive organs of a plant.
10. Why are the plants produced from vegetative propagation identical to the parent plant?
because the new plants have identical genes to the parent plant

**Test Tip**
Make sure that you understand what a test question is asking. Read a question twice before answering.

**3.** When pollen from the stamen of a flower moves to the pistil of the same flower.

**4.** When pollen from the stamen of a flower moves to the pistil of a flower on another similar plant or a different plant.

**7.** They would have to compete with each other for water, sunlight, and nutrients if they grew right under the parent plant.

—SC*L*INKS
Go online to www.scilinks.org. Enter the code **PMB118** to research **medicine from plants**.

**Research Project**

Plants with flowers are pleasant to look at. These types of plants belong to the division called anthophyta. Many important products, including medicines, come from these types of plants. Use the Internet and other resources to find out what kinds of products come from flowering plants. Share your list with the class.

See the *Classroom Resource Binder* for a scoring rubric for the Research Project.

# Unit 3 Review

Choose the letter for the correct answer to each question.

Use Figure U3-2 to answer Questions 1 and 2.

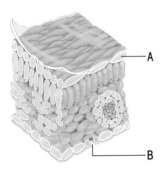

Figure U3-2 *The parts of a leaf*

**1.** Which part of the leaf is labeled *A*?

    A.  the palisade layer

    B.  the spongy layer

    C.  the epidermis

    D.  the stoma

**2.** Which part of the leaf is labeled *B*?

    A.  the stoma

    B.  the guard cell

    C.  xylem tissue

    D.  phloem tissue

**3.** The best type of soil for most plants to grow in is

    A.  loam.

    B.  clay.

    C.  sand.

    D.  gravel.

**4.** In most plants, which plant parts carry out photosynthesis?

    A.  leaves

    B.  roots

    C.  bark

    D.  flowers

**5.** Which of the following are examples of nonvascular plants?

    A.  roses and oak trees

    B.  mosses and ferns

    C.  mosses and liverworts

    D.  ferns and horsetails

**6.** A plant's growth in response to a stimulus is called

    A.  vegetative propagation.

    B.  fertilization.

    C.  pollination.

    D.  a tropism.

**7.** Which of the following is NOT an example of a fruit?

    A.  apple

    B.  lettuce

    C.  orange

    D.  tomato

### Critical Thinking

Blue-green bacteria, euglena, and some types of algae have chlorophyll, but they are not considered plants. Why not?

# Unit 4 ▶ The Animal Kingdom

Chapter 11  **Invertebrates**

Chapter 12  **Coldblooded Vertebrates**

Chapter 13  **Warmblooded Vertebrates**

Figure U4-1  *A sloth is a slow-moving mammal. It lives most of its life hanging upside down from trees. This warmblooded vertebrate has grayish-brown hair. However, algae that grows in its hair gives it a green appearance.*

## Biology Journal

In this unit, you will learn about animals. In your journal, write a list of questions that you have about animals. As you read each chapter, go back and answer your questions. You can also use the Internet or other references to help answer your questions.

**Biology Journal**  The journal activity can be an alternative assessment, a portfolio project, or an enrichment exercise. Have students write at least five questions. As they read the chapters, have them answer the questions.

**Figure 11-1** *This sea slug can be found in the ocean. It has a soft body, and tentacles on top of its head. What do you think tentacles do for a sea slug?*

Tentacles help it to find food.

## Learning Objectives

- Define *vertebrate* and *invertebrate*.
- Compare the body plans of different invertebrates.
- Describe and give examples of each main group into which invertebrates are classified.
- **LAB ACTIVITY:** Compare worms.
- **ON-THE-JOB BIOLOGY:** Learn how a clammer grows clams.

## Words to Know

| | |
|---|---|
| **vertebrate** | an animal with a backbone |
| **invertebrate** | an animal without a backbone |
| **organ system** | a group of organs and tissues that work together to perform a certain job |
| **symmetry** | a body plan that forms mirror images along a line |
| **sponge** | an invertebrate that lives in water and has many pores in its body |
| **cnidarian** | (ni-DARE-ee-uhn) an invertebrate with a hollow body and stinging cells |
| **parasite** | an organism that lives on or inside another organism, usually causing it harm |
| **host** | an organism that a parasite lives on or in |
| **mollusk** | an invertebrate with a soft body that may or may not have a shell |
| **echinoderm** | (ee-KY-noh-duhrm) an invertebrate that has an inner skeleton and rays that extend from a central point |
| **regeneration** | the ability to regrow lost body parts |
| **arthropod** | an invertebrate with an exoskeleton and jointed appendages |
| **exoskeleton** | a tough, stiff covering around the body of an organism |
| **appendage** | a movable part that extends out from the body, such as arms, legs, wings, and claws |
| **thorax** | the middle segment of an anthropod's body |
| **abdomen** | the rear segment of an arthropod's body |
| **metamorphosis** | the process of change and development that an insect goes through as it becomes an adult insect |

**Linking Prior Knowledge** Students should recall the characteristics of living things (Chapter 2) and the kingdoms of life (Chapter 6).

## Animal Classification

The animal kingdom is divided into two main groups, vertebrates and invertebrates. **Vertebrates** are animals with backbones. Fishes, birds, horses, snakes, and humans are vertebrates. **Invertebrates** are animals that do not have backbones. Invertebrates can be divided into groups that include jellyfish, worms, sea stars, clams, and insects.

## Invertebrate Bodies

Invertebrates have complex body structures. Their bodies have specialized tissues that make up organs. Groups of organs work together to do a certain job. Together they make up an **organ system**. Figure 11-2 shows the function of some organ systems in the bodies of invertebrate animals.

**Butterfly**
Bilateral symmetry

**Sea urchin**
Radial symmetry

**Sponge**
Asymmetry

Figure 11-3 *Different invertebrates have different body plans.*

| Some Invertebrate Body Systems | |
|---|---|
| **Organ System** | **Function of the System** |
| nervous | controls the rest of the body |
| digestive | breaks down food and absorbs nutrients for the body |
| respiratory | gets oxygen into the body; gets rid of carbon dioxide |
| circulatory | moves nutrients and gases throughout the body in blood |
| reproductive | produces offspring |

Figure 11-2 *Groups of organs work together to form organ systems.*

Different invertebrates have different body plans. **Symmetry** is a body plan that allows mirror images to form along a line. Two types of symmetry are *bilateral* and *radial*. An invertebrate with no symmetry is said to have *asymmetry*.

# Sponges

**Sponges** are invertebrates that live in water. They are usually found attached to objects on the ocean floor. Sponges have many pores in their bodies. Vase sponges and finger sponges are examples of sponges.

Sponges are the simplest kind of animal. Some biologists believe they were one of the first animals on Earth. They have few specialized cells. The body of a sponge is shaped like a hollow sack. There is a large opening at the top. The bottom of a sponge is closed.

Water enters through the pores in a sponge's body. Water leaves through the large opening at the top of the sponge's body. The sponge gets oxygen and food from this water. The water also carries away wastes.

A few specialized cells carry out simple functions for a sponge. *Collar cells* have flagella. They keep water flowing through the sponge. Stiff, sharp fibers called *spicules* help hold up the sponge. *Amoebocytes* carry food to other cells. Figure 11-4 shows the parts of a sponge.

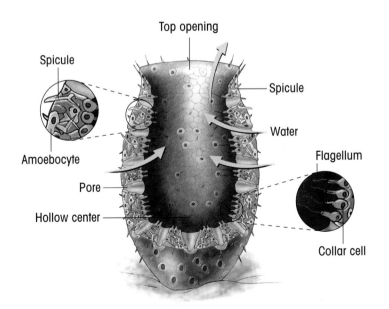

Figure 11-4 *The basic parts of a sponge*

Figure 11-5 *A sponge is shaped like a hollow sack.*

# Cnidarians

**Cnidarians** are invertebrate animals with hollow bodies and stinging cells. Sea anemones, corals, and jellyfish are cnidarians that live in the ocean. Hydras are cnidarians that live in lakes, ponds, and streams.

The mouth of this type of invertebrate is the only body opening. All the body parts are arranged around its mouth. Cnidarian bodies have radial symmetry.

### Polyp Body Form

Cnidarians can have two types of body forms. One form, called a *polyp*, is tubelike. The mouth is found at the top of the polyp. It is surrounded by tentacles. A polyp usually lives attached to objects in water. A hydra is an example of a polyp.

### Medusa Body Form

The other body form is called a *medusa*. A medusa has an umbrella-shaped body. Tentacles hang down from the edge of its body. The mouth of a medusa is at the center of its bottom surface. A medusa can swim or float in water. A jellyfish is an example of a medusa.

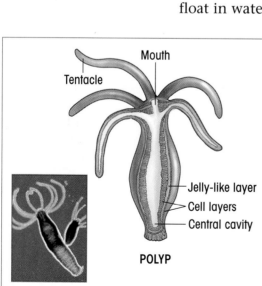

POLYP

Mouth
Tentacle
Jelly-like layer
Cell layers
Central cavity

Figure 11-6 *The hydra in the photograph has a polyp body form.*

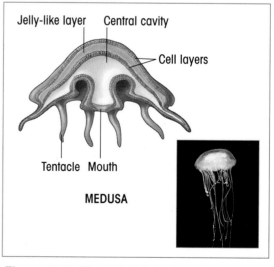

Jelly-like layer   Central cavity
Cell layers
Tentacle  Mouth

MEDUSA

Figure 11-7 *The jellyfish in the photograph has a medusa body form.*

## Stinging Cells

There are stinging cells on the tentacles of these invertebrates. These stingers are used for defense and to catch organisms for food. Once stung, poison in the stingers makes the animals unable to move. The cnidarian catches them with its tentacles and eats them.

### ✓ Check Your Understanding

Write your answers in complete sentences.

**1.** What is an invertebrate?

**2.** Compare the body forms of a polyp and a medusa.

**3.** CRITICAL THINKING  What type of symmetry does a sponge have?

**Check Your Understanding**
**1.** an animal without a backbone
**2.** A polyp is tubelike. A medusa is umbrella-shaped.
**3.** asymmetry

## Worms

Worms are more complex than sponges and cnidarians. They have cells, organs, and organ systems. All worms have bilateral symmetry. Worms can be classified into smaller groups. Three of these groups are: flatworms, roundworms, and segmented worms.

### Flatworms

The simplest type of worm is the flatworm. Flatworms have flat, ribbonlike bodies. They also have only one body opening called a pharynx. This opening is part of a simple digestive system. Food and wastes enter and leave through this opening.

A planarian is a flatworm that lives in ponds and streams. Most other types of flatworms are **parasites**. A parasite is an organism that lives on or in another organism, called its **host**. Parasites feed on and usually harm their hosts. A tapeworm is a flatworm parasite. It has a head with hooks and suckers on it. The head attaches to the host's intestine. A tapeworm absorbs nutrients through its body.

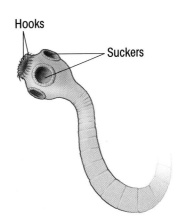

Hooks

Suckers

**Figure 11-8** *A tapeworm is a flatworm parasite.*

## PARASITIC WORMS

Humans can be hosts to parasitic worms. Sometimes it is hard to tell if a person has a worm inside his or her body. There are not always signs of them. Some worms can live in a person for years if they are not found and treated.

There are different types of parasitic worms, such as pinworms, hookworms, and tapeworms. The most familiar is the tapeworm. A tapeworm is a flatworm that lives in a person's intestines. The worm can get there if a person eats meat or fish that contains the eggs of a tapeworm. Not washing fruits and vegetables well enough or direct contact with a person who is infected can also cause a person to get a tapeworm. Tapeworms also can affect other animals. Certain types of tapeworms can be found in horses, sheep, pigs, dogs, and cats.

Figure 11-9 *A tapeworm can grow to several feet in length.*

Problems that can be caused by these parasites include stomach pain and nausea. If a person is not treated for the problem, other organs of the body can also be affected.

CRITICAL THINKING How can you avoid getting tapeworms?

Make sure the meat and fish you eat are cooked thoroughly and that the fruits and vegetables you are going to eat are carefully washed. If you do develop symptoms that last more than a few days, see a doctor.

## Roundworms

Roundworms have long, threadlike bodies that are pointed at the ends. They have two body openings. They can live in soil, fresh water, and salt water. They are also parasites of plants and animals. Hookworms are roundworms that can harm humans.

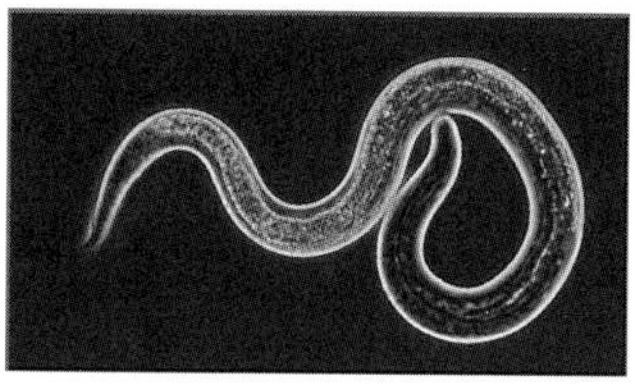

Roundworms have a simple digestive system. They have a mouth where food enters. They have an opening called an anus where wastes leave. A tube connects the mouth to the anus.

Figure 11-10 *A hookworm is a roundworm.*

## Segmented Worms

Segmented worms are more complex than flatworms and roundworms. They have many ringlike sections or segments. Each segment is separated by a wall of tissue. Segmented worms can live in water or in soil. Earthworms are segmented worms that live in soil. Leeches are segmented worms that live in water.

Segmented worms have more highly developed organ systems than the other kinds of worms. They have two body openings, a mouth and an anus. A tube, the intestine, connects the mouth and the anus. These parts make up the worm's digestive system. Segmented worms also have a simple nervous system, including a brain. They have a circulatory system of blood vessels and five simple hearts. They move by using strong muscles. They also have tiny bristles called setae that grip the soil as they move. Figure 11-11 shows the parts of a segmented worm.

**Everyday Science**

Have you ever seen an earthworm move through soil? They move through soil looking for food. They make holes in soil and loosen it. Why do you think a garden full of earthworms is considered a healthy garden?

As earthworms move through soil, the holes that they make and the loosening of the soil allow water and air to enter the soil. Earthworms make soil more fertile.

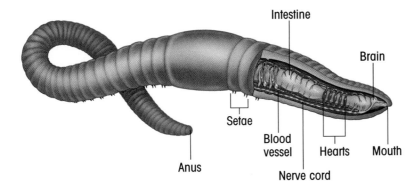

**Figure 11-11** *An earthworm has highly developed organ systems.*

### ✓ Check Your Understanding

Write your answers in complete sentences.

1. What type of symmetry do worms have?

2. What is a parasite?

3. **CRITICAL THINKING** Explain why segmented worms are more complex than roundworms.

1. bilateral symmetry
2. an organism that lives on or in another organism, usually causing it harm
3. Segmented worms have more highly developed organ systems than roundworms.

# Mollusks

**Mollusks** are invertebrates with soft bodies. Most mollusks are covered by hard shells. Mollusks live in water and on land.

Mollusks have three main body parts. These parts are the foot, the soft inner tissue, and a thin membrane that covers the body. The body contains well-developed organ systems. The thin membrane is called the *mantle*. It makes the hard shell on some mollusks. Mollusks can be divided into different classes. These classes include gastropods, bivalves, and cephalopods.

## Science Fact

The sea slug in Figure 11-1 at the beginning of this chapter is also called a nudibranch (NOO-dih-brank).

### Gastropods

*Gastropods* are the most common type of mollusks. These mollusks have one shell or no shell at all. The foot is used for very slow movement. Snails, land slugs, and sea slugs are gastropods.

Gastropods have a stomach and an intestine between their mouth and anus. These parts make up their digestive system. They also have a simple heart in their circulatory system. They have simple lungs or gills in their respiratory system. Figure 11-12 shows the parts of a sea snail.

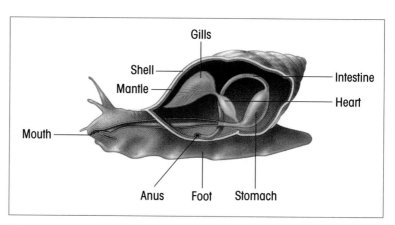

Figure 11-12 *A sea snail is a gastropod.*

### Bivalves

*Bivalves* are mollusks that have two shells. The shells are held together by strong muscles. Oysters, clams, scallops, and mussels are bivalves. They have similar body parts and organ systems to gastropods.

Figure 11-13 *A scallop is a bivalve. Each dark spot is an eye.*

### Cephalopods

*Cephalopods* are the most highly developed of all mollusks. They have a well-developed brain and nervous system. They may have an outer shell, an inner shell, or no shell at all. They also have tentacles. Octopuses and squids are cephalopods.

### ✓ Check Your Understanding

Write your answers in complete sentences.

**1.** What are three characteristics of mollusks?

**2.** What are the three main body parts of a mollusk?

**Check Your Understanding**

1. Possible answer: They have soft bodies. Most are covered by a hard shell. They live in water or on land.

2. the foot, the soft inner tissue, and a thin membrane that covers the body

Figure 11-14 *The scientists are studying a tentacle of a giant squid. A giant squid is a very large cephalopod.*

## Echinoderms

**Figure 11-15**
*Some sea urchins have long needle-like spines.*

Another group of invertebrates is the **echinoderms**. These invertebrates live only in the ocean. Sea stars, sand dollars, sea brittles, sea cucumbers, and sea urchins are all echinoderms.

These types of invertebrates have an inner skeleton made of bumps or spines. These spines can be rounded lumps or they can be long and needle-like. Sea stars have rounded spines. Sea urchins have long needle-like spines.

Echinoderms do not have a brain. However, they do have a nervous system. They also have a water-vascular system. This system of tubes moves water throughout the invertebrate's body. This system helps them in respiration, feeding, and moving.

Echinoderms have rays or arms that extend out from a central point. They have radial symmetry. If any of the arms is cut off, it can grow again. The ability to regrow lost body parts is called **regeneration**.

### Science Fact

Sea stars have tube feet. The tube feet act like suction cups. Sea stars use them to walk on the ocean floor. They also use them to feed on mollusks. The tube feet help pull open the shells of mollusks.

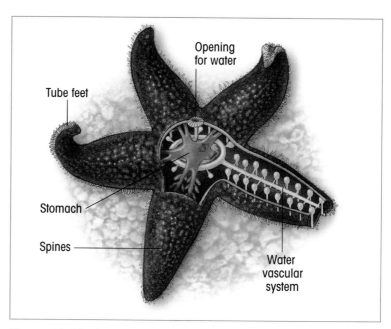

Tube feet

Opening for water

Stomach

Spines

Water vascular system

**Figure 11-16** *A sea star is an echinoderm.*

# Arthropods

**Arthropods** are invertebrates with outer skeletons and jointed legs. Arthropods make up the largest phylum of animals. They can live in water and on land. Lobsters, shrimp, ants, butterflies, spiders, scorpions, and centipedes are examples of arthropods.

Arthropods have some common characteristics. They have a hard outer covering called an **exoskeleton**. The exoskeleton protects and supports the arthropod's body. It is made of a hard material called *chitin* (KY-tin). An exoskeleton cannot grow. So, as the arthropod grows, it must shed its exoskeleton. This shedding is called *molting*.

Arthropods have movable parts that extend out from their bodies. These parts are called **appendages**. Legs, wings, arms, and claws are appendages.

Arthropods also have segmented bodies and an open circulatory system. Some arthropods, such as millipedes and centipedes, have many segments. The open circulatory system of arthropods moves blood through open spaces in their bodies. Blood does not flow through a system of tubes, like it does in the human body.

Arthropods are divided into several classes. The major classes are centipedes and millipedes, crustaceans, arachnids, and insects.

## Centipedes and Millipedes

Centipedes have flat bodies. They have a pair of legs attached to each body segment. They also have a pair of poisonous claws attached to their first body segment. Centipedes eat insects and worms. Millipedes have round bodies. A pair of legs are attached to each of the first four segments of their bodies. The other segments have two pairs of legs attached to each of them. Millipedes eat dead plants.

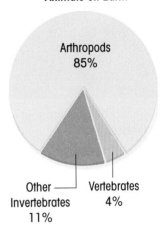

**Relative Percentages of Animals on Earth**

Arthropods 85%

Other Invertebrates 11%

Vertebrates 4%

**Figure 11-17** *Arthropods make up more than three-quarters of all animal species on Earth.*

**Figure 11-18** *A centipede has a flat body.*

Crustaceans

Lobsters, crabs, shrimp, and crayfish are all crustaceans. These arthropods share some common characteristics. They all have exoskeletons. They all have two pairs of *antennae* used to sense smell and touch. Crustaceans also have a type of jaw called a *mandible*. Most crustaceans use their jaws to tear and chew food.

**Figure 11-19** *A lobster is a crustacean.*

A crayfish has a typical crustacean body. It has three body sections. These sections are the head, the thorax, and the abdomen. The **thorax** is the middle section of an arthropod's body. In a crustacean, the head and the thorax are tightly joined to form the *cephalothorax*. The **abdomen** is the rear section of an arthropod's body. A crayfish has different appendages for different functions. For example, the front pincer claws are for defense. Look at the structure of a crayfish in Figure 11-20.

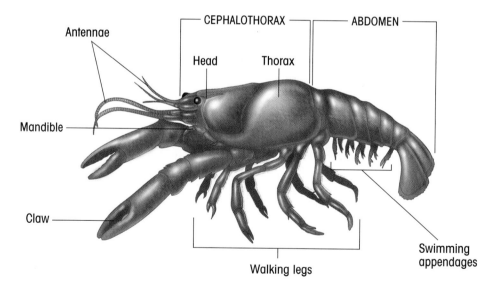

**Figure 11-20** *The basic parts of a crayfish*

## Arachnids

Spiders, ticks, scorpions, and mites are all arachnids. These types of arthropods have two main body sections. These parts are the head-chest section and the abdomen. They have six pairs of appendages. These appendages are all attached to the head-chest section. The first pair of appendages is used for feeding. The second pair may be used for touching and smelling. The last four pairs are used for walking.

**Figure 11-21** *A spider is an arachnid.*

## Insects

Ants, butterflies, mosquitoes, grasshoppers, and beetles are all insects. Insects make up the largest class of arthropods. In fact, there are many more different kinds of insects than there are of any other kind of animal. All insects have three body sections. These sections are the head, the thorax, and the abdomen. Insects have three pairs of legs. Most insects have a pair of antennae and wings. The grasshopper has a tympanic membrane that it uses to hear. It breathes through spiracles. Figure 11-23 shows the basic body parts of an insect.

**Figure 11-22** *A Colorado potato beetle is an insect.*

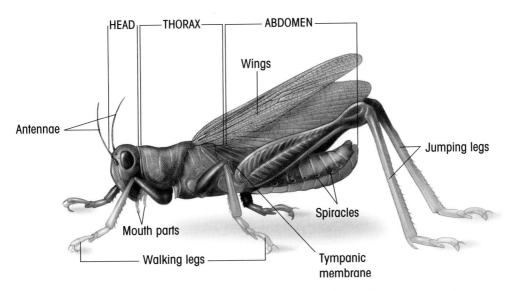

**Figure 11-23** *The basic parts of a grasshopper are shared by most insects.*

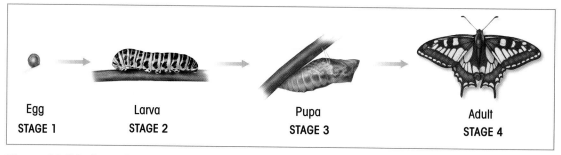

## Science Fact

During the pupa stage, many insects spin a cocoon around themselves. Inside the cocoon, the insect develops into an adult.

# Insect Metamorphosis

Insects go through different stages as they develop into adult insects. They change in form and size. These stages of development are called **metamorphosis**.

*Complete metamorphosis* has four stages: *egg, larva, pupa,* and *adult.* The egg hatches into a larva. The larva usually looks different from the adult. The larva eats a lot of food. After a while, the larva goes into a resting stage. This stage is called the pupa. The pupa develops into an adult. Butterflies and moths are some insects that go through complete metamorphosis.

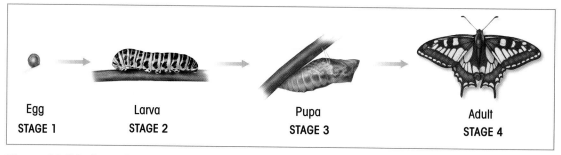

| Egg | Larva | Pupa | Adult |
| STAGE 1 | STAGE 2 | STAGE 3 | STAGE 4 |

Figure 11-24 *Complete metamorphosis has four stages.*

*Incomplete metamorphosis* has only three stages: *egg, nymph,* and *adult.* The egg hatches into a nymph. The nymph looks like the adult insect, but it does not have wings or reproductive organs. After a while, the nymph develops into an adult. The nymph will molt several times before it becomes an adult. Grasshoppers and crickets go through incomplete metamorphosis.

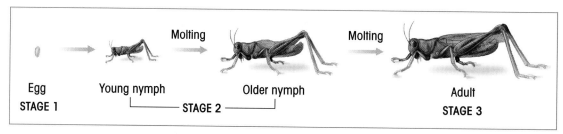

Figure 11-25 *Incomplete metamorphosis has three stages.*

✓ **Check Your Understanding**

Write your answers in complete sentences.

1. What does the water-vascular system do in an echinoderm?

2. What are some common characteristics of arthropods?

3. What are the stages of complete metamorphosis?

4. **CRITICAL THINKING** Compare the main body sections of an insect and an arachnid.

## A Closer Look

### REGENERATION

There are certain animals that can regenerate lost parts. Some of these animals can even develop a whole new organism from parts of their bodies. Some animals use regeneration as a defense against predators.

Lobsters and crabs can regenerate claws that have fallen off. Sea stars can regenerate rays that have fallen off. A sea star can regenerate all of its rays as long as its central body is healthy.

**Figure 11-26** *This sea star is regenerating three of its rays.*

Planarians and sponges can reproduce through regeneration. A planarian can stretch its body until it splits into two planarians. A sponge can be cut up into smaller pieces. Each smaller piece can grow into a new sponge. The new animals produced by regeneration are identical to the original animals.

**CRITICAL THINKING** How can regeneration increase the population of sponges? If a sponge is cut up, each piece can grow into a new sponge.

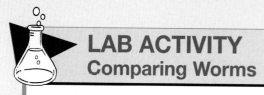

# LAB ACTIVITY
## Comparing Worms

### BACKGROUND
Flatworms, roundworms, and segmented worms are three groups of worms.

### PURPOSE
In this activity, you will compare the features of two different worms.

| Comparing Worms | | |
|---|---|---|
| Observations | Planaria | Vinegar eels |
| Appearance | | |
| Response to Light | | |
| Movement | | |

Figure 11-27 *Copy this chart onto a separate sheet of paper.*

### MATERIALS
safety goggles, dropper, planaria, vinegar eels, hand lens, black construction paper, microscope, microscope depression slides

### WHAT TO DO

1. Copy the chart in Figure 11-27 onto a separate sheet of paper.

2. Put on safety goggles. Remove them whenever you make observations using a hand lens or a microscope.

3. Use a dropper to place some planaria on a microscope depression slide. Observe the planaria using a hand lens.

4. Observe how they respond to light. Place a small piece of construction paper over half of the hand lens. Observe the planaria's response. Record your observations in your chart. Also, record how the worms move.

5. Use a dropper to place some vinegar eels on another microscope slide. Place the slide under the microscope and locate the vinegar eels under high power. Repeat Step 4, but place the construction paper under the microscope stage so that it blocks out half of the microscope field.

### DRAW CONCLUSIONS

- What were some of the major differences between the two types of worms?   their size and the number of openings in their bodies

- What similarities did they share?  they have no legs

# ON-THE-JOB BIOLOGY
## Clammer

Clams live on the bottom of bays, oceans, and salty waters. Some clammers use boats to get the clams. Ed White is a clammer. However, he is more like a clam farmer because he "grows" his own clams.

Figure 11-28 *Clammers use bull rakes to harvest clams.*

Ed plants tiny clams, called clam seeds, in seedbeds. The seedbeds are found in the bay where his farm is located. The bottom of the beds must be raked to remove crabs. Ed harvests the clams with a bull rake. This rake has a T-shaped handle, a long pole, a square wire basket on the front, and long teeth. The rake's teeth scoop clams out of the mud and into the basket.

Ed raises eastern hard-shell clams. These types of clams have different names, depending on their size. Look at the chart and answer the questions below. Write your answers on a separate sheet of paper.

| Types of Eastern Hard-Shell Clams | |
|---|---|
| **Type** | **Size** |
| Little neck | 2 to 2.5 in. (5 to 6.25 cm) |
| Topneck | 2.5 to 3 in. (6.25 to 7.5 cm) |
| Cherrystone | 3 to 4 in. (7.5 to 10 cm) |
| Chowder | Over 4 in. (over 10 cm) |

Figure 11-29 *Eastern hard-shell clam sizes*

**Critical Thinking**

A crab uses its claws to open the shell of a clam.

1. Which clams are the smallest? Which are the largest? little neck clams; chowder clams

2. If you were to buy 3.5 in. (8.75 cm) clams from Ed, what kinds of clams would they be? cherrystone clams

**Critical Thinking**

Crabs are predators of clams. How do you think a crab eats a clam?

## Summary

- The animal kingdom is divided into two main groups. Vertebrates have backbones. Invertebrates do not have backbones.

- Sponges are the simplest animals. They live in water. They get their food and oxygen from water as it filters through them.

- Cnidarians are invertebrates with hollow bodies and stinging cells. Jellyfish, hydras, and corals are examples of cnidarians.

- Three kinds of worms are: flatworms, roundworms, and segmented worms. Tapeworms, hookworms, and earthworms are part of this group.

- Mollusks have soft bodies. They have three body parts, the foot, the soft inner tissue, and the mantle. Oysters, clams, and scallops are mollusks.

- Echinoderms have an inner skeleton and a water-vascular system. Sea stars and sea urchins are echinoderms. Echinoderms can regenerate.

- Arthropods are invertebrates with exoskeletons, jointed appendages, and an open circulatory system. Arthropods are divided into different classes: centipedes and millipedes, crustaceans, arachnids, and insects.

| |
|---|
| abdomen |
| host |
| parasite |
| regeneration |
| vertebrate |

## Vocabulary Review

**Complete each sentence with a term from the list.**

1. The rear segment of an arthropod's body is called the _____.  abdomen

2. _____ is the ability to regrow lost body parts. Regeneration

3. An animal with a backbone is classified as a _____.  vertebrate

4. A _____ is an organism that lives on or inside another organism usually causing it harm.  parasite

5. An animal in which a parasite lives on or in is called a _____.  host

# Chapter Quiz

## Write your answers in complete sentences.

1. What is the difference between invertebrates and vertebrates? Invertebrates do not have backbones. Vertebrates have backbones.
2. What are some examples of cnidarians? Possible answer: hydras, jellyfish, and corals
3. Which type of worm is the most complex? earthworm
4. How does a parasite affect its host? It usually harms its host.
5. What part of a mollusk can make a hard shell? the mantle
6. What is an exoskeleton? Which invertebrates have exoskeletons? a tough, stiff covering around the body of an organism; arthropods
7. Which type of arthropod has a cephalothorax? crustaceans
8. What are the two types of metamorphoses called? complete metamorphosis and incomplete metamorphosis

## CRITICAL THINKING

9. What are the three different types of body plans? For each type, name an animal that has that type of body plan. bilateral: worm; radial: sea star; asymmetry: sponge
10. What would happen if a sea star lost one of its arms? It could regenerate a new one.

**Test Tip**
Before taking a chapter test, use the Vocabulary Review and the questions in the Chapter Quiz to help you review the chapter.

**Research Project** Students should state which invertebrate they are reporting on. They should include information about the invertebrate's classification, what the invertebrate looks like, how it moves, and how it gets food.

SC*L*INKS
Go online to www.scilinks.org. Enter the code PMB120 to research **invertebrates**.

## Research Project

Choose one of the invertebrate animals discussed in this chapter. Research the animal on the Internet or in other resources. Write a short report on the animal's characteristics. Be sure to include the following information: What classification group does the animal belong to? What does the animal look like? How does the animal move? How does the animal get food?

See the *Classroom Resource Binder* for a scoring rubric for the Research Project.

**Figure 12-1** *A great white shark is a vertebrate. How does the structure of its backbone help it to survive life in the ocean?*

The backbone is strong and flexible, helping it to swim fast and turn quickly.

## Learning Objectives

- Compare the body systems of vertebrates.

- Explain the difference between warmblooded and coldblooded animals.

- Describe two important features of fish, amphibians, and reptiles.

- **LAB ACTIVITY:** Make a model of a fish swim bladder.

- **BIOLOGY IN YOUR LIFE:** Compare the needs of different fish as pets.

# Chapter 12 ▷ Coldblooded Vertebrates

## Words to Know

| | |
|---|---|
| **notochord** | (NOHT-uh-kawrd) a strong but flexible support rod found below the nerve cord |
| **chordate** | (KAWR-dayt) organism that has a notochord at some stage of development |
| **skeleton** | a group of bones that give structure and support to an organism's body |
| **coldblooded** | having a body temperature that changes with the temperature of the surroundings |
| **warmblooded** | having a body temperature that remains about the same |
| **cartilage** | strong, flexible connective tissue |
| **gills** | organs used for getting oxygen from water |
| **amphibian** | a coldblooded vertebrate that lives part of its life in water and part on land |
| **reptile** | a coldblooded vertebrate that breathes air with lungs and has scaly skin |

## Vertebrates: Animals With a Backbone

What makes you different from jellyfish? A spider? A sponge? All these animals are invertebrates, or animals without a backbone. You, along with crocodiles, fish, birds, snakes, and whales, are part of the group of animals called vertebrates. You and all other vertebrates have backbones.

**Linking Prior Knowledge**
Students should recall the difference between vertebrates and invertebrates (Chapter 11).

# Chordates

Figure 12-2 *Lancelets, like the one shown here, are chordates but are not vertebrates. They do not develop a backbone.*

In the early stages of many vertebrates, a backbone does not exist. Instead, the nerve cord is supported by a strong, flexible rod called the **notochord**. In adult vertebrates, the notochord is usually replaced by a backbone. All organisms that have a notochord at some point during development are called **chordates**. There are a few chordates, such as the lancelet, that are not vertebrates. However, all vertebrates are classified as chordates.

## The Body Systems of Vertebrates

A backbone is part of the skeleton. A **skeleton** is a structure that supports an organism's body. The skeleton of a vertebrate is made up of bones.

The skeletal system is just one kind of system in vertebrate animals. Some organ systems in animals include the muscular, digestive, respiratory, circulatory, and reproductive systems. Figure 12-3 tells you the function of some of the systems in the bodies of vertebrate animals.

| Organ Systems in Animals | | |
|---|---|---|
| **System** | **Function of System** | **Organs** |
| Skeletal | Supports the body and protects organs | Bones |
| Muscular | Moves the body | Muscles |
| Digestive | Breaks down food into nutrients for the body | Stomach, intestines |
| Respiratory | Gets oxygen into the body; gets rid of carbon dioxide | Gills, lungs |
| Circulatory | Moves nutrients and gases throughout the body in blood | Heart, arteries, veins, capillaries |
| Reproductive | Makes offspring | Ovaries, testes |

Figure 12-3 *Different organ systems have different functions.*

# Coldblooded and Warmblooded

**Coldblooded** vertebrates do not have a constant body temperature. Their bodies take on the temperature of their surroundings. Fish, for example, are coldblooded. If they are swimming in cold water, their bodies are cold. If they are in warm water, their bodies are warm.

Coldblooded animals can use the outside environment to change their body temperature. Their thin body coverings do not hold in heat for long. For example, some reptiles will lie in a sunny spot to raise their body temperature. If they lie in the shade, their body temperature will lower.

**Figure 12-4**
*This Galápagos land iguana is coldblooded.*

| Coldblooded Animals | | |
|---|---|---|
| **Group** | **Body Covering** | **Examples** |
| Fish | Thin skin and/or scales | Bass, goldfish, shark |
| Amphibians | Thin, moist skin | Frog, toad, salamander |
| Reptiles | Thin, scaly skin | Snake, lizard, turtle, crocodile |

Figure 12-5  *The coverings of coldblooded animals are thin.*

You are **warmblooded**. This means that your body temperature stays about the same. Fat, feathers, and fur are thick coverings that insulate. They help warmblooded animals keep constant body temperatures. Warmblooded animals usually eat more than coldblooded ones. The extra food is needed for energy in order to stay warm. Warmblooded bodies control temperature from the inside.

**Think About It**

Do you think a mouse is coldblooded or warmblooded? What about a snake?

A mouse is covered with fur (mammal). It is warmblooded. A snake is covered with scaly skin (reptile). It is coldblooded.

| Warmblooded Animals | | |
|---|---|---|
| **Group** | **Body Covering** | **Examples** |
| Birds | Feathers | Cardinal, pelican, hawk, penguin |
| Mammals | Fat, fur, and/or hair | Mouse, dog, whale, human |

Figure 12-6  *The coverings of warmblooded vertebrates are thick.*

# Fish

**Everyday Science**

27

Fish are an important part of many people's diets. Make a list of fish that people eat. Put a check mark next to the types of fish you have eaten.

Fish are coldblooded vertebrates that live in water. Some fish live in fresh water, such as streams and lakes. Other fish, called saltwater fish, live in the oceans. Shark, salmon, bass, trout, cod, red snapper, and eels are all examples of fish.

## Groups of Fish

There are three main types of fish. The first group are the *jawless fish*. Fish in this group do not have jaws. Lampreys and hagfish are examples of jawless fish. Jawless fish are closely related to fish that lived on Earth hundreds of millions of years ago.

The second group of fish are called the *cartilaginous fish*. They have a skeleton that is made of cartilage instead of bone. **Cartilage** is a strong, flexible, connective material. Sharks and stingrays are examples of this type of fish.

The last group of fish are the *bony fish*. These fish have skeletons made of bone. Most fish that you are familiar with are bony fish. Goldfish, trout, bass, surgeonfish, and tuna are all examples of bony fish.

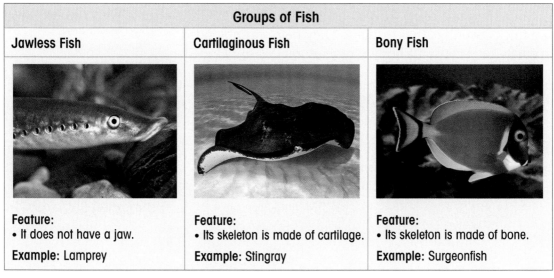

| Groups of Fish | | |
| --- | --- | --- |
| **Jawless Fish** | **Cartilaginous Fish** | **Bony Fish** |
| **Feature:** • It does not have a jaw. **Example:** Lamprey | **Feature:** • Its skeleton is made of cartilage. **Example:** Stingray | **Feature:** • Its skeleton is made of bone. **Example:** Surgeonfish |

Figure 12-7 *There are three main groups of fish.*

## Body Structure of Fish

All fish, except for the jawless fish, have scales covering their bodies. These scales have rings on them. Each ring stands for one year of growth. You can count the number of rings on a fish scale. The number of rings will tell you how old a fish is. Fish have eyes to see and nostrils to smell. They have fins and a tail for swimming. Figure 12-8 shows the parts of a fish from the outside.

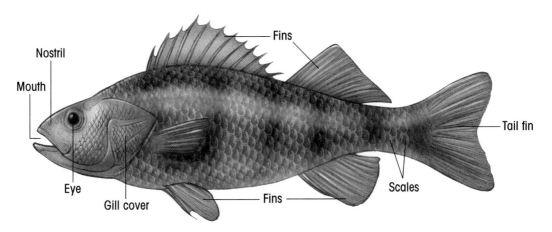

Figure 12-8 *The outside parts of a fish*

### Body Systems

Like other vertebrates, fish have many different organ systems. Fish have a small brain and a spinal cord. They have a skeleton made of bone and cartilage. The skeleton gives them strength and support. Fish also have muscles that help them to swim.

Although fish live in water, they still need oxygen to survive. Fish get their oxygen from the water. The organs responsible for absorbing oxygen are the gills. **Gills** are a major organ of their respiratory system. They are located behind the mouth, under a flap of skin and bone. This structure is called the *gill cover*. The gill cover protects the gills.

As fish swim, water passes into their mouths and across the gills. Oxygen passes through the gills into the blood. Fish have a well-developed circulatory system, including a heart with two chambers. The circulatory system gets oxygen to cells. It also carries away the wastes, ammonia and carbon dioxide. The ammonia is removed from the blood by the kidney. The carbon dioxide passes out through the gills into the water.

Some of the gases in the blood are used by the swim bladder. This organ controls the fish's position in water. It fills with gas to make the fish float.

Fish take in food through the mouth. Then, the food passes through the esophagus into the stomach. Chemicals in the stomach break down the food. Food then passes into the intestine. Other digestive organs include the pancreas, liver, and gallbladder. Waste products leave through the anus.

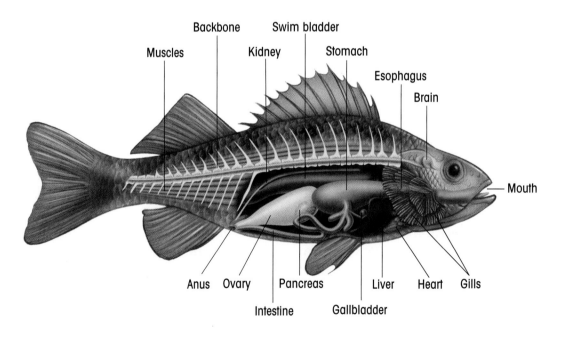

Figure 12-9 *The inside parts of a fish*

# Fish Reproduction and Development

Most fish reproduce by laying eggs made in the ovary. A female fish lays millions of eggs. A male fish swims over the eggs. It releases a substance that has sperm in it. The sperm fertilize only some of the eggs. Most fish eggs do not have shells to protect them. They can be eaten by other fish. Many of them never hatch. The young fish that do hatch have all of the same structures as an adult, only smaller.

**Figure 12-10** *These salmon eggs do not have shells.*

# Amphibians

**Amphibians** are coldblooded vertebrates. Amphibians can live both in water and on land. Amphibians usually live the first part of their lives in water and the second part of their lives on land. Amphibians generally have moist, slimy skin. Frogs, toads, and salamanders are all amphibians.

### Groups of Amphibians

The frog is an amphibian without a tail. Most frogs live in and near water. Toads are also amphibians without tails but they spend less time in the water than frogs do. Their skin is drier and more bumpy. They also have less webbing between their toes. Newts and salamanders are amphibians with tails. They often live under rocks and leaves or in streams.

### Body Structure of Amphibians

Amphibians have skeletons made of bone, not cartilage. They have muscles for movement. Many frogs have strong leg muscles used for jumping. Their body systems are adapted to their watery environment.

**Figure 12-11** *These red-spotted newts are amphibians with tails.*

### Body Systems

Amphibians have similar body systems to fish. They have a nervous system with a brain and a spinal cord.

Amphibians have a special structure used for hearing called a *tympanic membrane*. Figure 12-13 shows this structure.

**Figure 12-12** *Frogs capture insects with their sticky tongue.*

Amphibians have a digestive system that is similar to that of fish. The intestine of the amphibian is fairly short. A frog's tongue is attached to the front of its mouth instead of the back. It is also very sticky. Both the position and the stickiness of the tongue help the frog capture insects.

Amphibians have a more complex circulatory system than fish do. The heart of an amphibian has three chambers instead of two. Amphibians have kidneys to remove certain wastes from the blood. The wastes collect in the urinary bladder where they are released.

The respiratory system of amphibians is different from the respiratory system of fish. Amphibians have the ability to breathe in three ways. Young amphibians breathe using gills. Adult amphibians can breathe using lungs. They also can absorb oxygen through their skin.

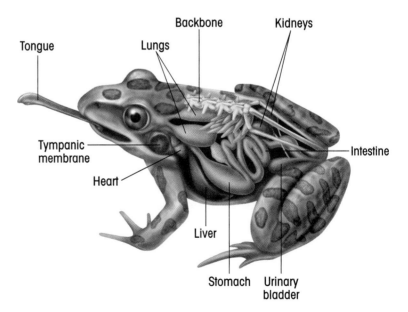

**Figure 12-13** *The parts of a frog*

Both male frogs and female frogs have vocal cords. However, the vocal cords in male frogs are more developed. Frogs make noise by passing air over their vocal cords. Their song is used to attract a mate.

## Amphibian Reproduction and Development

Many amphibians go through metamorphosis. Figure 12-15 shows the metamorphosis of a frog. First, the female lays eggs. The eggs are usually laid in water. Because the eggs have no shells, they would dry out if they were laid on land. Then, the male frog fertilizes the eggs by releasing sperm on them. The fertilized eggs develop into small organisms called *tadpoles*.

Tadpoles have a tail and live in the water. They have gills for getting oxygen from the water. As tadpoles develop, they lose their gills and develop lungs. Tadpoles also gradually lose their tails and grow legs. They are then considered adult frogs. A female adult frog can lay eggs. The cycle will then start over.

**Remember**
The term *metamorphosis* means "to change in form as an organism develops."

Figure 12-14 *Tadpoles lose their tails as they develop.*

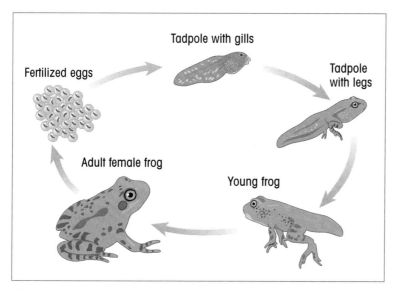

Figure 12-15 *Frogs go through complete metamorphosis.*

**Think About It** An amphibian spends part of its life in water and the other part on land.

4. Sharks have skeletons that are made of cartilage. Cartilage is very flexible. They also have backbones that allow them to bend and twist along their spines.

✓ **Check Your Understanding**

Write your answers on a separate sheet of paper. Use complete sentences.

1. What is similar about the way that fish and amphibians get oxygen? Both fish and young amphibians breathe using gills.

2. What is similar about the way fish and amphibians fertilize egg cells? In both fish and amphibians, the female lays eggs in water and then the male fertilizes the eggs.

3. What is different about the way that young fish and amphibians develop? Young fish do not go through metamorphosis like amphibians do.

4. **CRITICAL THINKING** Many sharks can bend and twist their bodies. What do you think makes this possible?

## A Closer Look

### THE PINE BARRENS TREE FROG

Many amphibians have become endangered over the past several years. *Endangered* means "close to becoming extinct." An animal can become endangered when the habitat it lives in is changed. A bio-indicator is a species of animal whose physical appearance and number in a given area tell you how well an ecosystem is doing. Amphibians make good bio-indicators because they are affected easily by pollution or other changes in the ecosystem.

Figure 12-16 *The number of Pine Barrens tree frogs is declining. At some point, they may become extinct.*

The Pine Barrens tree frog is an example of an endangered amphibian. Its natural habitat is an area of New Jersey called the Pine Barrens. It became an endangered species in 1982. This was a result of two things. First, part of the wetlands that it lives in was either drained or covered with dirt in order to build roads and buildings. Second, some of the wetlands were polluted with a chemical called DDT. Many people are fighting to save the Pine Barrens tree frog. One of the ways they can do so is by protecting the wetlands that the frog lives in.

**CRITICAL THINKING** Explain why it is important to monitor the number of Pine Barrens tree frogs. because it is a bio-indicator, which means it tells you about the health of an ecosystem

# Reptiles

Lizards, snakes, turtles, tortoises, crocodiles, and alligators are all **reptiles**. These are coldblooded, vertebrate animals. They breathe air with their lungs and live mostly on land.

The skin of reptiles is made of dry scales. These scales help reptiles to keep the water in their bodies. If an animal loses too much of the water in its body, it will die. The scales also protect the animals from rough surfaces such as sand and rocks.

## Groups of Reptiles

Many reptiles eat other animals, such as insects, mammals, and fish. Other reptiles eat only plants. Figure 12-17 lists the different groups of reptiles and their characteristics.

**Think About It**

Lizards and salamanders look alike. However, they are classified differently from each other. What are some differences between lizards and salamanders?

Salamanders are amphibians and lizards are reptiles. Salamanders have moist skin; lizards have dry scaly skin. Salamanders lay eggs in water; lizards lay leathery eggs on land or give birth to live offspring.

| Groups of Reptiles | |
|---|---|
| **Group** | **Characteristics** |
| Lizard | Four legs; usually small; can live in many environments |
| Turtle | Four legs; a shell; lives in water and on land; may have flippers for swimming |
| Tortoise | Four legs; a shell; claws; lives only on land |
| Crocodile | Four legs; low to the ground; a long, narrow snout; can live in fresh water or salt water |
| Alligator | Four legs; low to the ground; a wide, rounded snout; lives in fresh water only |
| Snake | No legs; can be small or very large; lives in many environments |

Figure 12-17 *Reptiles are grouped based on characteristics.*

### Lizards

Lizards also have four legs and a tail. Their backbone forms the tail. They are usually smaller than other reptiles. Lizards are the most common type of reptile. They can be found living in many different areas.

### Turtles and Tortoises

Turtles and tortoises are different from other reptiles because they have a shell. The shell is attached to the backbone of the animal. Sometimes turtles pull their head into their shell to protect themselves from predators. Most turtles spend a lot of time in or near water. Sea turtles have front legs that are shaped like flippers. These help the turtle swim. Tortoises are different from turtles because they live only on land.

### Crocodiles and Alligators

Crocodiles and alligators are both reptiles with four legs and a tail. Both crocodiles and alligators live in water. They have eyes close to the tops of their heads. This helps them to see animals they might want to eat, while keeping the rest of their bodies hidden underwater.

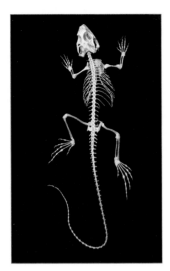

**Figure 12-18** *Lizards have skeletons made of bones.*

**Figure 12-19** *Alligators have powerful jaws.*

## Snakes

Snakes are reptiles without legs. Snakes, like all reptiles, have a backbone. Some snakes have fangs. These are hollow teeth. Venom flows through the fangs. The venom of some snakes can paralyze an animal. Then, the snake can swallow the animal whole.

Figure 12-20 *Some snakes have as many as 400 bones in their bodies.*

Figure 12-21 *The skeleton of a snake includes a skull, a backbone, and many ribs.*

## Reptile Reproduction and Development

In reproduction, reptiles fertilize eggs inside the female. In most cases, development of the young happens outside the female's body. Some reptiles lay leathery eggs on land. The soft shells protect the eggs. A shell keeps an egg from drying out. Some snakes and lizards produce offspring without laying eggs. The common garter snake is an example of a reptile that gives birth to live offspring.

Figure 12-22 *This Nile crocodile is hatching from an egg.*

### ✓ Check Your Understanding

Write your answers in complete sentences.

1. What are four examples of reptiles?

2. How are scales helpful to the snake?

3. How do most reptiles reproduce?

4. **CRITICAL THINKING** Is a snake more closely related to an earthworm or a frog? Explain your answer.

**Check Your Understanding**

1. turtles, snakes, crocodiles, and alligators

2. Scales prevent a snake from drying out and protect it from rough surfaces.

3. They lay leathery eggs on land.

4. Snakes are more closely related to frogs. Both have backbones and are vertebrates. An earthworm is an invertebrate.

# LAB ACTIVITY
## Investigating Swim Bladders

### BACKGROUND
The swim bladder fills with gases when a fish needs to float up or float in one position. It empties when the fish needs to sink.

### PURPOSE
In this activity, you will make a model of a fish swim bladder.

### MATERIALS
safety goggles, 0.5- to 1-liter glass beaker or jar, clean rubber tubing, small balloon, marble, thick rubber band, water

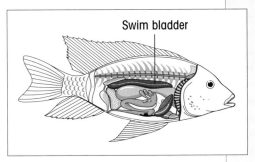

Figure 12-23 *Fish have a swim bladder.*

### WHAT TO DO
1. Place the marble inside the balloon. Fit the rubber tubing into the balloon. Use the rubber band to secure the balloon to the tubing. Press all of the air out of the balloon.

2. Fill the large jar or beaker with water. Sink the balloon with the marble in the water.

3. Blow into the tubing. What happens to the balloon? Write your observation.

4. Predict what will happen if you suck the air out. Record your prediction.

5. Suck the air back out and watch the balloon. Write down what you notice.

6. Try to make the balloon hover in the middle of the jar, so that it does not sink to the bottom or rise to the top. This is called *neutral buoyancy.*

### DRAW CONCLUSIONS
- What did it take to keep the balloon in the middle of the water?
filling the balloon with just enough air to bring the marble and balloon to neutral buoyancy, so it floats in one position
- If the balloon represents a fish swim bladder, what do fish need to do to rise to the surface? They need to have enough gas in their swim bladder to make them float up.

# BIOLOGY IN YOUR LIFE
## Keeping Fish as Pets

Millions of households in the United States have fish as pets. Fish make good pets, but you must know how to care for them.

Before you buy fish, you need a tank, a filter, and other items. A filter cleans the water so fish can thrive. An aerator (AIR-ay-tuhr) pumps air into the water and helps the water circulate.

Figure 12-24 *A freshwater aquarium*

It is important to understand that different types of fish have different needs. Some fish get along well with other species; others do not. Fish also require different types of water. For a group tank, you need fish that like the same water temperature, pH (acidity level), and mineral content.

**Figure 12-25 shows some fish and the best water for them. Use Figure 12-25 to answer the questions below.**

| Needs of Some Aquarium Fish | | |
|---|---|---|
| **Type of Fish** | **Temperature** | **pH** |
| Guppies | 62°–75°F | 6.8–7.6 (best in 7.0–7.2) |
| Swordtails | 75°–82°F | 7.0–8.3 |
| Platies | 68°–74°F | 7.0–8.3 |
| Mollies | 70°–80°F | 7.5–8.2 |

Figure 12-25 *Different types of fish have different needs.*

1. Which type of fish could live in a tank that is 65°F with a pH of 7? guppies

2. Which type of fish likes the warmest water? swordtails

3. What would be a good temperature and pH for a tank with guppies and mollies? any temperature between 70°F and 75°F with a pH between 7.5 and 7.6

**Critical Thinking**
Possible answer: What does it eat? How large does it grow? Can it live with other fish?

**Critical Thinking**
Other factors besides water conditions go into choosing fish. List three questions that you would ask about a species before buying it.

## Summary

- Vertebrates are animals with backbones. Vertebrates have several body systems: skeletal, muscular, digestive, respiratory, circulatory, and reproductive.

- A coldblooded vertebrate's body temperature changes with its environment. The three main classes of coldblooded vertebrates are fish, amphibians, and reptiles.

- Fish live in water. They have bodies suited for swimming. Fish have fins that help them to swim. They have scales covering their bodies. Fish use gills for getting oxygen from the water.

- Amphibians spend the early parts of their lives in water. When they develop into adults, they move onto land. Amphibians eat insects and worms.

- Reptiles live on land. They breathe air and have lungs. Their skin is made of dry scales.

| |
| --- |
| chordate |
| coldblooded |
| gills |
| notochord |
| skeleton |
| warmblooded |

## Vocabulary Review

**Write the term from the list that matches each definition.**

1. organism that has a notochord at some stage of development   chordate

2. a strong but flexible support rod found below the nerve cord   notochord

3. a group of bones that give structure and support to an organism's body   skeleton

4. having a body temperature that changes with the temperature of the environment   coldblooded

5. having a body temperature that remains about the same   warmblooded

6. organs used for getting oxygen from water   gills

# Chapter Quiz

**Write your answers on a separate sheet of paper. Use complete sentences.**

**Test Tip**
Reread your answers to all the questions. Be sure that you have answered both parts of a two-part question.

1. Are you coldblooded or warmblooded? Explain what that means. warmblooded; you maintain a constant body temperature

2. What characteristics help warmblooded animals stay warm? fat, feathers, and fur

3. How do coldblooded animals control their body temperature? by moving to sunny or shady areas to regulate their body temperatures

4. Compare how fish and lizards get oxygen into their bodies. Fish use gills; lizards use lungs.

5. Why do fish lay so many eggs? Many of the eggs never get fertilized. Some eggs may be eaten by other animals.

6. What are the three main types of fish? jawless, cartilaginous, bony

7. What parts of a young frog's body change as it grows older? The gills become lungs; it loses its tail and grows legs.

8. Why is a salamander classified as an amphibian and not a reptile? Salamanders have moist skin and lay their eggs in water.

## CRITICAL THINKING

9. How is reproduction different in fish, amphibians, and reptiles?

10. Why might reptiles be better suited to living in hot, dry conditions than amphibians would be?

**Critical Thinking**

9. Fish and amphibians both lay eggs in water; most reptiles lay eggs on land. Amphibians go through metamorphosis; fish and reptiles do not.

10. because they have thick, dry scaly skin that holds in moisture and because they have eggs with shells, so they do not have to lay their eggs in water

**Research Project** Check students' reports for completeness. You may also want to have students give an oral report about the organism.

## Research Project

Species that are in danger of becoming extinct are called endangered. Research an endangered, coldblooded animal using the Internet or other references. Write a report about the animal. In your report, explain what could be causing the number of animals of that species to decline.

*SciLINKS*

Go online to www.scilinks.org. Enter the code **PMB122** to research **endangered species**.

See the *Classroom Resource Binder* for a scoring rubric for the Research Project.

**Figure 13-1** *Eagles have long, sharp claws and sharp beaks. How do you think these traits help them?*

Possible answer: Sharp claws help eagles capture small animals. Sharp beaks help them to tear meat.

## Learning Objectives

- Identify the characteristics of bird body systems.
- Name the five characteristics of mammals.
- Describe the main groups of mammals.
- **LAB ACTIVITY:** Relate body structures to the behaviors of some birds.
- **ON-THE-JOB BIOLOGY:** Learn about the kinds of animals a large-animal veterinarian treats.

# Chapter 13

# Warmblooded Vertebrates

## Words to Know

| | |
|---|---|
| **migration** | the traveling over long distances of some animals from season to season for feeding, nesting, and warmth |
| **mammal** | a warmblooded vertebrate that is highly developed, fed milk from its mother, and is covered with fur or hair |
| **monotremes** | a group of mammals that lays eggs |
| **marsupials** | a group of mammals that has pouches |
| **placentals** | a group of mammals that is fully developed at birth |
| **primates** | a group of mammals that includes apes, monkeys, and humans |

## Keeping a Constant Body Temperature

Warmblooded vertebrates use a lot of energy to keep a constant body temperature. You are a warmblooded vertebrate. You have a backbone. Also, your body works to keep your temperature the same even when the temperature around you changes. When it is hot, you sweat to lower your body temperature. When it is cold, you shiver to raise your body temperature. Sweating and shivering are ways to keep your body at a constant temperature.

In this chapter, you will read about two classes of warmblooded vertebrates: birds and mammals.

**Linking Prior Knowledge**
Students should recall information that they learned about invertebrates (Chapter 11) and coldblooded vertebrates (Chapter 12).

# Birds

Birds are warmblooded vertebrates. They have a backbone. Their bodies are covered with feathers. Birds are the only animals that have feathers. Feathers help birds to keep constant body temperatures. Birds have lungs for breathing air. They also have four appendages—two wings and two legs.

**Remember**
Appendages are movable parts that extend from the body, such as arms, legs, and wings.

### Why Birds Can Fly

Most types of birds can fly. These birds have several characteristics that help them to fly. They have wings. They also have strong breast muscles. These muscles help them flap their wings. Birds have special skeletal systems. They have hollow bones. These bones keep their bodies very light. Birds have many air sacs inside their bodies. These air sacs help birds get lots of oxygen into their lungs. The large supply of oxygen helps birds to maintain the high energy needed for flight. Feathers also help birds move easily through air.

### Reproduction and Development

Fertilization takes place inside the female bird. The female then lays hard-shelled eggs. A female bird can lay one egg or many eggs at a time. A group of eggs that a female bird lays is called a clutch. It can take several weeks for the eggs to hatch. Once born, young birds need great care or they will die. The parents feed and guard them.

Because birds are warmblooded, their eggs must be kept warm. Either the male or female bird sits on the eggs to keep them warm until they hatch. For example, the female emperor penguin lays only one egg at a time. The egg must be kept warm in the cold climate of Antarctica. After the egg is laid, the male emperor penguin places it on the tops of its feet. The skin from the male emperor's stomach hangs down over the egg to keep it warm.

Figure 13-2 *This emperor penguin keeps an egg warm under its stomach.*

## Migration

Some birds, such as the bobwhite and the cardinal, live in the same place all year. Others, such as the robin, fly to warmer areas in the fall and return in the spring. This regular travel pattern is called **migration**. Other animals, such as deer, fish, and turtles, migrate, too. However, most migrating birds travel longer distances than these other animals. These birds go north to feed and nest in the spring and summer. They return south in the winter to feed on other food sources.

Birds that migrate use the same routes every year. Biologists think that they use stars and the Sun as guides. The arctic tern flies south in the fall to Antarctica. In the spring, it flies all the way back to the Arctic Circle using a different route. The bird makes a round trip of about 22,000 miles (35,000 kilometers) every year.

## Science Fact

During summer in the Arctic Circle, the 24 hours of sunlight help many plants to grow. These plants are the reason why some migrating birds travel to the Arctic Circle in the spring and summer.

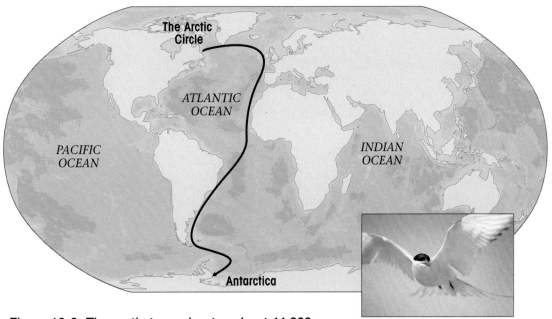

Figure 13-3 *The arctic tern migrates about 11,000 miles (17,600 kilometers) one way from the Arctic Circle in the north to Antarctica in the south.*

# Types of Birds

**Think About It**

Penguins cannot fly. In what other ways do they move?

They walk on land and they swim in water.

All birds are not the same. Many can fly, but some cannot. Penguins and emus are birds that cannot fly. Birds have other differences, too. They have different types of feet and beaks.

Birds that eat meat, such as eagles, have sharp, curved claws on their feet. They use them to grasp other animals. Birds such as crows have curved toes. These toes help them to grip tree branches while they perch. Ducks are swimming birds. They have webbed feet for paddling in water. Flamingoes have long legs that allow them to walk in shallow water.

Bird beaks come in many shapes. The different shapes help different types of birds to eat. For example, most ducks have broad, flat bills. These bills help ducks to strain water for food. Hawks have sharp, hooked beaks. These beaks are used for tearing meat. Other birds, such as parakeets, have strong beaks for cracking open seeds. The heron has a long pointed beak for fishing.

| Different Types of Birds | | | |
|---|---|---|---|
| **Birds That Eat Meat** | **Birds That Eat Seeds** | **Birds That Swim** | **Birds That Walk in Shallow Water** |
| Features: <br> • Sharp, curved claws <br> • Strong, sharp beak <br><br> Examples: <br> Hawks, owls, eagles, falcons | Features: <br> • Curved toes <br> • Small beaks <br><br> Examples: <br> Cardinals, sparrows, parakeets | Feature: <br> • Webbed feet <br><br> Examples: <br> Ducks, geese, swans, loons, gulls | Feature: <br> • Long legs <br><br> Examples: <br> Herons, flamingoes, sandpipers, cranes |

Figure 13-4 *The feet and beaks of these birds help them to behave and eat in certain ways.*

## Check Your Understanding

Write your answers in complete sentences.

**1.** Why are birds warmblooded vertebrates?

**2.** What are three characteristics that help birds to fly?

**3.** How do some birds care for their eggs?

**4.** CRITICAL THINKING What features make birds different from each other?

**1.** Birds have backbones and their body temperatures remain constant.

**2.** Accept any three of the following: They have wings, their breast muscles are very strong, they have hollow bones, they have many air sacs.

**3.** Possible answer: They sit on them to keep them warm until they hatch.

**4.** Possible answer: The shapes of their feet and beaks make them different from each other.

## Great Moments in Biology

### FLYING HOME

Some birds can find their way home from places they have never been before. This ability is called a homing instinct.

In 1953, scientists took a bird called a manx shearwater from its nest in the United Kingdom. They took the bird to Massachusetts, in the United States. There, the bird was let go. After 12½ days, the bird returned to its nest in the United Kingdom. It had flown across the Atlantic Ocean, on a route it had never traveled before. It even found its own nest—a single nest in a single tree—across a distance of about 3,000 miles (4,800 kilometers)!

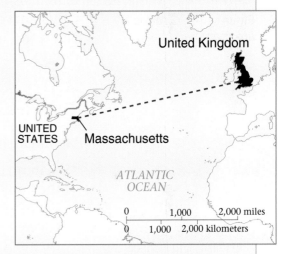

Figure 13-5 *A manx shearwater flew from Massachusetts back to its home nest in the United Kingdom.*

Scientists are still not sure how the bird found its way home. Migrating birds may use the Sun and stars to guide them. Some birds may use landmarks to find their way home. However, scientists cannot explain how a bird can fly home over a route it has never taken before.

CRITICAL THINKING How did the manx shearwater demonstrate its homing instinct? It flew from Massachusetts back to its nest in the United Kingdom.

## Mammals

**Figure 13-6** *Giraffes are mammals.*

**Mammals** are highly developed, warmblooded vertebrates that have hair on their bodies. Mammals feed on their mother's milk when young. Humans, bears, giraffes, mice, cats, dogs, whales, dolphins, walruses, and deer are all mammals. Most mammals live on land. Some mammals live in water.

Mammals have two sets of appendages. Mammals that live in water have flippers. Mammals that walk on land have legs. Many mammals have tails. All mammals breathe air through lungs. Many mammals also have well-developed lips and teeth. Some of these mammals have special teeth for ripping and tearing. Others have special teeth for cutting and gnawing. Flat teeth, like the ones you have in the back of your mouth, are for grinding food.

## The Five Characteristics of Mammals

Mammals all share five important characteristics.

1) Mammals have hair on their bodies. Hair helps keep the body temperature of mammals constant. Dogs, tigers, antelopes, and many other mammals are covered with fur. Humans only have a little hair on their bodies. Whales and dolphins also have very little hair. However, these sea mammals are born covered with fuzz.

2) Most mammals give birth to live young. These mammals do not lay eggs. At birth, the babies of most mammals are fully developed infant mammals.

3) Mammals feed their young with milk from their mother. All female mammals have mammary glands. *Mammary glands* is where the word *mammal* comes from. These glands produce milk. All mammals feed their young with this milk.

4) Mammals care for their young until they can care for themselves. Many species of animals do not take care of their young. Mammals, however, are different. They feed and protect their young. Many mammals teach their young how to hunt, clean, and protect themselves.

5) Mammals have well-developed brains. Mammals are able to think, understand, and learn behaviors that help them to live. They are the smartest group of animals on Earth.

**Think About It**

How do you know that a chicken is not a mammal?

Possible answer: A chicken does not give birth to live young and it does not feed its young milk. It has no hair or fur.

## Classification of Mammals

There are about 5,000 different kinds of mammals on Earth. Scientists classify mammals into three different groups. A mammal is placed into a group based on how it develops from birth.

Most mammals give birth to live young. However, a small group of mammals lays eggs. Mammals in this group are called **monotremes**. These mammals lay eggs covered with hard, leathery shells. There are two types of these egg-laying mammals. They are the duckbilled platypus and the spiny anteater.

**Remember**

Classification is a system used to group organisms based on their characteristics.

Figure 13-7 *A duckbilled platypus is an egg-laying mammal.*

**Figure 13-8** *Koalas are mammals with pouches.*

The second group of mammals has pouches. These mammals are called **marsupials**. They give birth to live young that are not very well developed. After they are born, the young must crawl into a pouch in their mother's body. Inside the pouch, these young mammals drink their mother's milk. There, they finish developing. Kangaroos, koalas, and opossums are mammals with pouches.

The third group of mammals is fully developed at birth. These mammals are called **placentals**. After they are born, the parents of these young mammals take care of them until the young mammals are ready to take care of themselves. Most mammals belong to this group. They include deer, mice, elephants, raccoons, dolphins, and humans.

## Groups of Placentals

There are many mammals that belong to the placentals group. These mammals can be further divided into smaller groups. Placental mammals can be placed into smaller groups based on how and what they eat, how they move about, or where they live.

### Rodents

The largest group of mammals is the rodent group. There are more species of rodents than any other mammal group. Rats, mice, squirrels, and beavers are rodents. They have sharp teeth for gnawing and cutting. Some rodents reproduce very quickly.

### Flying Mammals

The only true flying mammal is the bat. The wings of bats are made from their long finger bones. Skin stretches across these bones to make wings. Bats usually sleep hanging upside down in caves during the day. They become active at night.

## Meat-Eating Mammals

Meat-eating mammals have special body structures to help them capture and eat other animals. Some of these structures include special teeth and sharp claws. Wolves, cats, dogs, bears, seals, and walruses are meat eaters.

## Trunk-Nosed Mammals

The trunk-nosed mammals are all elephants. An elephant's trunk is really a nose and upper lip. An elephant uses its trunk to drink, to lift objects, and to scratch. An elephant's trunk contains many muscles.

### Science Fact

Today, there are three kinds of elephants on Earth: two types of African elephants and the Asian elephant.

## On the Cutting Edge

### TRACKING ANIMALS

Elephants are in danger. Many have been killed for their tusks. People make jewelry and other objects from elephant tusks. In order to save elephants, scientists record their movements in different parts of the world.

Tiny radio transmitters are attached to collars that the elephants wear. Information about the animals, such as their movements and their location, is sent to a satellite that orbits Earth. The information is then sent to a ground station. Scientists use computers to read the information. Knowing where the elephants are and where they travel can help scientists set up places of protection for them.

**Figure 13-9** *Elephants are tracked using radio transmitters, satellites, and computers.*

Many different kinds of animals, including birds and fish, are tracked using satellites. Scientists and researchers examine the information to find out more about the tracked animals.

**CRITICAL THINKING** Why do you think satellite-tracking collars are useful? The collars can give information about the movement and location of an animal.

## Science Fact

There are more groups, or orders, of placentals than we have covered in this chapter. See Appendix C on pages 414 and 415 for a listing of some of these orders.

## Insect-Eating Mammals

Insect-eating mammals include hedgehogs, moles, and shrews. These types of mammals have special features that help them to find and eat insects. For example, a star-nosed mole has tentacles around its nose to help it find insects in the ground.

Figure 13-10 *A star-nosed mole is an insect-eating mammal.*

## Toothless Mammals

Toothless mammals have no front teeth. They include armadillos, sloths, and giant anteaters. Armadillos and sloths do have back teeth. The anteaters in this group are different from spiny anteaters because they do not lay eggs. Giant anteaters and armadillos eat insects. Sloths eat plants.

## Science Fact

Many hoofed mammals have horns or antlers. These structures are used for protection. Antlers fall off each year. The animals grow a new set every spring. Horns do not fall off.

## Hoofed Mammals

There are two groups of hoofed mammals—even-toed and odd-toed. Even-toed include sheep, camels, cows, and deer. Odd-toed include horses and zebras. The hooves are made of a material that is similar to a human fingernail. Many of these types of mammals are plant eaters. They have special body parts that help them to eat plant matter. Some of these body parts include special teeth and special organs for digestion.

## Water Mammals

Whales, dolphins, and porpoises are mammals that live in water. Although these mammals seem a lot like fish, they are not fish. For one thing, they breathe with lungs. They do not have gills. These mammals must come to the surface to get air. They breathe in air through a hole on the top of their heads. They also give birth to live young. The young mammals are fed milk from their mothers. They are warmblooded. Thick layers of fat on their bodies help them stay warm in the cold seas.

## Primates

You are a member of the most highly developed group of mammals. In fact, you are the most intelligent kind of organism on Earth. Scientists classify humans as **primates**. Primates also include monkeys, chimpanzees, gibbons, orangutans, and gorillas.

Primates have several characteristics that set them apart from other mammals. They have hands with movable fingers. They also have thumbs that can touch all the other fingers on the same hand. The thumbs of primates allow them to pick up things easily. Many animals, such as horses, fish, and many types of birds, have eyes on the sides of their face. Each eye sees something a little different. However, primates and some other mammals have both eyes on the front of their face. Both eyes are looking forward. The two eyes have views that overlap almost completely. This characteristic gives primates very good depth perception. They can tell if objects are near or far from them.

More than 200 species of mammals are classified as primates. Most primates live in warm parts of the world. Some of these animals, such as baboons and gorillas, spend a lot of time on the ground. Others, such as gibbons and orangutans, live in trees.

Figure 13-11 *Dolphins are water mammals.*

### Everyday Science

Your thumb is very important. It allows you to grasp and hold things. All primates have thumbs. Because the thumb bends the opposite way of other fingers, it is called the opposing digit. Try picking up objects without using your thumb. Is it difficult?

Students may find it difficult to pick up objects without using their thumbs.

Nearly all primates, including humans, live in groups. Within these groups, they depend on each other for protection. Groups also work together to gather food. Different members may have different roles, such as babysitter or lookout. Many groups also have leaders. Some members of the groups compete with each other to be leader.

## Modern Leaders in Biology

**JANE GOODALL, Primate Biologist**

Much of the information that we have on chimpanzees comes from the work of Dr. Jane Goodall. Dr. Goodall is a researcher. She spent many years studying chimpanzees in an area of East Africa called Tanzania.

While studying chimpanzees, she learned that they are very smart primates. They live in groups. They can make and use simple tools. Dr. Goodall also found that chimpanzees have feelings very similar to humans. They can love and fight. During her research, Dr. Goodall set up a feeding station for some of them. She was even accepted as one of them by a group of chimpanzees.

Figure 13-12 *Dr. Jane Goodall studied chimpanzees in East Africa.*

Dr. Jane Goodall has written books and articles about her experiences with chimpanzees. She speaks to groups of people about the importance of chimpanzees and other wild creatures. She has also set up areas of protection for chimpanzees in Africa.

CRITICAL THINKING What characteristics of primates did Dr. Goodall observe? Possible answer: She found that chimpanzees can love and fight.

# Humans

Humans are the most highly developed primates. Our skeletons allow us to walk completely upright. We have very good control over our fingers and thumbs. Humans have the most developed brains. We can communicate with one another using spoken language. We are also the only primates that can make and use complex tools.

Figure 13-13 *Humans can make and use complex tools.*

## ✓ Check Your Understanding

Write your answers in complete sentences.

1. What are the five characteristics of mammals?

2. What are the three main classification groups of mammals?

3. List four smaller groups of placental mammals.

4. CRITICAL THINKING What evidence is there that humans are more developed than primates?

1. Mammals have hair on their bodies. Most mammals give birth to live young. Mammals feed their young milk from their mother. Mammals care for their young. Mammals have well-developed brains.

2. monotremes, marsupials, and placentals

3. Accept any four of the following: rodents, flying mammals, meat-eating mammals, trunk-nosed mammals, insect-eating mammals, toothless mammals, hoofed mammals, water mammals, primates

4. Possible answer: Humans are more developed because they can use complex tools and they can communicate with one another using spoken language.

# Comparing Digestion in Mammals

Mammals need a lot of energy to maintain a constant body temperature. As a result, mammals must eat more food than other animals. The digestive systems of different kinds of mammals are adapted to the food they eat.

### Specialized Teeth

Most mammals have specialized teeth to help them chew food. There are four types of teeth: incisors, canines, molars, and premolars. *Incisors* are like a chisel. They are used to cut food. *Canines* are pointed teeth. They are used for tearing. *Molars* and *premolars* are used to crush and grind food.

Look at Figures 13-14 and 13-15. The wolf on the left is a meat eater. The cow on the right is a plant eater. Notice how their teeth are different. The wolf has molars that can lock together while eating. The wolf's canines are very sharp. This sharpness allows meat eaters, like the wolf, to cut and tear their food more easily. The cow, however, has wide, flat molars. These teeth help plant eaters, like the cow, grind tough plants.

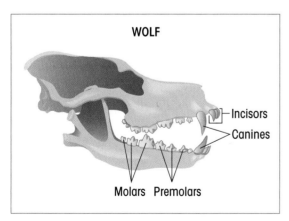

Figure 13-14 *A wolf has specialized teeth that help it to eat meat.*

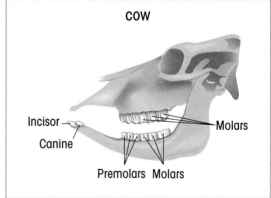

Figure 13-15 *A cow has specialized teeth that help it to eat plants.*

## Specialized Stomachs

Meat eaters, such as wolves have a digestive system with one simple stomach. Plants, however, can be more difficult to digest than meat. Some plant eaters, such as cows, have complex stomachs with four chambers.

After a cow chews and swallows food, the food enters the first stomach chamber. Bacteria in the chamber help digest plant matter. The food is then brought back up to the cow's mouth. The food that is brought back up to the mouth is known as cud. The cud is chewed and swallowed again. It then travels to the other three stomach chambers and digestion continues. Figure 13-16 shows the digestive system of a cow.

**Think About It**

Hoofed animals are often said to be "chewing the cud." What do you think this statement means?

It refers to the fact that these animals bring chewed food, called *cud*, back up and then chew it again.

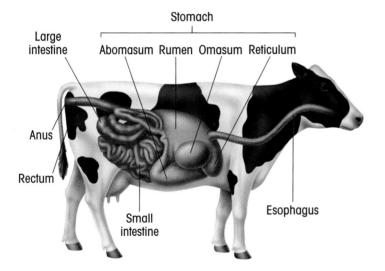

Stomach

Large intestine  Abomasum  Rumen  Omasum  Reticulum

Anus

Rectum

Small intestine

Esophagus

**Figure 13-16** *Cows have specialized stomachs that help them digest plant matter.*

**Science Fact**

It takes longer to digest plants than it does to digest meat. This difference is the reason why meat eaters have shorter intestines than plant eaters do.

## ✓ Check Your Understanding

Write your answers in complete sentences.

1. Why do mammals eat more food than other kinds of animals?

2. What are the four special kinds of teeth used for chewing food?

1. because they need a lot of energy to maintain a constant body temperature

2. incisors, canines, molars, and premolars

# LAB ACTIVITY
## Relating Bird Body Structures to Behavior

### BACKGROUND
The structures of beaks and feet are designed to help birds get food and to behave in certain ways.

### PURPOSE
In this activity, you will relate the structures of the beaks and feet of some birds to their behavior.

### MATERIALS
pencil, index cards

### WHAT TO DO
1. Write the names *Hawk*, *Heron*, and *Duck* on separate index cards.

2. Read the following descriptions of each type of bird.

**Bird Body Structures**

A  B  C

D  E  F

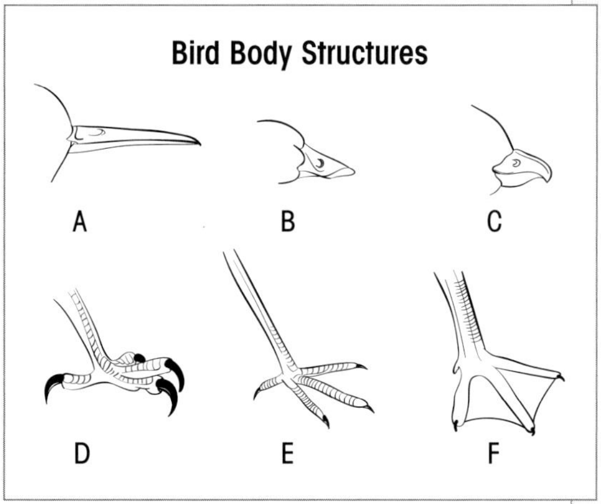

Figure 13-17 *The structures of its beak and feet are related to a bird's behavior.*

**Hawks:** Hawks are hunting birds. They use their beaks for tearing meat. Their powerful claws help them to kill and to carry their prey.

**Herons:** Herons live in nests by water. They walk into shallow bodies of water to eat fish, eels, small mammals, and frogs.

**Ducks:** Ducks live by water. Their legs and feet are very good for paddling in water. Some ducks have broad, flat bills that are used to strain plants from water.

3. Look at Figure 13-17. Decide which beak and foot goes with each bird. Draw and label the beak and foot for each bird on the index cards.

### DRAW CONCLUSIONS
- If you saw a bird with a broad, flat bill, what would you guess it eats?  Possible answer: plants from water

- Explain how body structure relates to a bird's behavior.
  Possible answer: Powerful claws allow hawks to kill and carry their prey.

# ON-THE-JOB BIOLOGY
## Large-Animal Veterinarian

Dr. Anthony Suarez is a veterinarian, or vet. A vet is a doctor that treats animals. Some vets treat pets such as dogs, cats, and birds. Dr. Suarez is a large-animal vet. He travels to farms and ranches to take care of animals such as horses, cows, pigs, and chickens.

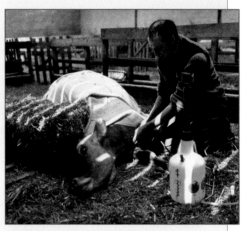

Figure 13-18 *A large-animal vet takes care of animals that live on farms.*

Dr. Suarez helps farmers keep their animals healthy. He tests the animals for diseases. He gives them medicines that help keep them from getting diseases. He also treats animals when they are sick or injured. Sometimes a large animal has trouble giving birth. Dr. Suarez may be called in the middle of the night to help the animal deliver its baby. Dr. Suarez brings medical tools with him so he can bandage wounds, take X-rays, and even do surgery.

Being a large-animal vet has dangers. Dr. Suarez works outdoors in all kinds of weather. He might be kicked or bitten by an animal that is scared or in pain. However, Dr. Suarez's job also has rewards. He loves taking care of animals.

**Here are some farm animals that a large-animal vet might treat. Use this list to answer the questions.**

| | | |
|---|---|---|
| geese | pigs | chickens |
| cows | horses | sheep |

**1.** Which animals are mammals?
pigs, cows, horses, sheep
**2.** Which animals produce milk?
pigs, cows, horses, sheep
**3.** Name an animal that is not a mammal. What food does it produce?
Possible answer: chicken; meat, eggs

### Critical Thinking
The meat, eggs, and milk we eat come from farm animals. By vaccinating and treating food animals, vets help to protect our food supply from disease.

### Critical Thinking
How do large-animal vets protect the health of humans?

## Summary

- Birds and mammals are two classes of warmblooded vertebrates. They have a backbone and their body temperature remains constant.

- Birds are able to fly because of wings, strong breast muscles, hollow bones, and air sacs inside their bodies. Birds' feet and beaks are suited to where they live and what they eat. Many birds migrate.

- Mammals share five characteristics: 1) mammals have hair on their bodies; 2) most mammals give birth to live young; 3) mammals feed their young with milk from their mother; 4) mammals care for their young until the young can care for themselves; and 5) mammals have well-developed brains. Mammals are classified into three different groups: monotremes, marsupials, and placentals.

- People are classified into a placental group called primates. People are the most highly-developed mammals.

- The digestive systems of meat-eating mammals and plant-eating mammals are different.

mammal

marsupial

migration

monotreme

placental

primates

## Vocabulary Review

**Complete each sentence with a term from the list.**

1. The traveling of some animals over long distances from season to season is called _____. migration

2. The group of mammals that includes apes, monkeys, and humans is called _____. primates

3. A mammal with a pouch is called a _____. marsupial

4. A mammal that lays eggs is called a _____. monotreme

5. A mammal that is born fully developed is a _____. placental

6. A warmblooded vertebrate that is fed milk from its mother and is covered with fur is called a _____.
mammal

# Chapter Quiz

**Write your answers in complete sentences.**

1. Why must birds sit on their eggs?
The eggs must be kept warm.
2. Why are a bird's bones hollow?
It helps them to fly by making them light.
3. Why do you think some birds migrate?
Some birds migrate to find food and to nest.
4. How do biologists think migrating birds find their way?
by using the Sun and the stars as guides
5. Why do hawks have sharp hooked beaks and claws?
These traits help them to capture and eat meat.
6. Why do ducks have webbed feet?
Webbed feet help them to swim in water.
7. How do biologists know that whales are mammals
and not fish? Whales are born fully developed, they feed their
young with milk, and they use lungs to breathe.
8. What is unusual about kangaroos as mammals?
Kangaroos have pouches.
9. What is the difference between a monotreme and
other types of mammals?  A monotreme lays eggs.
10. What is the largest group of mammals? Name four
kinds of animals in this group.

**CRITICAL THINKING**

11. How is the migration of a fish different from the
migration of a bird?

12. If bats can fly, why are they not classified as birds?

**Test Tip**
Reread the summaries for
each chapter before taking
the quiz. The summaries
usually cover the main points
and tie the pieces of the
chapter together.

10. placentals; Accept four
kinds of placental mammals,
such as whales, deer, rats,
elephants, bats, or humans.

11. Possible answer:  A fish
swims instead of flying and
a fish may not travel as far
as a bird.

12. Possible answer: Bats
are not classified as birds
because they do not have
feathers and they do not
lay eggs.

*SCi**LINKS*

Go online to www.scilinks.org.
Enter the code **PMB124** to
research **flightless birds**.

## Research Project

Ostriches cannot fly. They are flightless birds. Research
the body structure of ostriches. Find out how their
bodies are different from the bodies of birds that can
fly. Write a short report that explains why ostriches cannot fly. Include
pictures of a bird that can fly and an ostrich. Point out the differences
between both types of birds. Present your report to the class.

See the *Classroom Resource Binder* for a scoring rubric for the Research Project.

# Unit 4 **Review**

**Standardized Test Preparation** This Unit Review follows the format of many standardized tests. A Scantron® sheet is provided in the *Classroom Resource Binder*.

**Choose the letter for the correct answer to each question.**

Use the diagram to answer Questions 1 and 2.

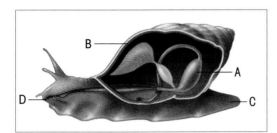

Figure U4-2 *Illustration of an invertebrate*

**1.** What type of animal is shown above?  A (p. 172)

A. mollusk

B. sponge

C. reptile

D. arthropod

**2.** What is the name of the structure labeled *A*?  C (p. 172)

A. mouth

B. foot

C. stomach

D. shell

**3.** A hard, outer covering that protects an arthropod is called  B (p. 175)

A. an endoskeleton.

B. an exoskeleton.

C. a backbone.

D. a notochord.

**4.** Which of the following is NOT an invertebrate?  C (p. 166)

A. jellyfish

B. earthworms

C. turtles

D. spiders

**5.** What do all vertebrates have at some point during their development?  D (p. 186)

A. lungs

B. scales

C. shells

D. notochord

**6.** An animal's backbone is part of its

A. respiratory system.  C (p. 186)

B. reproductive system.

C. skeletal system.

D. digestive system.

**7.** The warmblooded vertebrates are

A. birds and mammals.  A (p. 203)

B. birds and reptiles.

C. fish and amphibians.

D. amphibians and mammals.

**Critical Thinking**

How is reproduction in reptiles similar to reproduction in birds? How is it different?

**Critical Thinking** Both reptiles and birds reproduce by internal fertilization, but the development of the embryo takes place outside the mother's body in eggs. Reptiles lay leathery eggs. Birds lay eggs with hard shells. (Chs. 12–13)

# Unit 5 ▷ The Human Body

Chapter 14 **Support and Movement**

Chapter 15 **Circulation, Respiration, and Excretion**

Chapter 16 **Digestion**

Chapter 17 **Regulating the Body**

Chapter 18 **The Sense Organs**

Chapter 19 **Human Health**

Chapter 20 **Reproduction and Development**

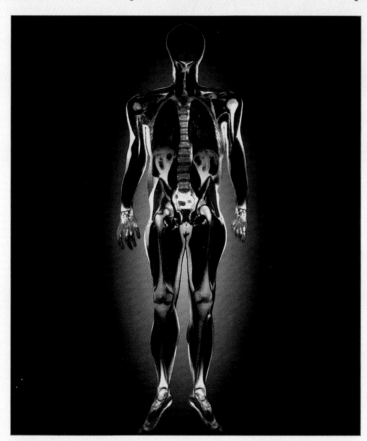

Figure U5-1 *The human body is an important part of scientific research. This photograph was taken through a PET scan. Pictures like these help doctors "see" the body.*

## Biology Journal

In this unit, you will learn about the human body. In your journal, write a list of questions that you have about how the body works. As you read each chapter, go back and answer your questions. You also can use the Internet or other references to help answer your questions.

**Biology Journal** The journal activity can be an alternative assessment, a portfolio project, or an enrichment exercise. Have students write at least five questions. As they read the chapters, have them answer the questions.

**Figure 14-1** *Athletes have greater bone mass and muscle mass than people who do not exercise regularly. Why do you think exercise is important for your bones and muscles?*

Possible answer: because your bones and muscles support your body and help you to move

## Learning Objectives

- Name the five jobs of bones.
- Explain what bones are made of.
- Describe the structure of bones.
- List and give examples of bone joints.
- Describe the three kinds of muscles.
- Explain how muscles work.
- **LAB ACTIVITY:** Identify joints.
- **BIOLOGY IN YOUR LIFE:** Explain how to treat a pulled muscle.

# Chapter 14

# Support and Movement

## Words to Know

| | |
|---|---|
| **marrow** | the soft tissue inside bones that makes blood cells |
| **fracture** | a break or crack in a bone |
| **joint** | a place in the body where two or more bones meet |
| **ligament** | a type of tissue that holds bones together at a joint |
| **tendon** | a type of tissue that connects muscles to bones or other muscles |
| **voluntary muscle** | a muscle that moves because you control it |
| **involuntary muscle** | a muscle that moves without your control |
| **cardiac muscle** | the muscle tissue that makes up the heart |

## Bones and Muscles

You have 206 bones in your body. Your bones have different sizes and different shapes. Some bones are long and flat. Others look like small pebbles. All of these bones combine to make up your skeleton. Without a skeleton, you would not be able to stand. Your body would be a soft bag of skin and organs. Even your head would be soft. Also, your organs would not be well protected.

You have more than 600 muscles in your body. Muscles help your body to move. You can control some muscles. Other muscles move without your control.

This chapter is about bones and muscles. Your bones make up your skeletal system. Your muscles make up your muscular system. Bones and muscles support your body and help you to move.

**Linking Prior Knowledge**
Students should recall that humans are vertebrates (Chapter 13).

# The Skeletal System

## Science Fact

The longest bone in your body is your thigh bone or femur. The smallest bones are found in your ears.

Bones have five jobs.

**1)** They support your body and give your body shape.

**2)** They protect many organs inside your body. For example, your skull protects your brain.

**3)** Many bones work with muscles to help your body move.

**4)** Some bones make blood cells. These bones have soft tissue called **marrow** inside them that makes blood cells.

**5)** Bones store minerals such as calcium and phosphorus.

## Think About It

Look at Figure 14-2. Which bones are you familiar with? What are their common names?

Possible answers: The cranium is the skull. The clavicle is the collarbone. The femur is the thigh bone.

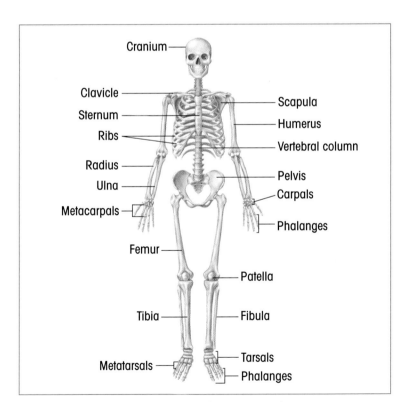

Figure 14-2 *The human skeletal system makes a framework that gives the body its shape.*

# What Bones Are Made Of

Bones are made of both living and nonliving materials. The living materials are bone cells, blood cells, and nerve cells. These cells make up about one-third of bones. About two-thirds of bones are made up of nonliving materials, including minerals.

### Cartilage

The skeleton of an unborn baby is made of cartilage. This tissue is tough but flexible. As an unborn baby develops, most of its cartilage skeleton changes to hard bone. The cartilage that is left has many functions. For example, it prevents the ends of bones from rubbing against each other.

Your outer ear is also made of cartilage.

## The Structure of Bone

A bone is covered with a strong membrane. This membrane is called the *periosteum*. This membrane has many blood vessels that pass through it. Blood supplies living bone cells with nutrients and oxygen.

Under the periosteum is *compact bone*. It is the hardest part of a bone. It is made up of living bone cells. It is also made up of the mineral calcium. Calcium in compact bone makes it hard and gives it strength. A network of tubes also runs through compact bone. These tubes contain blood vessels and nerves.

The centers of flat bones and the ends of long bones are not as hard as compact bone. These softer parts of bone are called *spongy bone*. They even look like sponges because they are filled with many spaces. The spaces are filled with connective tissue and blood vessels. Although it is softer than compact bone, spongy bone is quite strong. The spaces form a support structure that is lightweight yet strong. Spongy bone provides support to areas in bones where there is great pressure and stress.

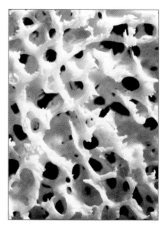

Figure 14-3 *Spongy bone is filled with many spaces.*

The centers of your bones are filled with marrow. You read that marrow makes blood cells. Marrow is red or yellow in color. Red marrow actively makes red blood cells. Yellow marrow is not active. It is found in the center of long bones. Yellow marrow is made mostly of fat cells.

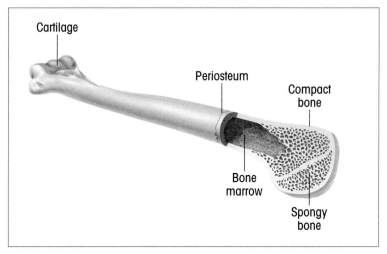

Figure 14-4 *The center of a long bone can contain red marrow and yellow marrow.*

## Bones Change

Bones grow and change. Some cells help build bone tissue. Some cells break down bone tissue. When bone tissue is broken down, some of the minerals inside bones are released into the body for use. However, your body normally replaces those minerals. The building up and breaking down of your bones changes the size and shape of your bones as you age.

### Broken Bones

Bones can be broken. A break or crack in a bone is called a **fracture**. The living material in a bone is made of cells that reproduce. These cells form new bone tissue to heal a broken bone.

There are different types of broken bones. Compare a closed fracture and an open fracture in Figure 14-5.

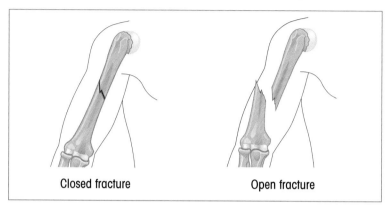

Closed fracture          Open fracture

Figure 14-5 *In a closed fracture, the skin is not opened. In an open fracture, the broken bone breaks through the skin.*

✓ **Check Your Understanding**

Write your answers in complete sentences.

1. What are the living materials that bones are made of?

2. What is a fracture?

3. **CRITICAL THINKING** What is the difference between red marrow and yellow marrow?

## Joints

The places where two or more bones meet are called **joints**. Bones are held together by strong bands of tissue called **ligaments**.

### Immovable Joints

There are several kinds of joints in your body. The joints in the skull are an example of *immovable joints*. The curved bones in a skull join together in what look like cracks. These bones do not move. Another example of an immovable joint is where a rib meets a vertebra in your backbone.

**Check Your Understanding**
1. bone cells, blood cells, and nerve cells
2. a break or crack in a bone
3. Possible answer: Red marrow makes blood cells. Yellow marrow contains fat.

## Movable Joints

Move your head from side to side. The joint between your skull and neck is a *pivotal joint*. This type of joint also allows your head to nod up and down. It also allows your head to move from side to side.

Your hip and shoulder joints are examples of *ball-and-socket joints*. These joints allow bones to move in almost any direction. You can even swing your arms or legs in a circle.

The joint at your elbow is a *hinge joint*. It can only move in two directions—back and forth. Your knees, toes, and fingers also have hinge joints.

There are *gliding joints* in your neck, wrists, and feet. These joints allow bones to glide over one another. You use gliding joints in your wrist when you unscrew the lid of a jar.

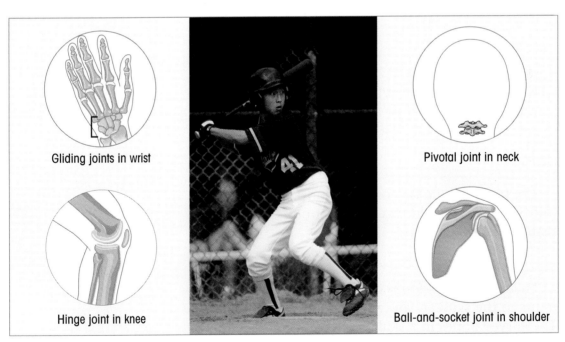

Gliding joints in wrist

Hinge joint in knee

Pivotal joint in neck

Ball-and-socket joint in shoulder

Figure 14-6 *Different kinds of joints allow this batter to hit a ball.*

# The Muscular System

Your bones cannot move on their own. They move because they are attached to muscles. **Tendons** are a type of tissue that connect muscles to bones or to other muscles. Your body has over 600 muscles. Almost half your body weight is muscle weight.

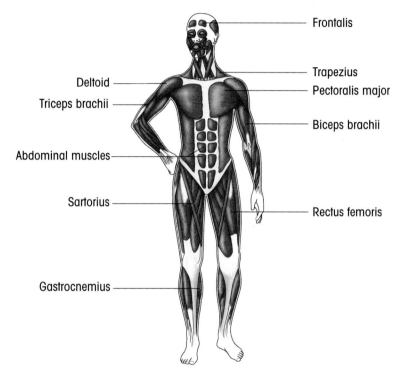

Frontalis

Trapezius

Deltoid

Pectoralis major

Triceps brachii

Biceps brachii

Abdominal muscles

Sartorius

Rectus femoris

Gastrocnemius

Figure 14-7 *The human muscular system has more than 600 muscles.*

## Different Kinds of Muscle Tissue

There are three kinds of muscle tissue in your body: skeletal muscle, smooth muscle, and cardiac muscle. *Skeletal muscles* are attached to your bones. They are **voluntary muscles**. You can control when these muscles move. This type of muscle moves your skeleton by pulling on tendons attached to bones.

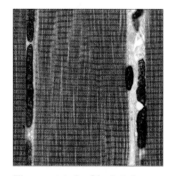

Figure 14-8 *Skeletal muscle is also called striated muscle. Seen under a microscope, it has a striped, or striated, appearance.*

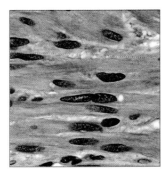

Figure 14-9 *Seen under a microscope, smooth muscle cells are long and pointed at the ends. They are not striated.*

Figure 14-10 *Seen under a microscope, cardiac muscle is striated.*

*Smooth muscles* are **involuntary muscles**. They work whether you want them to or not. They work even while you are sleeping. Smooth muscle is found in the walls of your blood vessels and intestines. Smooth muscles are not attached to bones.

**Cardiac muscle** is a type of involuntary muscle found in your heart. Cardiac muscle works all the time. Your heart is always beating.

## How Muscles Work

When a muscle moves, it contracts, or shortens. When a muscle relaxes, it lengthens. Muscles can only pull. They cannot push. For this reason, skeletal muscles work in pairs.

A muscle that extends a limb is called the *extensor*. The triceps in your arms are extensors. A muscle that bends the limb at the joint is called a *flexor*. The biceps, also in your arms, are flexors. The extensor and the flexor work opposite to each other. They balance each other. For movement to be controlled, one muscle must contract and the other must be relaxed.

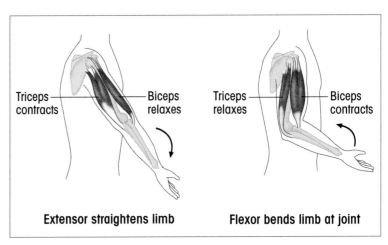

Triceps contracts — Biceps relaxes

Extensor straightens limb

Triceps relaxes — Biceps contracts

Flexor bends limb at joint

Figure 14-11 *The extensor and the flexor balance each other. This balance allows you to move smoothly.*

## ✓ Check Your Understanding

Write your answers in complete sentences.

1. What are ligaments?
2. What is a joint?
3. What are the three types of muscles?
4. **CRITICAL THINKING** What is the difference between a ligament and a tendon?

## A Closer Look

### MUSCULAR DYSTROPHY

Muscular dystrophy is an inherited disease. It is a disease that affects the skeletal muscles of the body. As time goes by, the muscles are destroyed. A person with the disease may no longer be able to use his or her arms and legs.

There are different types of muscular dystrophy. In one type, muscle problems start when a person with the disease is very young. With another type, the disease does not appear until adulthood.

Figure 14-12 *Muscular dystrophy is an inherited disease. This woman and her son both have the disease.*

Researchers are looking for ways to fight the disease. Some researchers are inserting certain genes into muscle cells. They hope that the cells will produce the right proteins to keep muscles healthy. Scientists are also trying to make chemicals that will act like the proteins the body uses to produce healthy muscles and membranes.

**CRITICAL THINKING** Could muscular dystrophy affect a person's heart? Explain your answer. No, muscular dystrophy affects skeletal muscles, not cardiac muscles.

# LAB ACTIVITY
## Identifying Joints

**BACKGROUND**

The four types of movable joints are the pivotal joint, ball-and-socket joint, hinge joint, and gliding joint.

**PURPOSE**

In this activity, you will classify the types of joints that help you to move in certain ways.

**MATERIALS**

book, pencil, paper

**WHAT TO DO**

1. Copy the chart in Figure 14-13 onto a separate sheet of paper.

2. Turn your hand at the wrist. What type of motion occurs at the joint? What kind of joint is used? Record your answer in your chart.

3. Continue with the other six motions. Complete the chart.

| Motion and Joints | | |
|---|---|---|
| **Motion** | **Motion of Bones** | **Type of Joint** |
| Turn your hand | gliding | gliding |
| Open and close your fist | back and forth | hinge |
| Tap your foot | gliding | gliding |
| Look behind yourself | circular | pivotal |
| Lift a book | up and down | hinge |
| Pretend to throw a ball | circular | ball-and-socket |
| Pretend to kick a ball | back and forth | hinge |

Figure 14-13 *Copy this chart onto a separate sheet of paper.*

**DRAW CONCLUSIONS**

• How are the motions of turning your hand and opening and closing your fist different? What types of joints control these motions?
Turning the hand is a gliding motion. Opening and closing a fist is a back and forth motion.
• How are the elbow and the knee similar?           Gliding joint and hinge joint.
They are both hinge joints.

# BIOLOGY IN YOUR LIFE
## Treating a Pulled Muscle

Playing sports and exercising are good for your health. Without care, though, you can get hurt. You can pull a muscle. A pulled muscle is actually small tears in the muscle. You feel some pain, but you can still move fairly well. A second-degree tear is more severe. It is painful and restricts movement. A third-degree tear is the most severe. The muscle fibers rip apart completely. It is very painful, and you cannot move the muscle at all. A very severe tear may need surgery. Tendons can also be torn.

If you tear a muscle or tendon, proper care is important. You must take steps to limit swelling. You also need to protect the injured muscle. An easy way to remember these steps is to think of R.I.C.E. Figure 14-14 explains R.I.C.E.

**The R.I.C.E Treatment Method**

| Letter | Treatment | Application | Effect |
|---|---|---|---|
| R | Rest | Immediately | Prevents more damage; gives the body energy to heal |
| I | Ice | Apply ice or cold packs. Leave on 15–20 minutes, then off 20 minutes. | Limits swelling |
| C | Compression | Wrap an elastic bandage over the area. | Limits swelling |
| E | Elevation | Raise the injury above the level of the heart. | Reduces swelling |

**Figure 14-14** ***R.I.C.E. is used to treat a pulled muscle.***

**Use the table to answer the questions.**

1. Why should you rest after an injury? To prevent more damage and to give the body energy to heal
2. Why should you put ice on a pulled muscle? to limit swelling
3. What does the *E* in R.I.C.E. stand for? elevation

**Critical Thinking**

After a day or two of R.I.C.E., many injuries begin to heal. What would you do if the pain and swelling did not decrease after two days? see a doctor

## Summary

- There are 206 bones in the human skeletal system. These bones have five jobs: 1) they support the body and give it shape; 2) they protect organs inside the body; 3) they work with muscles to help the body move; 4) some bones help make blood cells; and 5) they store some minerals that the body needs.

- Bones are made of bone cells, blood cells, and nerve cells. They also contain minerals that make bones hard. The structure of a bone includes the periosteum, compact bone, spongy bone, and marrow.

- The places where bones meet are called joints. Ligaments hold bones together at the joint. The types of joints are immovable joints, pivotal joints, ball-and-socket joints, hinge joints, and gliding joints.

- Tendons connect muscles to bones and other muscles. There are three kinds of muscles: skeletal muscles, smooth muscles, and cardiac muscles.

- When a muscle moves, it contracts, or shortens. When a muscle relaxes, it lengthens. Muscles can only pull. Skeletal muscles work in pairs.

cardiac muscle

fracture

involuntary muscle

joint

ligament

marrow

## Vocabulary Review

**Complete each sentence with a term from the box.**

1. A _____ is a type of tissue that holds bones together at a joint. ligament

2. The muscle tissue that makes up the heart is called _____. cardiac muscle

3. The soft tissue inside bones that makes blood cells is called _____. marrow

4. A muscle that moves without your control is called an _____. involuntary muscle

5. A _____ is a place in the body where two or more bones meet. joint

6. A break or crack in a bone is called a _____. fracture

# Chapter Quiz

**Write your answers on a separate sheet of paper. Use complete sentences.**

1. What are the jobs of bones and muscles?
   to support the body and to help it move
2. What tissue found in bones makes blood cells?
   red marrow
3. What minerals are found in bone? What do these minerals do for your bones?
   calcium and phosphorus; they make bones hard
4. What part of a bone has many blood vessels in it?
   the periosteum
5. What kind of tissue holds bones together?
   ligaments
6. What kind of joint allows you to bend your fingers?   hinge joint

7. Give two examples of involuntary muscles.
   muscles found in blood vessels and the stomach
8. What kind of joints are found in your skull?
   immovable joints
9. What kind of tissue connects muscles to bones?
   tendon
10. What kind of muscles can you control?
    voluntary muscles

## CRITICAL THINKING

11. Why do skeletal muscles work in pairs?
    because muscles can only pull, they cannot push
12. How can bones change in size and shape?
    The building up and breaking down of bones can change their size and shape.

**Research Project**

Students should find that the clavicle is the collarbone. It supports the shoulder and arm bones. The sternum is the breastbone. It protects the heart and supports the rib cage. The cranium is the skull. It protects the brain. The vertebrae are backbones. They support the body and protect the spinal cord.

*SCiLINKS*

Go online to www.scilinks.org. Enter the code **PMB126** to research **bones and joints**.

## Research Project

Bones support your body and protect your organs. Find out the jobs of some specific bones such as the clavicle, the sternum, the cranium, and the vertebrae. Use the Internet, encyclopedias, or other resources to research the common names of these bones and what their jobs are.

See the *Classroom Resource Binder* for a scoring rubric for the Research Project.

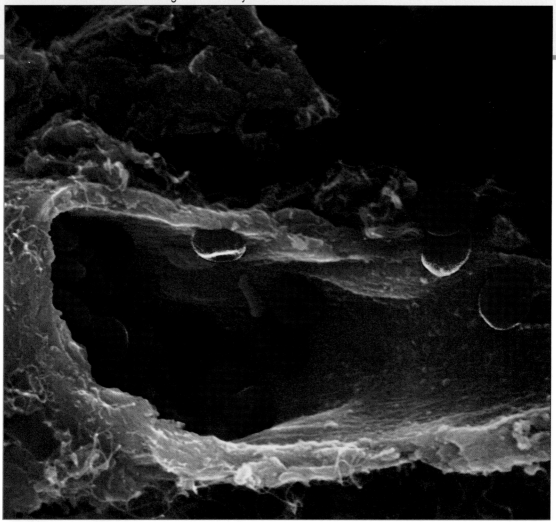

**Figure 15-1** *The shape of red blood cells help them to move through blood vessels. Red blood cells are flexible. Why do you think it is important for red blood cells to be flexible?*

It helps them to move through blood vessels easily.

## Learning Objectives

- Explain and list the parts of the circulatory system.
- Explain and list the parts of the lymphatic system.
- Explain and list the parts of the respiratory system.
- Explain and list the parts of the excretory system.
- **LAB ACTIVITY:** Compare blood cells.
- **ON-THE-JOB BIOLOGY:** Learn how a personal trainer creates a fitness program.

# Chapter 15

# Circulation, Respiration, and Excretion

## Words to Know

| | |
|---|---|
| **circulatory system** | the organ system that moves blood throughout the body |
| **vein** | a blood vessel that carries blood toward the heart |
| **artery** | a blood vessel that carries blood away from the heart |
| **capillary** | a tiny blood vessel that connects arteries to veins |
| **plasma** | the liquid part of blood that carries nutrients and wastes throughout the circulatory system |
| **platelet** | a piece of a cell found in blood that helps blood cells clump together |
| **lymphatic system** | the organ system that returns fluid lost by blood to the circulatory system |
| **respiratory system** | the organ system that moves oxygen into the body and carbon dioxide out of the body |
| **excretory system** | the organ system that removes wastes from the body |
| **nephron** | a network of tubes and blood vessels in a kidney that filters wastes from blood |

## The Needs of Cells

All cells need nutrients and oxygen. These materials help cells to carry out life functions. Cells also produce wastes. Cells need to get rid of these wastes.

In this chapter, you will read about the circulatory system, the lymphatic system, the respiratory system, and the excretory system. These organ systems help your cells get the materials that they need. They also help your cells get rid of wastes.

**Linking Prior Knowledge**
Students should recall the seven characteristics of life (Chapter 2), cellular respiration (Chapter 4), organ systems (Chapter 11), and the formation of blood cells in bone marrow (Chapter 14).

# The Circulatory System

**Remember**
Nutrients come from food that has been broken down into molecules.

Your cells need certain materials. These materials are brought to your cells by blood. Blood flows throughout your body. It carries oxygen and nutrients to all your body cells.

When cells carry out life functions, they also produce wastes. Some of these wastes include carbon dioxide and water. Blood carries these wastes away from body cells.

The **circulatory system** moves blood throughout the body. This organ system is made up of the heart, blood vessels, and blood. Figure 15-2 shows some parts of the circulatory system.

**Think About It**

When you exercise, more blood flows to the muscles in your arms and legs. When you eat, more blood flows to the muscles of your digestive system. How do you think eating a big meal before exercising might affect you?

One of two things can happen. Your muscles can tire easily because more blood is concentrated in your digestive system. Or, you can get an upset stomach because blood is flowing to your muscles instead of your digestive system.

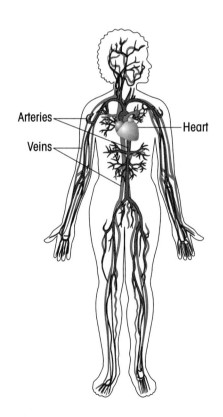

Arteries

Heart

Veins

Figure 15-2 *The circulatory system moves blood throughout the body.*

# The Heart

The heart is the main organ of the circulatory system. It is made up of cardiac muscle. The heart works by pumping and relaxing. These actions keep blood flowing throughout the body. The heart is divided into four parts called chambers. The two top chambers are called *atria* (singular *atrium*). The two bottom chambers are called *ventricles*. *Valves* open and close to control the direction of blood flow.

Figure 15-3 shows the path of blood through the heart. First, blood flows into the right atrium. From there, it is pumped into the right ventricle. Then, arteries carry the blood to the lungs. In the lungs, the blood picks up oxygen. Now, the oxygen-rich blood returns to the left atrium through large veins. The heart pumps the blood into the left ventricle. From there, it moves through the aorta and out to the body. The blood delivers nutrients and oxygen to all cells.

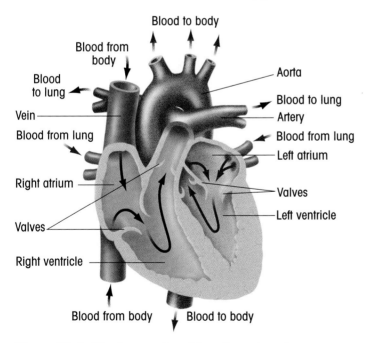

Figure 15-3 *The human heart has four chambers and four valves.*

## Blood Vessels

**Veins** are blood vessels that carry blood toward the heart. **Arteries** are blood vessels that carry blood away from the heart. **Capillaries** are tiny blood vessels that connect arteries to veins. These blood vessels are only one cell thick. Figure 15-4 shows capillaries that connect an artery to a vein.

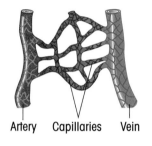

Artery    Capillaries    Vein

Figure 15-4 *Capillaries connect arteries to veins.*

Oxygen molecules and nutrient molecules pass through capillary membranes into body cells. At the same time, waste molecules from the body cells pass through the capillary membranes into the blood. These wastes are carried through veins all the way back to the heart. From there, the blood containing the wastes is pumped into the lungs. You exhale some of the wastes into the air.

## Great Moments in Biology

### BLOOD CIRCULATES IN THE BODY

Dr. William Harvey was an English doctor who lived in the 1600s. Dr. Harvey studied blood circulation. He found that blood moved in a continuous circular pattern.

Many people did not believe in Dr. Harvey's observations. At the time, most people thought that blood was produced in the liver and absorbed by the body. However, the discovery of capillaries about 30 years later supported Dr. Harvey's theory.

This discovery showed that large blood vessels and smaller blood vessels are connected by capillaries. This network of blood vessels allows blood to move in the circular pattern that Dr. Harvey described.

**CRITICAL THINKING** Why was the discovery of capillaries important to Dr. Harvey's theory?

Figure 15-5 *Dr. William Harvey explained that blood moves in a continuous circular pattern in the body.*

This discovery showed that large blood vessels and smaller blood vessels are connected by capillaries. This network of blood vessels allows blood to move in the circular pattern that Dr. Harvey described.

Write your answers in complete sentences.

**1.** What does the circulatory system do?

**2.** What parts make up the circulatory system?

**3.** What are the three types of blood vessels called?

**1.** It moves blood throughout the body.

**2.** the heart, blood vessels, and blood

**3.** veins, arteries, and capillaries

## Blood

Blood has many important jobs. One job is to transport nutrients, oxygen, and wastes. Another job is to fight infection.

Blood is a liquid tissue. It has a liquid part and a solid part. About 50 percent of blood is a liquid called **plasma.** This liquid is made up mostly of water. It carries nutrients and wastes throughout the circulatory system.

There are three kinds of solids in blood: platelets, red blood cells, and white blood cells. **Platelets** are actually pieces of cells. They help blood cells to clump together. This action helps wounds to stop bleeding.

### Red Blood Cells

Red blood cells deliver oxygen to other body cells. A protein in these cells carries the oxygen. This protein, called *hemoglobin,* contains the mineral iron. It gives blood cells their red color. The more oxygen the protein carries, the brighter the color of the cells. Red blood cells usually live about four months.

### White Blood Cells

White blood cells fight infection. They destroy harmful bacteria. They also defend the body against other harmful substances that enter the body. You have fewer white blood cells than red blood cells. There is about one white blood cell for every 1,000 red blood cells. However, when your body is fighting an infection, the number of white blood cells in your body can quickly increase.

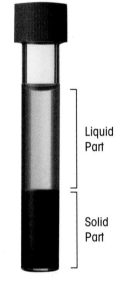

Liquid Part

Solid Part

**Figure 15-6** *This test tube contains blood that has been separated into its liquid part and solid part.*

# The Lymphatic System

As blood moves through the body, some fluid leaks into body tissues. As much as three quarts (three liters) of fluid can leak into body tissues per day.

The **lymphatic system** collects this fluid and returns it to the circulatory system. The clear fluid, called *lymph,* contains special white blood cells. These white blood cells, called *lymphocytes,* fight disease. Lymph moves through vessels and special organs, including the tonsils, thymus, spleen, and lymph nodes shown in Figure 15-7.

Lymph is filtered by the *lymph nodes* and *tonsils.* When you have an illness, your lymph nodes or tonsils can become larger. They become filled with a large number of white blood cells that are fighting the bacteria or virus causing your illness. The thymus gland helps certain white blood cells to grow. The spleen destroys damaged red blood cells.

## Science Fact

The lymphatic system does not have a pump like the heart. Pressure from blood and the squeezing of skeletal muscles moves lymph through lymph vessels.

## Everyday Science

Have you ever heard a doctor use the term *swollen glands*? What do you think the doctor is referring to?

the swelling of lymph nodes

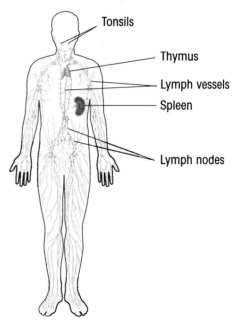

Tonsils

Thymus

Lymph vessels

Spleen

Lymph nodes

**Figure 15-7** *The lymphatic system returns fluid back to the circulatory system.*

# The Respiratory System

The **respiratory system** moves oxygen into the body and carbon dioxide out of the body. This system is made up of the lungs and the pathways that air moves through.

### Breathing Air In and Out

Breathe in through your nose. The air you breathe contains oxygen. It goes into your nose. The small hairs and mucus in your nose filter out dust and dirt. You can also breathe in air through your mouth. But your mouth cannot clean the air as well as your nose can.

As you breathe out, air rushes past your vocal cords. *Vocal cords* are stretched across the larynx as shown in Figure 15-8. The *larynx* is found at the top of the *trachea* (TRA-kee-uh) or windpipe. Vocal cords can vibrate and produce sound.

### The Diaphragm

As you breathe in, the muscles in your chest lift your rib cage up and outward. At the same time, a large skeletal muscle, called the *diaphragm* (DY-uh-fram), pulls the bottom part of your chest cavity lower. Both actions expand the volume of your chest cavity. Air moves into your lungs. When you breathe out, the rib muscles and diaphragm relax. The size of your chest cavity gets smaller and air is forced out of your lungs.

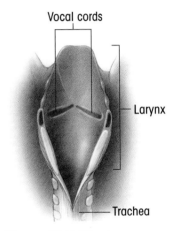

Figure 15-8 *The larynx contains vocal cords.*

### Tubes and Pathways

Inhaled air moves into your throat, or pharynx. The *pharynx* is a tube. Both air and food pass through it. Food continues through it to another tube that leads to the stomach.

From the pharynx, air moves into the trachea. You read that the larynx is found at the top of the trachea. When you swallow, a flap of tissue called the *epiglottis,* closes the larynx. It prevents food and liquids from getting into the trachea. Figure 15-9 on page 246 shows the location of these parts.

The trachea is a tube that divides into two smaller tubes. These smaller tubes are called *bronchi* (BRAHN-ky). Each bronchus goes into a lung. Air moves through these small tubes and into your lungs. Figure 15-9 also shows the parts that air passes through as it makes its way into the lungs.

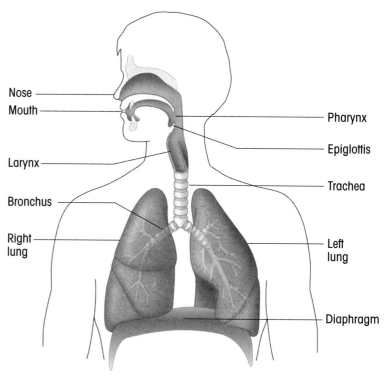

Nose
Mouth
Larynx
Bronchus
Right lung

Pharynx
Epiglottis
Trachea
Left lung
Diaphragm

Figure 15-9 *The respiratory system is made up of the pathways and the lungs that air moves through.*

## Science Fact

Asthma is a disease of the respiratory system. During an asthma attack, smooth muscles reduce the size of air pathways in the lungs. Breathing becomes very difficult. People who suffer from asthma can take certain medications to help relax the muscles in their air pathways.

### Lungs

You have two lungs. They are big, spongy organs. In each lung, the bronchus branches into smaller and smaller tubes. These tubes end in millions of tiny air sacs. These air sacs are called *alveoli* (al-VEE-uh-ly).

The air sacs are covered with capillaries. Like capillaries, the walls of each air sac are only one cell thick. Oxygen molecules pass from the air sac into the capillary. Blood vessels carry the oxygen-rich blood to the heart and then to all parts of the body.

At the same time, blood in capillaries passes water and carbon dioxide into the air sacs. You get rid of these wastes when you exhale.

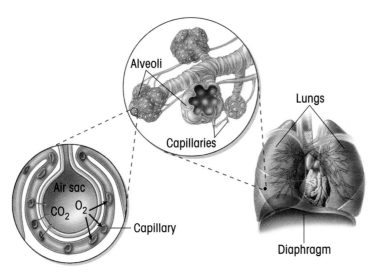

Figure 15-10 *Each lung is filled with air sacs that exchange gases with air.*

## The Dangers of Smoking

Smoking can seriously damage the respiratory system. Tar and nicotine are substances found in cigarettes. These substances can clog the air sacs in the lungs. Oxygen cannot pass into the blood. Without oxygen, your cells cannot carry out life functions. Wastes also get trapped in the lungs. Clogged lungs damage the heart. The heart has to pump harder to move blood throughout the body.

✓ **Check Your Understanding**

Write your answers in complete sentences.

**1.** What are the three kinds of solids in blood?

**2.** What does the respiratory system do?

**3.** Where are vocal cords found?

**4.** Describe what the lungs look like.

**Check Your Understanding**

1. platelets, red blood cells, and white blood cells

2. It provides oxygen to cells and removes carbon dioxide from cells.

3. in the larynx at the top of the trachea

4. They are big spongy organs filled with smaller branching tubes.

# The Excretory System

Wastes are produced as a cell carries out life functions. The body needs to get rid of these wastes in order to stay healthy. The **excretory system** removes wastes from the body. The lungs, the kidneys, the intestines, and the skin are a part of this system.

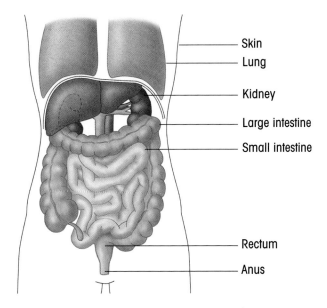

Figure 15-11 *The excretory system removes wastes from the body.*

## Wastes From Respiration

The lungs get rid of carbon dioxide and water. As you breathe out, you are releasing these waste products into the air.

## Solid Waste From Digestion

During digestion, material that cannot be digested and absorbed by the body moves from the small intestine to the large intestine. In the large intestine, water is removed from the material and it becomes solid. The solid waste moves into the rectum. From there, the solid waste material leaves the body through the anus.

## Liquid Waste From Digestion

Wastes leave the kidneys as a liquid called *urine*. Urine is made up mostly of water. It also contains salts and other substances. One of these substances is urea. *Urea* is a compound that forms when the body uses proteins. Urine leaves the kidneys through tubes called *ureters*. Then, it is collected in an organ called the *urinary bladder*. Urine leaves the body through another tube called the *urethra*.

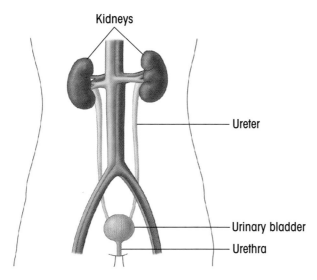

**Kidneys**

— Ureter

— Urinary bladder

— Urethra

**Figure 15-12** *The kidneys, along with the ureter, bladder, and urethra, are parts of the excretory system.*

## The Kidneys

The kidneys are the organs that remove waste products from the blood. You have two kidneys. They are located just above your waistline. These organs contain a network of small tubes called **nephrons**. The tubes remove extra water, salt, and other waste materials from the blood.

Each nephron has a cuplike structure at one end. This structure is called *Bowman's capsule*. Inside of this structure, blood flows through the many coiled capillaries. Water and other materials filter out of the blood.

### Science Fact

Dialysis is a procedure used to treat kidney failure. During a type of dialysis, waste products and extra fluid in blood are filtered through a machine. Clean blood flows back into the body.

## How a Nephron Works

The following steps explain how a nephron filters wastes and nutrients from blood. Look at Figure 15-13 as you read each step.

**STEP 1** Blood enters the nephron in an artery.

**STEP 2** The blood passes into the coiled capillaries in Bowman's capsule.

**STEP 3** Water, salts, urea, and other substances are filtered in Bowman's capsule. The filtered materials pass into a tubule.

**STEP 4** In the coiled part of the tubule, the nutrients pass back into the blood. Extra water, salts, and urea form urine.

**STEP 5** Urine is concentrated in an area called the *loop of Henle.*

**STEP 6** The urine then empties into the last part of the tube called the collecting duct. The filtered blood returns to the heart in veins. Urine flows through ureters to the bladder. It is stored there until it is released through the urethra.

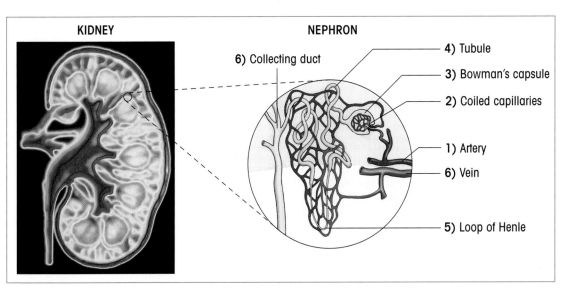

Figure 15-13 *The photograph on the left is a CT scan of a kidney. Nephrons in a kidney, on the right, filter blood.*

# The Skin

The skin is the largest organ in the human body. It covers and protects the body. It also helps rid the body of certain wastes.

The skin contains many sweat glands. Each sweat gland extends to a pore in the skin. Blood carries water and salts to the sweat glands. The sweat glands release these wastes as perspiration, or sweat, through the pores.

Perspiration cools down the body. When your body temperature rises, blood circulation increases. Your skin becomes warm. The sweat glands in your skin release sweat. As water in sweat evaporates from your skin, your body cools. Your body temperature lowers.

**Figure 15-15** *This athlete sweats in order to lower her body temperature.*

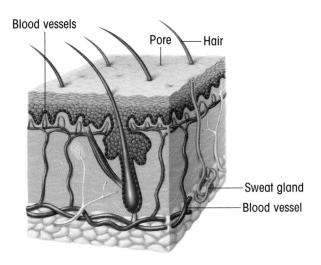

Blood vessels
Pore — Hair
Sweat gland
Blood vessel

**Figure 15-14** *The skin gets rid of wastes through sweat glands.*

## ✓ Check Your Understanding

Write your answers in complete sentences.

1. What organs are part of the excretory system?

2. What are the functions of the kidneys?

3. **CRITICAL THINKING** Which organ in the excretory system reacts to changes in temperature?

**Check Your Understanding,**
1. the lungs, the kidneys, and the skin and the intestines.

2. They remove waste products from the blood.

3. the skin

# LAB ACTIVITY
## Observing Blood Cells

### BACKGROUND
Red blood cells and white blood cells are two kinds of solids found in blood.

### PURPOSE
In this activity, you will compare red blood cells and white blood cells.

### MATERIALS
prepared slide of human blood, microscope, pencil, paper

### WHAT TO DO
1. Copy the chart in Figure 15-16 onto a separate sheet of paper.

2. Place the prepared slide on the stage of the microscope.

3. Focus the microscope using the low-power lens. Look for red blood cells. Draw or write your observations on your chart.

4. Switch to the high-power lens. Focus the lens, using the fine adjust knob, on the same cells. Draw or write your observations in your chart.

5. Switch back to the low-power lens. Try to find a white blood cell. Record your observations on your chart.

6. Repeat Step 4.

| Compare Blood Cells | | |
| --- | --- | --- |
| Type of Cell | Observations Under Low Magnification | Observations Under High Magnification |
| Red blood cell | | |
| White blood cell | | |

Figure 15-16 *Copy this chart onto a separate sheet of paper.*

### DRAW CONCLUSIONS
- Did the red blood cells have nuclei? Did the white blood cell have a nucleus? The red blood cells do not have nuclei. The white blood cells do.

- Which is smaller, a red blood cell or a white blood cell? A red blood cell is smaller than a white blood cell.
- How do you think the characteristics of a red blood cell are related to its job? The small size of a red blood cell allow it to fit through capillaries. Not having a nucleus allows a red blood cell to hold more oxygen.

# ON-THE-JOB BIOLOGY
## Personal Trainer

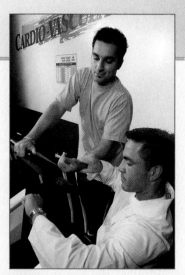

Ajay Singh is a personal trainer. He creates fitness programs for his clients based on each client's fitness level and goal. His clients walk, jog, bike, or swim to strengthen their hearts.

Ajay explains to his clients that the harder the body works, the faster the heart beats. Within a certain range of heartbeats, a person can increase his or her fitness. This range is the training zone. This zone is based on the number of heartbeats and age of a person. Ajay helps his clients create an exercise program that helps them reach their training zones.

**Figure 15-17** *A personal trainer helps people reach their training zones.*

Ajay took courses to learn to be a personal trainer. After he finished his courses, he took a test to become a certified personal trainer.

**Ajay created the table in Figure 15-18 to help Ann, one of his clients. Use it to answer the questions that follow.**

**Critical Thinking**
Possible answer: Your heart pumps harder to get blood to your muscles. Your muscles get more oxygen and nutrients.

**Critical Thinking**
Why does your heart beat faster as you exercise harder?

| Ann's Fitness Program | | |
|---|---|---|
| **Warm Up** | **Training Zone** | **Cool Down** |
| 5–10 minutes | 115–130 beats per minute for 40 minutes | 5–10 minutes |

**Figure 15-18** *A sample fitness program*

1. What is Ann's training zone?  115 to 130 beats per minute

2. How long does Ann need to exercise in the training zone?  40 minutes

3. How long should she warm up and cool down?  5 to 10 minutes each

# Chapter

## 15 ▶ Review

## Summary

- The circulatory system moves blood throughout the body. It is made up of the heart, blood vessels, and blood. The heart pumps blood throughout the body. Veins, arteries, and capillaries are the different types of blood vessels. Blood contains plasma, platelets, red blood cells, and white blood cells.

- The lymphatic system includes lymph, lymph vessels, and lymph nodes. Lymph contains white blood cells that fight disease.

- The respiratory system delivers oxygen to the blood. It is made up of the lungs, tubes, and pathways that air moves through. Air enters through the nose or mouth and travels to the air sacs in the lungs. Oxygen passes from the air sacs into the blood. Water vapor and carbon dioxide pass from blood into the air sacs.

- The excretory system removes wastes from the body. The lungs, kidneys, intestines, and skin are part of the excretory system. The lungs remove carbon dioxide and water from the body. Nephrons in kidneys remove wastes from the blood. Sweat glands in the skin release perspiration, or sweat.

circulatory system

excretory system

nephron

plasma

vein

## Vocabulary Review

**Complete each sentence with a term from the list.**

1. The liquid part of blood that carries nutrients and wastes throughout the circulatory system is called _____. plasma

2. The organ system that moves blood throughout the body is called the _____. circulatory system

3. A blood vessel that carries blood toward the heart is called a _____. vein

4. The organ system that includes the lungs, kidneys, and skin that removes wastes from the body is called the _____. excretory system

5. A tube in a kidney that filters wastes from blood is called a _____. nephron

# Chapter Quiz

**Write your answers in complete sentences.**

1. What are some materials that all cells need?
oxygen and nutrients

2. What are some wastes that blood carries away from body cells? carbon dioxide and water

3. What are the parts that make up the circulatory system? the heart, blood vessels, and blood

4. What are the two top chambers of the heart called? What are the two bottom chambers of the heart called? atria; ventricles

5. What are the three kinds of solids found in blood?
platelets, red blood cells, and white blood cells

6. What are some parts that make up the lymphatic system? lymph, lymph vessels, and lymph nodes

7. What parts make up the respiratory system?
the lungs, tubes, and pathways that air moves through

8. What is another name for the windpipe?
the trachea

9. What is the job of the excretory system?
It removes wastes from the body.

10. What part of the skin releases sweat?
the pores

**CRITICAL THINKING**

11. What is the difference between arteries and veins?

12. Why is breathing in and out important?

## Research Project

Narrow blood vessels lead to high blood pressure. High blood pressure can damage the heart and cause heart disease. A heart attack can occur if blood vessels become clogged. Find out what you can do to prevent heart disease and other diseases of the circulatory system. Use the Internet and other sources to research ways to keep your heart healthy.

See the *Classroom Resource Binder* for a scoring rubric for the Research Project.

---

**Test Tip**
Make sure that your answer sheet has the same number of answers as there are questions on the test.

**Research Project**
Students' research should include not smoking, not eating foods high in fat, getting plenty of exercise, and getting regular physical examinations by a doctor.

**Critical Thinking**
11. Arteries carry blood away from the heart. Veins carry blood toward the heart.
12. Breathing in brings oxygen into the body. Breathing out removes carbon dioxide and water vapor from the body.

*SCi*LINKS
Go online to www.scilinks.org. Enter the code **PMB128** to research **circulatory system**.

**ESL Note** To introduce the digestive system to your students, point out each organ on a model or large poster. Say the names of the organs of the digestive tract in the correct order. Then, ask students to do the same.

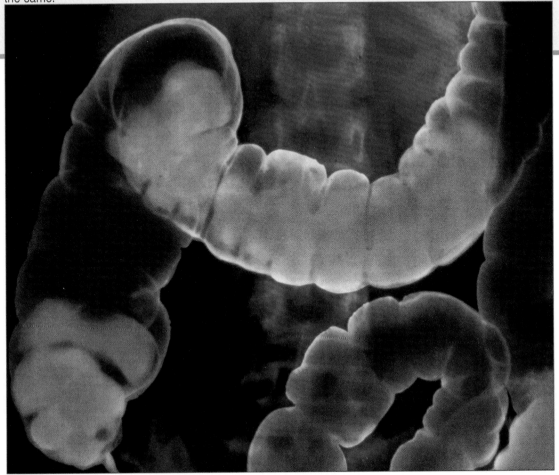

**Figure 16-1** *The X-ray image above shows a part of the large intestine. This organ is part of the digestive system. What do you think are some other organs of the digestive system?*

Possible answer: mouth, stomach, small intestine

## Learning Objectives

- Name the main organs of the digestive system.
- Describe what happens to food as it moves through the digestive system.
- Compare mechanical and chemical digestion.
- Describe the roles of the pancreas, liver, and gallbladder in digestion.
- **LAB ACTIVITY:** Investigate the role of enzymes in digestion.
- **BIOLOGY IN YOUR LIFE:** Describe some ways you can prevent tooth decay.

# Chapter 16 ▷ Digestion

## Words to Know

| | |
|---|---|
| **digestion** | the process by which the body breaks down food into nutrients that can be absorbed by cells |
| **enzyme** | a substance that helps to change chemical reaction rates in the body |
| **saliva** | a liquid in the mouth that helps digestion |
| **esophagus** | (ih-SAHF-uh-guhs) a tube behind the windpipe that carries food from the mouth to the stomach |
| **chyme** | (KYM) partially digested food material that leaves the stomach |
| **pancreas** | an organ that produces digestive enzymes and hormones |
| **liver** | a large organ that produces bile |
| **bile** | a green liquid produced by the liver that helps digest fats |
| **gallbladder** | an organ that stores bile |
| **villi** | tiny finger-shaped structures in the walls of the small intestine that absorb digested food into the blood—singular *villus* |

## Digesting Food

All living things need energy to survive. You get energy from the foods you eat. Your blood carries nutrients from the food you eat all over your body. First, the food must be broken down into simple substances. A whole tuna sandwich cannot move through your blood vessels. The process of breaking down food into nutrients is called **digestion**. This process is carried out by your digestive system.

**Linking Prior Knowledge**

Students should recall that systems of the human body are made of organs that work together (Chapter 14).

Students should also recall the parts of the digestive systems of other animals (Chapters 12 and 13).

# The Parts of the Digestive System

**Remember**
The digestive systems of all animals are not the same. Animals that eat different foods have slightly different digestive systems. For example, an animal that eats mostly plants usually has a longer intestine. This helps to digest the fibers of the plant material.

Digestion is carried out by many organs. These organs include the mouth, esophagus, stomach, small intestine, and large intestine. These organs are all connected. In fact, they form one long tube called the digestive tract. Figure 16-2 shows the parts of the digestive system.

In many of these organs, enzymes help with digestion. An **enzyme** is a substance that changes the rate of a chemical reaction. When a chemical reaction happens in the digestive system, the process is called chemical digestion. Other times, food is broken down into smaller pieces but not changed chemically. This is called mechanical digestion. Breaking, crushing, and mashing food are forms of mechanical digestion.

## Science Fact

Some structures exist in the human body that have no known purpose. The appendix is a small fingerlike structure at the end of the large intestine. It has no real function and can be removed if it gets infected.

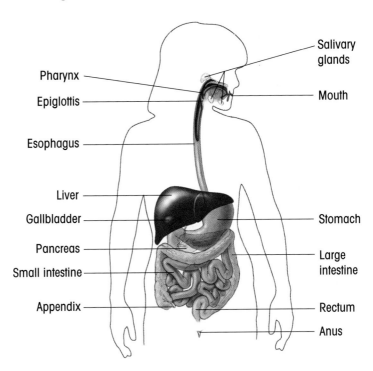

Figure 16–2 *The digestive system*

Pharynx

Epiglottis

Esophagus

Liver

Gallbladder

Pancreas

Small intestine

Appendix

Salivary glands

Mouth

Stomach

Large intestine

Rectum

Anus

# The Mouth and the Esophagus

Once you put food into your mouth, your teeth go to work. They bite and chew the food into smaller pieces. Humans have 32 permanent teeth. Different teeth are used to do different things. In the front of your mouth are teeth that bite, called incisors. Next to them are teeth that tear food, called canines. Behind the canines are premolars. In the back of your mouth are molars. The premolars and molars crush and grind food.

Teeth are covered in a hard material called *enamel*. They also have *roots*. The roots contain nerves and blood vessels.

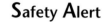
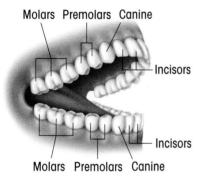

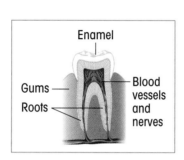

Figure 16-3 *Humans have four kinds of teeth. The shape of each tooth is related to its function.*

As you chew, a liquid called **saliva** is released. The saliva wets the food and helps you to swallow. Saliva also has an enzyme that breaks down starch into sugar. This is a form of chemical digestion.

There are two openings at the back of your mouth. One opening leads to your windpipe. The other opening leads to the esophagus. The **esophagus** connects the mouth to the stomach. Muscles in the esophagus squeeze together, pushing the food down. There is a flap of tissue called the epiglottis between these two openings. The epiglottis closes automatically when you swallow. This keeps food from going down the windpipe.

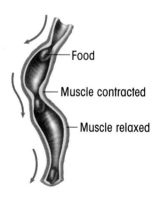

Figure 16-4
*The muscles of the esophagus squeeze food down into the stomach.*

✔ **Check Your Understanding**

Answer the questions on a separate sheet of paper.

1. What are the two types of digestion?
   mechanical and chemical
2. What substances affect the speed of chemical reactions? enzymes

3. What are the four types of teeth?
   incisors, canines, premolars, and molars

## The Stomach

Food passes from the esophagus into the stomach. It can be stored there for up to 6 hours. Both mechanical digestion and chemical digestion take place in the stomach. Muscles in the stomach move the food around. Enzymes and acids are released to break down the food chemically. A coating of mucus protects the stomach lining from the acid.

When the food leaves the stomach, it is considered chyme. **Chyme** is a partially digested form of food material. The chyme then goes into the small intestine. The small intestine is tightly coiled below the stomach. Most digestion takes place in the small intestine.

**Think About It**

The prefix *anti-* means "against." Why do you think some people take antacids for stomach problems?

**Think About It**

Antacids are taken to neutralize the acid in the stomach.

## The Pancreas, Liver, and Gallbladder

The pancreas and the liver are two organs that help in the digestion process. The **pancreas** produces enzymes that help digestion. It also makes a substance that neutralizes the acid in chyme. This is important because the small intestine could be damaged by the acid. The **liver** is a large organ that makes a substance called bile. **Bile** is a green liquid that helps digest fat. Bile is stored in the **gallbladder**. The bile is pushed out of the gallbladder and into the small intestine.

The pancreas, liver, and gallbladder are called accessory organs. This is because they help digestion occur but are not part of the digestive tract.

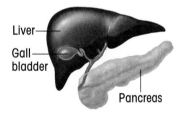

Liver
Gall bladder
Pancreas

Figure 16-5 *The pancreas, liver, and gallbladder take part in chemical digestion.*

## The Small Intestine and Large Intestine

Next, the mixture of chyme and digestive enzymes moves to the small intestine. The lining of the small intestine has many folds and tiny finger-shaped structures called **villi**. The villi increase the surface area of the small intestine. This means more nutrients can be absorbed. The villi are full of tiny blood vessels. They absorb nutrient molecules into the bloodstream.

You cannot digest every bit of the food you eat. Some leftovers pass into the large intestine. The large intestine absorbs water from the leftovers. The water returns to the body. Then, the large intestine forms solid waste. The solid waste is passed out of the body through the rectum.

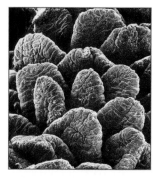

Figure 16-6 *Villi help the body absorb nutrients from food.*

✓ **Check Your Understanding**

How do the accessory organs help the stomach and intestines carry out digestion? Write your answer on a separate sheet of paper.

**Check Your Understanding**

The pancreas produces enzymes and a substance that neutralizes the acidity of chyme. The liver produces bile, which helps to digest fat. The gallbladder stores bile and releases it into the small intestine.

## A Closer Look

### LACTOSE INTOLERANCE

Many people cannot enjoy a glass of milk or slice of cheese pizza because they are lactose intolerant. A person is *lactose intolerant* if he or she cannot digest the sugar in milk products, called lactose. Milk cannot be digested without enough of the enzyme lactase. Lactase helps break down lactose. Symptoms of lactose intolerance include stomach pain, nausea, and cramps.

Figure 16-7 *Some people cannot digest the lactose found in dairy foods.*

**CRITICAL THINKING** Why should calcium from other sources be included in the diet of a person who is lactose intolerant?
Because they are not able to get calcium from dairy products.

# LAB ACTIVITY
## Investigating the Digestion of Starch

**BACKGROUND**
Living things use enzymes to change the rate of chemical reactions. Enzymes start the job of digestion from the minute we take food into our mouths.

**PURPOSE**
In this activity, you will test for starch. You will also observe the way the enzyme amylase turns starch into sugar.

**MATERIALS**
safety goggles, gloves, apron, plain crackers, iodine solution, dropper

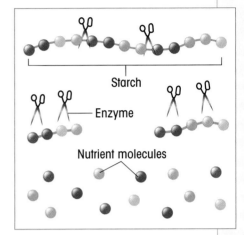

Figure 16-8 *Digestive enzymes act like scissors that cut starch chains into smaller molecules.*

**WHAT TO DO**
**Part A**

1. Use a dropper to place two drops of iodine onto a cracker.
   **Caution**: Be careful when using iodine because it will stain your skin and clothing.

2. Observe the cracker. Describe any color change on a separate sheet of paper. Note that iodine will turn a bluish-purple color when starch is present.

**Part B**

3. Obtain another cracker. Chew the cracker, but do not swallow it.
   Make sure students use unsalted crackers.
4. Let the cracker sit on your tongue and mix with your saliva for at least 1 full minute.

5. Notice the flavor change.

**DRAW CONCLUSIONS**
- Do crackers contain starch? How do you know?
  Yes, because they turned purple when iodine solution was added.
- What caused the cracker to change flavor? Look at Figure 16-8 for help. Enzymes broke down the starch into sugars.

# BIOLOGY IN YOUR LIFE
## Preventing Tooth Decay

Have you ever had a cavity? If you have, you are not alone. A cavity is a hole caused by tooth decay. By age 17, almost 80 percent of young people have had at least one.

After eating, a sticky film of bacteria can form on the teeth and gums. This film is called *plaque*. When you eat sugar or starch, the bacteria eat it, too. They create acids as a waste product. The acids can eat away at the enamel. This is the tooth's hard outer shell. Plaque helps the decay process by holding the acids against the tooth's surface.

**There are many ways to prevent tooth decay. Look at the following fact sheet. Then, answer the questions.**

Figure 16-9 *It is important to brush your teeth at least twice daily.*

### Ways to Prevent Tooth Decay
- Brush soon after eating meals and snacks to clean off plaque. If you cannot brush, rinse.
- Brush at least twice a day for 2 to 3 minutes to remove plaque from teeth.
- Floss at least once a day to remove plaque between teeth and below the gum line.
- Limit foods high in sugar. The less sugar for the bacteria to feed on, the less acid they produce.

**Critical Thinking**

Students could choose low-sugar snacks, such as vegetable sticks, cheese, or pretzels. They also should brush or rinse after eating.

**Critical Thinking**

Suppose you are going to be out of the house all day. What can you do to help reduce tooth decay throughout the day?

1. What is the minimum number of times per day you should brush your teeth?
   at least twice daily
2. Kaya does not like flossing because it takes too much time. How is this harmful to her teeth?
   It leaves plaque between teeth and below the gum line.

# Chapter

## 16 ▷ Review

### Summary

- Digestion is carried out by many organs, including the mouth, esophagus, stomach, small intestine, and large intestine.

- There are two forms of digestion: mechanical and chemical. Mechanical digestion tears and grinds food into smaller pieces. Chemical digestion changes the molecules of food through chemical reactions.

- The pancreas, liver, and gallbladder help the digestive process by making enzymes and other digestive fluids.

- Nutrients pass from the villi in the small intestine into the bloodstream.

- Undigested food passes from the small intestine into the large intestine. The large intestine absorbs water from the undigested food and forms waste. The waste then passes out of the body.

bile

chyme

digestion

enzyme

esophagus

saliva

## Vocabulary Review

**Write the term from the list that matches each definition.**

**1.** a tube behind the windpipe that carries food from the mouth to the stomach   esophagus

**2.** a substance that helps to change chemical reaction rates in the body   enzyme

**3.** the process by which the body breaks down food into nutrients that can be absorbed by the cells   digestion

**4.** partially digested food material that leaves the stomach   chyme

**5.** a liquid in the mouth that helps digestion   saliva

**6.** a green liquid produced by the liver that helps digest fats   bile

# Chapter Quiz

**Answer the questions in complete sentences. Write your answers on a separate sheet of paper.**

1. What must happen to a tuna sandwich before your body cells can use it? It must be digested.

2. What are five organs of the digestive system?
   mouth, esophagus, stomach, small intestine, large intestine

3. What are enzymes?
   substances that change the speed of a chemical reaction

4. What does saliva do?
   breaks down starch into sugar

5. What kind of teeth are used for grinding?
   molars

6. How do teeth help in digestion?
   They break up food into smaller particles.

7. How does food get down the esophagus?
   Muscles squeeze and push the food down.

8. What is food material called when it leaves the stomach? chyme

9. What is the role of the liver and the pancreas in digestion? The liver and the pancreas make substances like enzymes and bile that help digest food.

10. From which organ is food absorbed into the bloodstream? the small intestine

## CRITICAL THINKING

11. Why are villi in the small intestine so important to digestion?

12. Compare mechanical and chemical digestion.

Answer the questions you are sure of first. Then, go back and answer those you need to think about.

**Research Project**
Students' posters should show how the two systems are related. Ask students to share their posters with the class. You may want to create a "Body Systems" bulletin board using the posters.

**Critical Thinking**
11. They increase the surface area of the small intestine, which increases the amount of nutrients absorbed.

12. Mechanical digestion is the grinding and crushing food into smaller pieces. Chemical digestion occurs when food is broken down into smaller molecules by a chemical reaction.

*SCLINKS*
Go online to www.scilinks.org. Enter the code **PMB130** to research **body systems**.

---

## Research Project

The digestive system is closely linked to the circulatory system. This is because blood delivers nutrients to all of the cells of the body. Find out more about how the digestive system and the circulatory system are related. Use library resources or the Internet. Present your information in a poster.

See the *Classroom Resource Binder* for a scoring rubric for the Research Project.

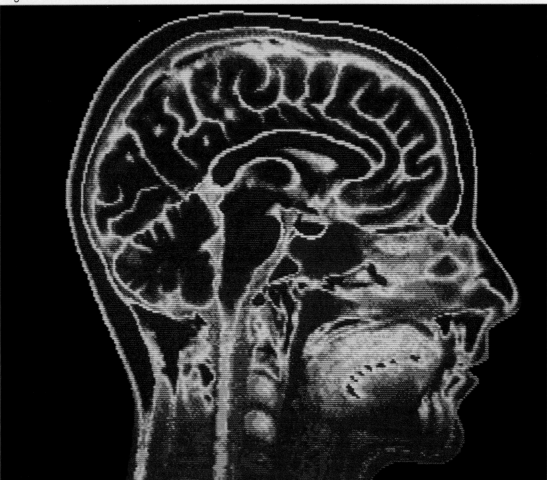

Figure 17-1 *Your brain receives and interprets messages from other parts of your body. Why is "control center" a good name for your brain?*

Possible answer: "Control center" is a good name because the brain controls the activities of the rest of the body.

## Learning Objectives

- Define neurons and explain their job in the body.
- Describe the functions of the main parts of the brain.
- Explain how a reflex works.
- Compare the central and peripheral nervous systems.
- Explain the job of the endocrine system.
- **LAB ACTIVITY:** Investigate how the brain handles coordinated and competing movements.
- **ON-THE-JOB BIOLOGY:** Describe what a nuclear medicine technologist does.

# Chapter 17 ▷ Regulating the Body

## Words to Know

| | |
|---|---|
| **neuron** | (NOOR-ahn) a nerve cell |
| **dendrite** | the fiber on neurons that carries messages into the cell |
| **axon** | the fiber on neurons that carries messages away from the cell |
| **gland** | an organ that makes a chemical substance that is used or released by the body |
| **synapse** | the space between the axon of one neuron and the dendrite of another |
| **reflex** | an automatic and involuntary response to an outside stimulus |
| **cerebrum** | the part of the brain that controls voluntary muscle movements, thinking, learning, memory, speech, and the senses |
| **cerebellum** | the part of the brain that controls balance and coordination |
| **medulla** | the part of the brain that controls involuntary functions |
| **endocrine gland** | organ that produces chemical substances called hormones that are released into the bloodstream |
| **hormone** | a type of chemical messenger in the body that is produced by endocrine glands |

## The Nervous System

Have you ever answered a ringing telephone, thrown a ball, or solved a difficult math problem? These activities are all controlled by your nervous system. Your nervous system allows you to think, speak, see, move, and feel.

**Linking Prior Knowledge**
Students should recall that the skull protects the brain and the backbone protects the spinal cord (Chapter 14).

## Neurons: Cells that Carry Messages

A **neuron** is a nerve cell. Neurons are the basic units of structure that make up the nervous system.

A typical neuron has three main parts. The largest part of the neuron is the cell body. The cell body contains the nucleus and most of the cytoplasm. There are long fibers that branch from the cell body. Some of these fibers carry messages from other neurons into the cell body. These fibers are called **dendrites**. Other fibers carry messages away from the cell body to other neurons. These fibers are called **axons**. Look at Figure 17-2 to see the main parts of a neuron.

### Science Fact

Neurons can be either large or small. In fact, one neuron in your leg can be more than 3 feet long!

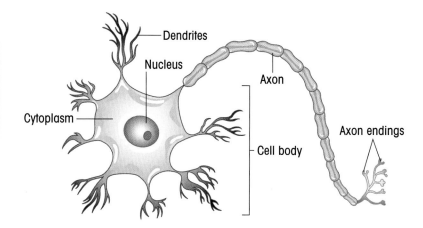

Figure 17-2 *A neuron is a nerve cell.*

### Three Kinds of Neurons

Neurons come in many sizes and shapes. However, there are three basic kinds of neurons. They can be classified by the direction in which they carry impulses. *Sensory neurons* are connected to your sense organs. The eyes, ears, nose, tongue, and skin are your sense organs. Sensory neurons take messages away from the sense organs to the central nervous system. The *central nervous system* is made up of the brain and the spinal cord.

*Motor neurons* carry messages away from the central nervous system to muscles and glands. **Glands** are organs that make chemical substances that are used or released by the body. Muscles and glands respond to information sent from the brain.

*Interneurons* connect other neurons together. They are found in the central nervous system. Interneurons process impulses from the sensory neurons and send the response impulses to the motor neurons. Then, the motor neurons respond. This process is called a response circuit.

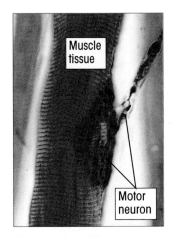

Figure 17-3 *Motor neurons attach to muscles.*

Neurons at Work

An impulse travels through your body in a series of steps. Look at Figure 17-4 to learn how an impulse travels. Suppose you are crossing a street and you hear a car horn. What do you do?

The first step is that your ears pick up the sound. In order to respond to the sound, your brain has to receive and then interpret the information. Step two is the brain receiving the information. Step three is the brain interpreting the information. Next, the brain has to send an impulse to your body telling it how to react. This is step four. Step five is the actions that your body does in response to this message. Even though this process involves many different organs and types of cells, it happens very fast.

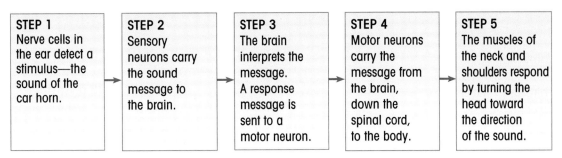

| STEP 1 | STEP 2 | STEP 3 | STEP 4 | STEP 5 |
|--------|--------|--------|--------|--------|
| Nerve cells in the ear detect a stimulus—the sound of the car horn. | Sensory neurons carry the sound message to the brain. | The brain interprets the message. A response message is sent to a motor neuron. | Motor neurons carry the message from the brain, down the spinal cord, to the body. | The muscles of the neck and shoulders respond by turning the head toward the direction of the sound. |

Figure 17-4 *Impulses travel in a series of steps.*

## Synapses: Connections Between Neurons

When an impulse travels along a neuron, the axon of one neuron usually does not actually touch the dendrite of the next neuron. The spaces between neurons are called **synapses**. Impulses must "jump" across this space. Chemicals released by the axon carry impulses across the synapse to the dendrite of the next neuron. Then, the impulse continues traveling along the next neuron.

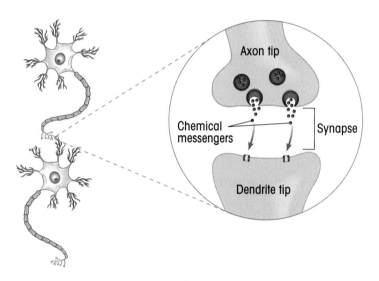

Figure 17-5 *Impulses travel from the axon of one neuron to the dendrite of the next neuron.*

## Reflexes

**Remember**
A stimulus is a change in environment that causes a response.

The nervous system is like the body's telephone service. It sends messages among the different organs of the body. These messages are signals called impulses. Sometimes an impulse is processed in your spinal cord rather than in your brain. This is called a reflex. **Reflexes** are automatic and involuntary responses to outside stimuli. This means they happen without your brain "thinking" about it.

Suppose you touch a very hot pan. Sensory receptors pick up an impulse from your finger. That impulse says that the finger is feeling heat and pain. This message travels along a sensory neuron to the spinal cord. Interneurons in the spinal cord pick up the message and deliver it to the motor neurons. The motor neurons send a message to your arm muscles.

Your arm muscles lift your hand away from the pan. This movement occurs before the pain message reaches the brain. The path that the impulse traveled is known as a *reflex arc*.

Possible answer: Reflexes can help an organism react quickly to dangerous situations such as an attack from a predator or an accident.

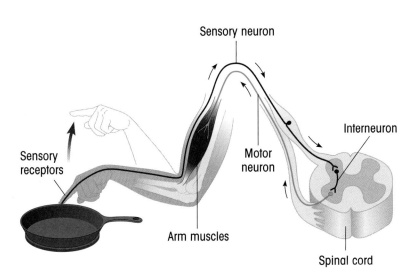

Figure 17–6 *In a reflex arc, impulses are processed in the spinal cord.*

## Two Divisions of the Nervous System

The nervous system is divided into two parts—the *central nervous system* and the *peripheral nervous system*. This is true of all vertebrates, including humans. The central nervous system acts as a control center. The peripheral nervous system links all the other parts of the body to the central nervous system.

# The Central Nervous System

The central nervous system is made up of the brain and the spinal cord. It receives and transmits messages to the peripheral nervous system. Look at Figure 17-7. It shows the position of the brain and spinal cord.

### The Brain

The main job of the brain is to receive, interpret, and send impulses to the rest of your body. The brain is made up of more than 160 billion neurons. There are three main parts to the brain. Figure 17-8 shows the parts of the brain.

The largest part of the brain is the cerebrum. In humans, it makes up about three-fourths of the brain. The cerebrum controls many muscle movements. It also controls thinking, learning, memory, and decision making. Another job of the cerebrum is to interpret messages from the sense organs.

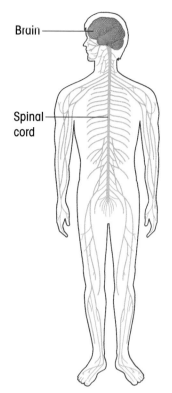

**Figure 17-7**
*The central nervous system includes the brain and spinal cord.*

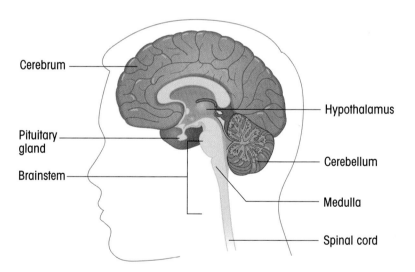

**Figure 17–8** *The brain has three major parts: the cerebrum, cerebellum, and brainstem. The inside of the brain has smaller parts, such as the pituitary gland and hypothalamus.*

The part of the brain that is located just below the back of the cerebrum is the cerebellum. The **cerebellum** helps to coordinate muscle movements. The cerebellum works with the cerebrum to enable you to perform such activities as walking and throwing a ball. The cerebellum also helps you to keep your balance.

The *brainstem* is the third part of the brain. It connects the brain to the spinal cord. In the lower part of the brainstem, there is a special organ called the medulla. The **medulla** controls involuntary body functions such as breathing, digestion, and heart rate. The medulla also controls the function of many glands. Look at Figure 17-9 to review the main parts of the brain.

| Parts of the Brain | |
|---|---|
| Cerebrum | • Largest part of the brain<br>• Controls voluntary movement, thinking, and memory |
| Cerebellum | • Located in the back of the head<br>• Helps with coordination and balance |
| Brainstem | • Connects the brain to the spinal cord<br>• Includes the medulla, which controls involuntary body function |

Figure 17-9 *The human brain is divided into three sections.*

### The Spinal Cord

The spinal cord is also part of the central nervous system. It is a ropelike structure that is made up of many neurons. The spinal cord runs from the base of the brain down to the lower part of your back. The spinal cord is protected by your backbone.

The spinal cord acts as the nervous system's main switchboard. Most of the messages between your brain and the rest of your body must travel along the spinal cord. Thirty-one pairs of nerves branch out from your spinal cord to all parts of your body.

**Safety Alert**

Like other parts of your body, your brain can be injured in an accident. Protect your brain by wearing a helmet when performing activities in which you could hurt your head. These activities include biking, skating, skiing, and football.

## ✓ Check Your Understanding

Answer the questions in complete sentences.

**1.** What is the central nervous system?

**2.** What is the function of an axon?

**3.** What are the three main parts of the brain?

**4.** CRITICAL THINKING  Which part of the brain are you using when you memorize a telephone number?

## On the Cutting Edge

### TREATING SPINAL CORD INJURIES

Spinal cord injury is damage to the spinal cord. It can result in some loss of feeling or function of body parts. These injuries can be caused by car accidents, falls, or disease. Spinal cord injuries can cause a person to lose function of just their lower body, or their entire body. Most spinal cord injury patients receive physical therapy, which includes various kinds of exercises. More than 450,000 people in the United States have some form of spinal cord injury.

In recent years, scientists have been doing a great deal of research on the treatment of spinal cord injury. One method that is currently being tested in the laboratory is nerve cell transplants. Scientists hope these transplanted neurons will help heal the spinal cord.

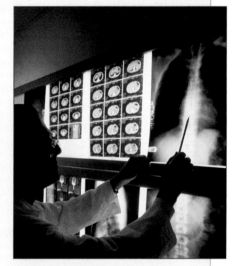

Figure 17-10 *Doctors use different techniques to diagnose spinal cord injuries.*

Another treatment that is being used is drug therapy. One drug has been shown to improve recovery from a spinal cord injury. However, this drug usually only works when given within eight hours of the injury. Many doctors and researchers feel that a combination of drug therapy, cell transplants, and physical therapy may be the best treatment for spinal cord injury.

CRITICAL THINKING  One problem with damaged neurons is that their axons no longer work. Why do you think nerve cell transplants could be used in treating spinal cord injury? because these new cells may be better able to regrow their axons

## The Peripheral Nervous System

Impulses leave the brain. They travel to different parts of your body. Then, other impulses travel back to the brain. Impulses travel along neurons. These include both sensory neurons and motor neurons. The neurons are bundled together into long fibers called nerves. These nerves make up the *peripheral nervous system.*

The peripheral nervous system contains twelve pairs of nerves that connect the brain to the head and neck. It also contains thirty-one pairs of nerves that connect the spinal cord with the rest of the body. Most impulses travel back and forth between the peripheral nervous system and the central nervous system.

## Hormones: Another Kind of Messenger

Some body functions and messages are controlled by nerve impulses. Others are controlled by chemical messengers. **Endocrine glands** produce chemical substances that are released directly into the bloodstream. These chemical substances are called **hormones**. Hormones regulate body functions. You may have heard of the hormone *adrenaline*. This hormone affects heart beat and blood pressure. Adrenaline is released when you are excited, angry, or frightened.

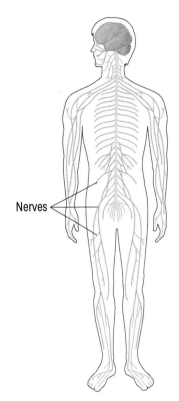

**Figure 17-11**
*The peripheral nervous system includes the nerves of the body.*

Nerves

Figure 17-12 *Extreme sports such as rafting can cause your levels of adrenaline to rise.*

### Science Fact

*Exocrine* glands are those that secrete a substance onto a surface either directly or by using ducts. Ducts are tubelike structures. Sweat glands and salivary glands are examples of exocrine glands.

# The Endocrine System

There are many different endocrine glands in the body. These glands are scattered throughout the body. Together, they make up the endocrine system. Figure 17-13 shows the parts of the female endocrine system. The job of the endocrine system is to help control body functions.

Endocrine glands may often control more than one organ at a time. However, hormones only react with certain cells. These are called target cells. Target cells are cells that have receptors for specific hormones. A hormone only binds to cells that have receptors for that particular hormone.

Different endocrine glands produce different hormones. Each gland has different functions. Look at Figure 17-14 to find out more.

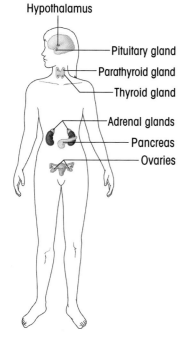

Figure 17–13 *The female endocrine system*

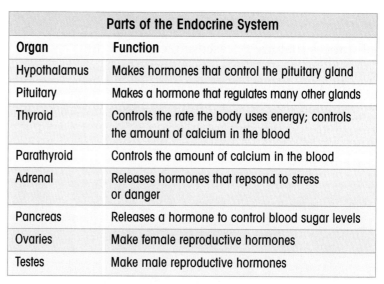

| Parts of the Endocrine System | |
|---|---|
| **Organ** | **Function** |
| Hypothalamus | Makes hormones that control the pituitary gland |
| Pituitary | Makes a hormone that regulates many other glands |
| Thyroid | Controls the rate the body uses energy; controls the amount of calcium in the blood |
| Parathyroid | Controls the amount of calcium in the blood |
| Adrenal | Releases hormones that repsond to stress or danger |
| Pancreas | Releases a hormone to control blood sugar levels |
| Ovaries | Make female reproductive hormones |
| Testes | Make male reproductive hormones |

Figure 17-14 *The endocrine system is made up of many different glands. Each gland may have one or more functions.*

**Remember**
Diffusion is the movement of molecules from an area that is more crowded to an area that is less crowded.

Endocrine glands release their hormones into the space that surrounds them. Then, they pass into capillaries by diffusion. The blood carries the hormones throughout the body to the correct cells.

## Hormones at Work

The endocrine system plays an important part in maintaining homeostasis. This is because the hormones produced by endocrine glands help to keep the body's cells and tissues in balance. The way endocrine glands accomplish this balance is by making sets of hormones that have opposite effects. For example, the thyroid gland and the parathyroid gland both release a hormone that controls the amount of calcium in the blood. However, one hormone increases the amount of calcium in the blood. The other hormone decreases the amount of calcium in the blood. In this way, the two hormones balance each other's effects. This type of hormone activity is called *negative feedback*. Look at Figure 17-15 to learn more about how hormones work.

### Everyday Science

27 The heating system in your home is a good example of negative feedback. Think about how a thermostat works and relate this to negative feedback.

Possible answer: The thermostat keeps the temperature stable. It turns the heat on when it gets too cool, and turns the heat off if it gets too hot.

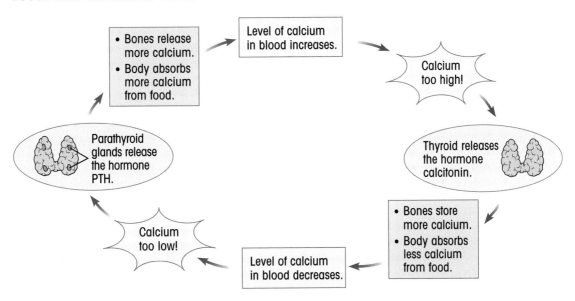

• Bones release more calcium.
• Body absorbs more calcium from food.

Level of calcium in blood increases.

Calcium too high!

Parathyroid glands release the hormone PTH.

Thyroid releases the hormone calcitonin.

• Bones store more calcium.
• Body absorbs less calcium from food.

Calcium too low!

Level of calcium in blood decreases.

Figure 17-15 *Calcium levels in the body are regulated by negative feedback between two hormones.*

### ✓ Check Your Understanding

On a separate sheet of paper, write a paragraph that explains how hormones help to maintain homeostasis.

### Check Your Understanding

Possible answer: Hormones help to maintain homeostasis by keeping the levels of certain chemicals in the body in balance. For example, the thyroid and parathyroid glands work together to control the amount of calcium in the body.

# LAB ACTIVITY
## Testing Coordination

### BACKGROUND
Human brains are good at doing many kinds of complex tasks, such as writing, talking, and walking. We can easily learn new movements that are coordinated. However, it takes practice to do tasks that require each hand to do its own thing. This is because the brain likes to focus on one coordinated task at a time.

### PURPOSE
In this activity, you will compare coordinated movements with competing movements.

**Figure 17-16** *Complex movements, such as drawing with both hands at once, can be difficult.*

### MATERIALS
2 pencils, paper, lab notebook

### WHAT TO DO
1. Try just tapping your head. Try just rubbing your stomach. Was it difficult to do?

2. Put one hand on your stomach and one hand on your head. Pat your head while rubbing your stomach. Keep trying until you are able to do it. Was it hard or easy? Did it get easier with practice? Write your observations on a separate sheet of paper.

3. Draw a triangle and a circle one at a time with each hand.

4. Now, set up a sheet of paper so you can draw on it with both hands. With one hand, draw a circle. At the same time, draw a triangle with the other hand. What do you notice?

### DRAW CONCLUSIONS
- Which movements were the easiest to do? Which ones were the hardest? Answers will vary. In general, tasks that require two competing movements will be harder to perform than single movements.
- Why do you think it is more difficult to do two different movements at the same time rather than one movement? because the brain is better at focusing on one action at a time

# ON-THE-JOB BIOLOGY
## Nuclear Medicine Technologist

Sue Jones is a nuclear medicine technologist. She helps conduct PET scans. *PET* stands for "positron emission tomography." A PET scan gives doctors a picture of the brain at work. This picture helps them diagnose diseases of the nervous system.

Sue begins by giving the patient a radioactive drug. The drug emits energy. The scanner in the PET machine detects the energy. It then creates images of the brain. Sue checks the images and the data it produces.

PET scans are used to study many diseases. Alzheimer's disease attacks the brain. People with the disease lose their memory over time.

**Look at the following graph. Then, answer the questions.**

**Figure 17-17**
*Sue Jones checks images created by the PET machine.*

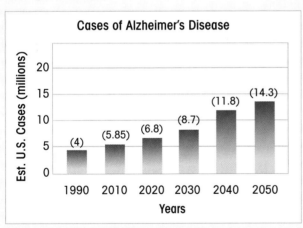

**Cases of Alzheimer's Disease**

Est. U.S. Cases (millions)

(4) (5.85) (6.8) (8.7) (11.8) (14.3)

Years: 1990 2010 2020 2030 2040 2050

Figure 17-18 *Predicted occurrence of Alzheimer's disease*

**Critical Thinking**
The number of Alzheimer's cases may start to decrease.

**Critical Thinking**
How might this chart change if a cure for the disease is found?

1. About how many people were diagnosed with Alzheimer's disease in 1990? about 4 million

2. What trend does the graph show? that the number of people diagnosed with Alzheimer's disease will increase

# Chapter

## 17 ▷ Review

## Summary

- The nervous system is made up of neurons. Neurons carry impulses to and from the brain and the spinal cord.

- Neurons are cells made of a cell body with a nucleus, dendrites, and axons. Between each neuron is a small gap, called a synapse. A special chemical carries an electrical impulse over this gap.

- A reflex is an automatic and involuntary response to outside stimulation. Many reflex messages are processed in the spinal cord rather than in the brain.

- The brain and the spinal cord make up the central nervous system. Other nerves make up the peripheral nervous system. The brain has three main parts: the cerebrum, the cerebellum, and the brainstem.

- Hormones are a different kind of body messenger. Hormones are released by the endocrine glands. Most hormones travel in the bloodstream. They control such things as body growth and behavior.

---

axon

cerebellum

cerebrum

dendrite

medulla

neuron

---

## Vocabulary Quiz

**Complete each sentence with a term from the list.**

1. The part of the brain that controls voluntary muscle movements, thinking, learning, memory, and the senses is the _____.   cerebrum

2. The part of the brain that controls body functions such as involuntary muscle movements is the _____.   medulla

3. A nerve cell that carries impulses to and from the brain is called a _____.   neuron

4. The fiber on neurons that carries messages into the cell is the _____.   dendrite

5. The fiber on neurons that carries messages away from the cell is the _____.   axon

6. The part of the brain that controls balance and the coordination of muscles is the _____.   cerebellum

Research Project
Make sure students' reports include current information on gene therapy of these diseases.
Also make sure the reports focus on a disease of the nervous system.

# Chapter Quiz

**Write your answers on a separate sheet of paper. Use complete sentences.**

1. What is the function of a neuron?
   to carry impulses to and from the brain
2. Explain the difference between axons and dendrites. Axons carry impulses away from a neuron; dendrites carry impulses to a neuron.
3. How do messages make it across synapses?
   Chemical messengers jump across the synapse.
4. What are the functions of the three main parts of the brain?

5. How do reflexes occur? An impulse travels to the spinal cord and back; it does not go to the brain.
6. What is the function of the spinal cord?
   acts as a switchboard for impulses going to and from the brain
7. Explain the differences between the central nervous system and the peripheral nervous system.

8. What are hormones? chemical messengers produced by endocrine glands that travel in the bloodstream

**CRITICAL THINKING**

9. How are nerve impulses and hormones similar? How are they different?

10. Why do you think it is important to protect your brain and spinal cord by using safety equipment such as seatbelts and helmets?

4. cerebrum – controls thinking and memory; cerebellum – controls coordination and balance; brainstem – controls involuntary body functions

7. The central nervous system is made up of the brain and spinal cord. The peripheral nervous system contains all the nerves in the rest of the body.

9. Nerves and hormones are similar because they both help regulate the body. They are different because neurons are cells that carry electrical impulses and hormones are chemical substances.

10. Injury to the brain or spinal cord can cause serious disease, paralysis, or death.

*SCiLINKS*

Go online to www.scilinks.org. Enter the code **PMB132** to research **noninfectious diseases**.

## Research Project

Over the past few years, scientists and doctors have been interested in using the study of genetics as a way of learning more about noninfectious diseases. Gene therapy may be used to treat diseases of the nervous system, such as Alzheimer's disease, Parkinson's disease, and Lou Gehrig's disease. Research one of these diseases. Find out what kind of genetic research is being done to study the disease.

See the *Classroom Resource Binder* for a scoring rubric for the Research Project.

**Figure 18-1** *Your senses take in information from your surroundings. What types of information do you think the people here are getting from their senses?*

Possible answer: Their senses could be sending them information about the noises, smells, and bright colors of the carnival.

## Learning Objectives

- Name the five senses and their organs.

- Describe the jobs of the skin, eyes, ears, tongue, and nose.

- Explain the relationship between the brain and the sense organs.

- **LAB ACTIVITY:** Investigate the action of the pupil.

- **BIOLOGY IN YOUR LIFE:** Test optical illusions.

## Words to Know

| | |
|---|---|
| **taste bud** | a group of sensory neurons on the tongue that detects taste |
| **epidermis** | the outermost layer of skin |
| **dermis** | a thick layer of skin under the epidermis |
| **cornea** | the clear, curved covering on the outer surface of the eye |
| **iris** | the colored part of the eye that controls the size of the pupil |
| **pupil** | the opening in the eye that lets light in |
| **lens** | the part of the eye that focuses light on the retina |
| **retina** | the layer of sensory neurons at the back of the eye that detects light |
| **optic nerve** | the bundle of nerves that carries information from the eye to the brain |
| **eardrum** | the tightly stretched layer of tissue inside the ear that vibrates when hit with sound waves |
| **cochlea** | (KAHK-lee-uh) snail-shaped organ in the ear that receives sound waves |
| **auditory nerve** | the bundle of nerves that carries information from the ear to the brain |

## What are the five senses?

People enjoy tasting food, listening to music, touching a puppy's fur, smelling flowers, and watching movies. These activities are possible because of five important organs: tongue, ears, skin, nose, and eyes.

However, these organs do more than help you enjoy the world. They send very important information to your brain every second. Information about your surroundings is first gathered by your senses. Then, it is sent to your brain. This information helps you to live.

**Linking Prior Knowledge**
Students should recall the term *neuron* (Chapter 17). They should also be familiar with the five sense organs: nose, tongue, skin, eyes, and ears.

The ear picks up the sound. Information is passed to the brain very quickly. This is an automatic response that occurs without the person thinking about it.

**Remember**

A cell is the basic unit of living things. There are many different types of cells. Each type of cell has a different job.

Each sense organ has special parts and cells that allow it to work. In this chapter, you will learn how each organ gathers information from the world around you.

## Gathering Information

The five senses are smell, taste, touch, sight, and hearing. Each sense begins with a special organ. The sense of smell begins with the nose. The sense of taste begins with the tongue. The sense of touch begins with the skin. The sense of sight begins with the eyes. The sense of hearing begins with the ears.

These sense organs collect information. This information is received by sensory neurons. Then, it is passed along to the brain. The brain decides what to do with the information. This figuring out what something means is called *interpreting*. The sensory neurons and brain cells work very fast. The whole process of collecting and interpreting information can happen in less than a second.

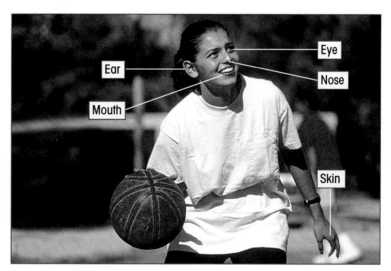

Figure 18-2 *Your sense organs help you to understand the world around you.*

## Nose and Smell

Much of your sense of taste is really smell. As you eat, scents from your food enter your nose. If you do not believe this, think about eating when you have a head cold. It is hard to taste the food. That is because your nose is clogged. You are not smelling the food.

Smells are picked up by sensory neurons in the nose. These cells have long hairlike cilia. The cilia are connected to nerves. The cilia pass the molecules that cause the smell along to cells. The nerves carry the smell message to the brain. The brain interprets the smell.

## Tongue and Taste

Your tongue is covered with sensory neurons that sense taste. These are contained in taste buds. You have four kinds of **taste buds**—some taste sour things, some taste sweet things, some taste salty things, and some taste bitter things. The "tongue map" in Figure 18-3 shows where each of these kinds of taste buds is often found.

**Remember**
Cilia are tiny hairlike objects that can help to move objects.

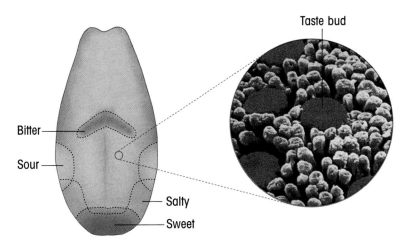

Bitter
Sour
Salty
Sweet
Taste bud

Figure 18-3 *The tongue senses four types of tastes. Each taste is sensed by a different part of the tongue. The photo on the right shows a close-up picture of taste buds on the tongue.*

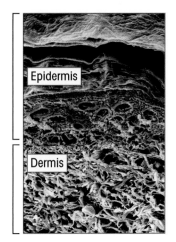

Epidermis

Dermis

Figure 18-4 *Skin is made up of many layers of cells. Old skin cells are constantly being replaced with new skin cells.*

## Skin and Touch

Skin is the sense organ of touch. Some parts of your skin can sense touch better than others. For example, your fingertips have many more sensory neurons than your arms do. The skin is the largest human organ. It covers the whole body.

The outermost layer of skin is called the **epidermis**. It is made up of living and dead cells. You shed millions of dead cells every day. The next layer of skin is called the **dermis**. This is a thick layer. It contains blood vessels, nerves, hair follicles, oil glands, and sweat glands. Hair, fingernails, and toenails are made of protein. They are made by living cells contained in the dermis. Under the dermis, there is a layer of fat tissue.

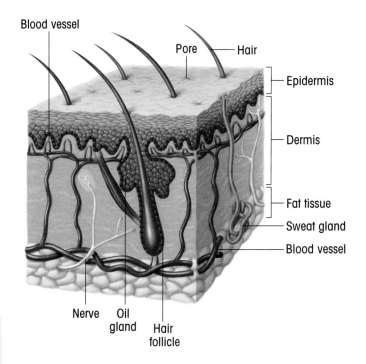

Figure 18-5 *Skin is the largest human organ. It is the sense organ of touch.*

### Safety Alert

Your skin can be damaged by too much sunlight. Wear sunscreen during all outdoor activities.

Skin has the following four important jobs:

- Skin is the sense organ of touch. Skin can feel heat, cold, pressure, and pain.

- Skin keeps harmful microscopic organisms out of your body.

- Skin keeps the water in your body from drying up. The outer layer of skin is waterproof.

- Skin helps to control the temperature of your body. Layers of fat in the skin help to keep you warm. Sweat glands in the skin let the body give off moisture through the skin. This helps you to cool off.

**Think About It**

Your skin also can feel different textures. Texture is how smooth or bumpy something feels. What are some examples of objects with different textures?

## Great Moments in Biology

### THE INVENTION OF BRAILLE

Braille is a type of writing that can be used by people who are blind. This system was named after its inventor, Louis Braille (BRAYL).

As a child, Louis was blinded in an accident. He went to a school that helped people who were blind. The pages of books were made using wires formed into the shapes of the letters. They were very hard to read because many of the letters felt similar to other letters.

Louis tried to come up with a better way to write for blind people. He developed a system made up of raised dots that represented different letters or words. These dots could be felt with the fingertips and interpreted by the reader. In 1827, the first book written completely in Braille was published. Today, Braille is used all over the world.

Figure 18-6 *People who are blind use their sense of touch to read. They read books written in Braille.*

**Critical Thinking** Braille has allowed people with vision impairments to learn and to enjoy books the same way that people with sight can. Braille also allows people to carry out daily activities, such as using an elevator.

**CRITICAL THINKING** Why do you think Braille is an important invention?

## Eyes and Sight

Everything you see is light. Suppose you are looking at a car. Light bounces off the car and enters your eye. Sensory neurons pick up the light. The "light information" is sent to your brain. There it is interpreted. Then, you know you are seeing a car.

Light enters the eye through a clear, curved covering called the **cornea**. Behind the cornea is a colored part called the **iris**. The iris surrounds an opening in the eye called the **pupil**. The pupil allows light to enter the eye. The iris has muscles. The muscles change the size of the pupil. This allows the correct amount of light to enter the pupil so that you can see better.

Once light enters your eye, it passes through the **lens**. The lens is the clear part of the eye. The lens focuses light. This means it bends the light rays so that they correctly hit the retina. The **retina** is the layer of sensory neurons at the back of your eye. The retina can sense brightness and color.

## Science Fact

Your eye works like a camera. When it is bright, your iris shrinks the pupil to let less light in. When it is dark, your iris widens the pupil to let more light in.

**Think About It**

Eyelids, eyebrows, eyelashes, and tears all protect your eyes. Tears come from a gland above each eye. What kinds of things do you think tears protect the eyes from?

Tears wash dust and other particles such as pollen from the eyes. They also contain a substance that fights harmful bacteria.

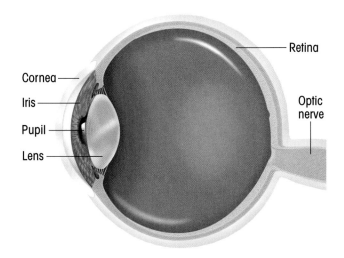

Cornea

Iris

Pupil

Lens

Retina

Optic nerve

**Figure 18-7** *The eye is the organ of sight. It captures and focuses light.*

The retina is attached to a thick bundle of nerves. This bundle of nerves is called the **optic nerve**. The optic nerve carries the information to your brain. Your brain makes sense of the information and you see a picture.

✓ **Check Your Understanding**

Write the main job or jobs of each item listed. Do your work on a separate sheet of paper.

**1.** sweat gland      **2.** epidermis

**3.** iris      **4.** sensory neuron

**5.** optic nerve      **6.** lens

**Check Your Understanding**

**1.** gives off moisture, regulates temperature

**2.** keeps bacteria out of the body, waterproofs the body

**3.** controls the size of the pupil

**4.** receives outside information

**5.** carries information from the eye to the brain

**6.** focuses light on the retina

## On the Cutting Edge

**ADVANCES IN EYE SURGERY**

Laser eye surgery is becoming a popular way of correcting vision problems. A laser is a narrow but powerful beam of light. One type of laser surgery is called Lasik surgery (LAY-sihk). During Lasik surgery, a laser is used to reshape the cornea. This helps a person to see better.

Surgery is also used to treat people with cataracts. A cataract is a disease that occurs when the lens becomes cloudy. A tiny rod is inserted into the eye. This device gives off sound waves that break up the lens. Then, a new lens can be placed in the eye.

Figure 18-8 *During laser eye surgery, a narrow but powerful beam of light is used to reshape the cornea.*

**CRITICAL THINKING** What steps might be taken to make sure new forms of surgery are safe and work correctly?

New procedures should be tested on many different volunteer patients first, before they are used on the general public. Researchers may also test procedures on laboratory animals to make sure they are effective.

## Ears and Hearing

The ears are the organs of hearing. They detect sound. The ear is divided into three sections: the outer ear, the middle ear, and the inner ear. As you read about the parts of the ear, look at Figure 18-9.

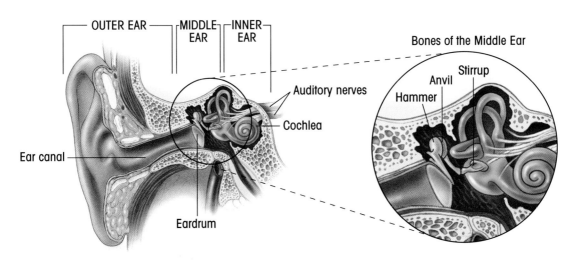

Figure 18-9 *The ear is the organ of hearing. It collects sound waves. Then, it sends messages to your brain to be interpreted.*

Sound waves enter the outer ear and travel through the ear canal. The waves pass into the middle ear through the **eardrum**. The eardrum is a thin layer of tissue. When sound hits the eardrum, it vibrates.

### Science Fact

The three ear bones are the smallest bones in the human body. All three bones could fit on a dime!

The middle ear contains three tiny bones. They are called the hammer, the anvil, and the stirrup. The first bone, the hammer, is attached to the eardrum. The hammer bone picks up the sound waves from the eardrum and passes them to the anvil bone. The anvil passes the sound waves to the stirrup bone. From there, the sound waves go to the inner ear.

In the inner ear, there is a snail-shaped structure called the **cochlea**. It receives the sound waves from the middle ear. The cochlea has tiny hairs and a liquid inside.

The liquid picks up the sound waves and passes them to the tiny hairs. These tiny hairs contain sensory neurons. They pass the sound waves to the **auditory nerves**. The auditory nerves carry the signal to the brain. The brain interprets the message as sound.

Your ears also help to maintain your balance. There are three looped tubes in your inner ear. They are filled with liquid. As the liquid moves, receptor cells send information about balance to your brain.

✓ **Check Your Understanding**

Explain how you hear a whistle blow. Describe the path of sound waves on a separate sheet of paper.

**Check Your Understanding**
First, sound enters your ear. The sound waves cause your ear to vibrate. This signal is passed along to the three ear bones and the cochlea. Tiny hairs in the cochlea pass the signal to the auditory nerves, which transfer it to the brain.

## A Closer Look

### SOUND WAVES
Sound waves occur because of vibrations. Vibrations are back and forth movements. Sound is often measured by the number of vibrations per second. This measurement is called the frequency of the sound. Frequency is measured in units called hertz.

Humans cannot hear all of the different sounds that exist in the world. We can only hear sounds that have a frequency of 20 to 20,000 hertz. Other animals can hear other frequencies of sound. For example, bats can hear sounds with a frequency of 100,000 hertz. The graph shown in Figure 18-10 shows the hearing abilities of different animals.

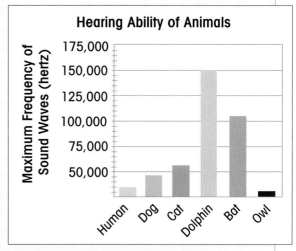

Figure 18-10 *Many animals can hear sounds that humans cannot.*

**CRITICAL THINKING** Which animal can hear better, a bat or an owl? How could this affect the way they get food?
A bat can hear much better than an owl. Bats find their prey using sound instead of sight.

# LAB ACTIVITY
## Investigating the Action of the Pupil

### BACKGROUND
Our eyes take in light through the pupil. The light is received by receptor cells. Nerve cells carry this light information to the brain. There it is interpreted as a picture.

**Figure 18-11** *Look at your partner's eye as light is shining on it.*

The sensory neurons are in a part of the eye called the retina. The retina is like the film in a camera. It needs the right amount of light to make a good picture. In bright light, the iris adjusts the size of the pupil to let in a little bit of light. In the dark, the iris opens the pupil wide to let in as much light as possible.

### PURPOSE
In this activity, you will observe the pupil changing size.

### MATERIALS
penlight, mirror, pencil, paper

### WHAT TO DO
1. Use the mirror to look very closely at one of your eyes.

2. Make a close-up drawing of your eye. Notice the dark opening in the middle of the colored iris of your eye. This is the pupil. It is the hole that light goes through to your retina.

3. Find a partner. Look at your partner's eye. Point a penlight at your partner's eye, but keep it turned off. When your partner is ready, switch on the light. Observe the pupil of your partner's eye.

**Teacher's note:** Warn students not to look directly into the light.

### DRAW CONCLUSIONS
- What did you notice in this activity? Students should notice that the pupil changes size.

- What might happen to the size of your pupil if you leave a dark movie theatre and go outside? Your pupil would get smaller when you step out into bright light.

Sometimes things are not as they appear to be. Near the horizon the Moon looks large. High in the sky it looks small. However, the size of the Moon has not changed. It is an optical illusion. An illusion is something that plays tricks on your brain. Optical illusions teach us about how the eyes and brain work together. In some illusions, your brain fills in pictures that are not there. Look at Figure 18-12. Is the white triangle really there?

In other optical illusions, our brains make mistakes about the sizes of objects. Two lines that are the same length may look different, depending on their position. Color can also change our vision. In Figure 18-13, the white goblet may be clearly visible against the black background. Look at the image a different way. Can you see two faces looking at each other?

There are many more optical illusions you can read about. They show that you cannot always believe what you see.

**Answer these questions about the two lines in Figure 18-14.**

**1.** Which of the horizontal lines appears longer? The bottom line appears longer.

**2.** Is one line really longer than the other? Both lines are actually the same length.
**3.** How could you test which line is longer? Measure both lines with a ruler or some other measure.

Figure 18-12 *Find the triangle.*

Figure 18-13 *Find the goblet and the faces.*

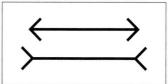

Figure 18-14 *Compare the lines.*

**Critical Thinking**

Why do you think one line appears longer than the other in Figure 18-14?

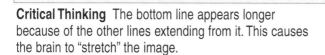

**Critical Thinking** The bottom line appears longer because of the other lines extending from it. This causes the brain to "stretch" the image.

## Summary

- The sense organs collect information about the world. The brain interprets the information.

- Smell, taste, touch, sight, and hearing are the five senses. The nose, tongue, skin, eyes, and ears are the five sense organs.

- The nose is the organ of smell. Sensory neurons inside the nose are connected to nerves. The nerves carry the smell messages to the brain.

- The tongue is the organ of taste. Taste buds on the tongue can pick up four different flavors: bitter, sweet, salty, and sour.

- Skin has four jobs. It keeps harmful bacteria out. It keeps the body from drying out. It regulates body temperature. It is the sense organ of touch.

- The eyes take in light through the pupil. Sensory neurons on the retina receive the light. Nerves carry this information to the brain. There, it is interpreted as a picture.

- The ear is the organ of hearing. It picks up sound waves and passes them along to special nerve cells. The nerves carry the message to the brain. The brain interprets the message as sound.

eardrum

epidermis

lens

optic nerve

taste bud

## Vocabulary Review

**Match each term to its definition. Write your answers on a separate sheet of paper.**

**1.** the tightly stretched layer of tissue inside the ear that vibrates when hit with sound waves   eardrum

**2.** the outermost layer of skin   epidermis

**3.** the part of the eye that focuses light on the retina   lens

**4.** the bundle of nerves that carries information from the eye to the brain   optic nerve

**5.** a group of sensory neurons on the tongue that detects taste   taste bud

# Chapter Quiz

Write your answers on a separate sheet of paper. Use complete sentences.

1. What are the five senses and their organs?
   The senses and their organs are touch/skin, sight/eye, hearing/ear, smell/nose, and taste/tongue.

2. Which human organ interprets most of the information taken in by the senses?
   The brain interprets most of the information.

3. What are the three bones of the ear called?
   the hammer, anvil, and stirrup

4. What is the outer layer of skin called?
   The outer layer of skin is called the epidermis.

5. Where does light enter the eye? How does enough light enter so that people can see at night?
   Light enters through the pupil. The pupil opens wider at night to allow more light in.

6. What protects someone's eyes?
   Eyelids, eyelashes, eyebrows, and tears protect eyes.

7. What does the eardrum do?
   It vibrates when hit by sound waves.

8. What four things can the tongue taste?
   The tongue can taste sweet, sour, bitter, and salty things.

9. What are the sensory neurons in the nose like?
   The sensory neurons in the nose are attached to long hairlike cilia.

10. Which of the sense organs is the largest?
    The skin is the largest sense organ.

**CRITICAL THINKING**

11. How do the senses of taste and smell work together?

12. What are the functions of the skin?

**Critical Thinking**

11. Sensory neurons in the nose help you to taste flavors other than salty, sweet, sour, and bitter.

12. It is the organ of touch; keeps harmful microscopic organisms out of the body; waterproofs the body, and controls the body's temperature.

SC*L*INKS
Go online to www.scilinks.org. Enter the code **PMB134** to research **the senses**.

## Research Project

The ears are a very complex sense organ. Find out more about how the ears function. Also find out about hearing problems. Write a report that explains some health problems that can damage someone's sense of hearing. Include the possible causes of the problem. Also, describe how the problem can be treated.

See the *Classroom Resource Binder* for a scoring rubric for the Research Project.

**ESL Note** Ask students to list five healthy foods from their native country. Have them write these foods in English and in their native language. Then, encourage them to place the foods in the correct category of the Food Guide Pyramid.

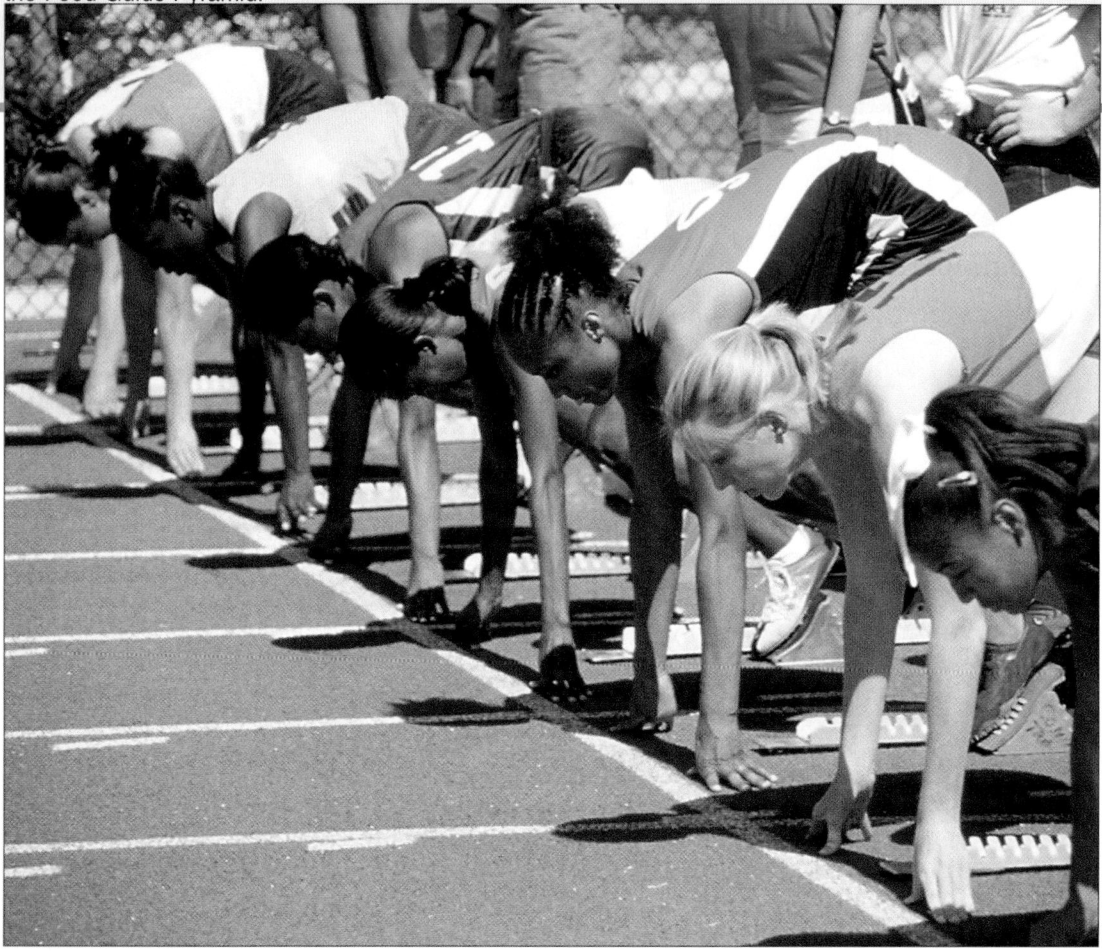

**Figure 19-1** ***Running takes a great deal of energy. These athletes get the energy they need by eating healthy foods. What other things do you think these athletes do to stay healthy?***

Possible answer: exercise regularly, drink enough water, stay away from drugs and alcohol

## Learning Objectives

- Explain how the body prevents disease.
- Describe the role of the immune system.
- Compare contagious and noncontagious diseases.
- Describe the six nutrients the body needs.
- Explain how to use the Food Guide Pyramid.
- Analyze the effects of drugs and alcohol on the body.
- LAB ACTIVITY: Test foods for vitamin C.
- ON-THE-JOB BIOLOGY: Describe the job of a dietitian.

## Words to Know

| | |
|---|---|
| **pathogen** | any substance that causes disease |
| **white blood cell** | a type of blood cell that helps to protect the body against disease |
| **antibody** | a molecule that attaches to a specific pathogen |
| **immunity** | resistance to a specific disease |
| **vaccine** | a substance made from dead or weakened pathogens, that stimulates immunity against a disease |
| **contagious** | can be spread from one person to another |
| **vitamins** | organic substances in foods that the body needs to function properly and to remain healthy |
| **minerals** | inorganic substances in foods that the body needs to function properly and to remain healthy |
| **Food Guide Pyramid** | a diagram that shows the kinds of foods and the numbers of servings of each that you should eat each day |

## Staying Healthy

Do you remember the last time you had a cold or other illness? While you were sick, your body was hard at work fighting the disease. Different parts of your body work together to help keep you healthy. In this chapter, you will learn about how the body fights disease. You will also learn how you can keep your body healthy.

You may not realize it, but your body is always working to prevent disease. Things that cause disease are called **pathogens**. Bacteria and viruses can both be pathogens. Other things such as fungi, protists, and worms can also be pathogens.

**Linking Prior Knowledge**

Students should recall the characteristics of viruses and bacteria (Chapter 6). They should also recall that the mouth and stomach contain enzymes and other digestive fluids (Chapter 16).

## The First Lines of Defense

Your body has many different ways to fight disease. The first lines of defense keep the pathogens from getting too far into your body. They act like a shield around your body. Different body parts have features that block or kill pathogens. Look at Figure 19-3 to learn more about these parts and how they work.

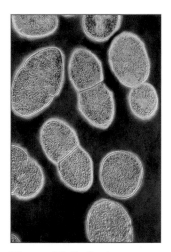

Figure 19-2 *Pathogens, such as these streptococci bacteria, are things that cause disease. What disease do you think these bacteria cause?* strep throat

| The Body's Defenses | |
|---|---|
| **Body Part** | **How It Fights Disease** |
| Skin | • Layers of dead cells at the surface of skin prevent pathogens from entering.<br>• Oils in skin help kill pathogens. |
| Mouth and stomach | • Enzymes in your mouth kill pathogens.<br>• Acids kill pathogens that reach the stomach. |
| Nose | • Mucus in the nose takes pathogens to the stomach, where they are killed. |

Figure 19-3 *Different parts of your body work together to prevent pathogens from entering.*

## The Immune System

The first lines of defense are not completely perfect. Sometimes a pathogen can get past these defenses. When this happens, the *immune system* starts working. This system defends against pathogens spreading inside the body. The immune system contains different kinds of cells and molecules that each fight against pathogens in a different way.

**Think About It**

Sometimes pathogens can get past the barrier of your skin. How do you think this happens?

Possible answer: Bacteria can get into the body through scrapes, burns, or punctures.

### White Blood Cells

One type of cell is the **white blood cell**. White blood cells travel in your bloodstream. However, they are not the same as red blood cells. White blood cells are much larger and can change shape. When a pathogen enters the body, the white blood cells find it. They enter the tissue that is infected and attack the pathogen.

## T Cells and B Cells

T cells and B cells are special kinds of white blood cells called *lymphocytes*. They are part of the lymphatic system. They are also part of the immune system because they help the body to fight disease. *T cells* identify pathogens. Then, they may attack the pathogens. T cells also attack the damaged body cells that the pathogens have already harmed. Other times, the T cells do not attack the pathogens directly. Instead, they call the B cells into action. The *B cells* produce substances called antibodies.

**Antibodies** find certain pathogens and attach to them like a flag. This attachment makes it easier for other cells of the immune system, such as white blood cells, to identify the pathogens. Then, the other cells find the pathogens and attack them.

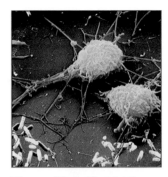

**Figure 19-4** *The white blood cells, shown in blue, are attacking E. coli bacteria, shown in yellow.*

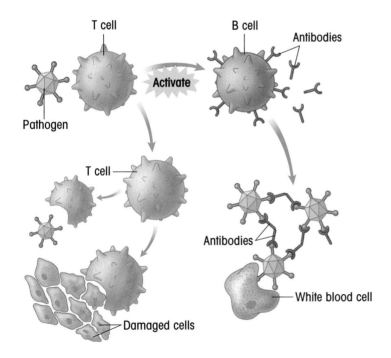

**Figure 19-5** *T cells and B cells are part of your immune system.*

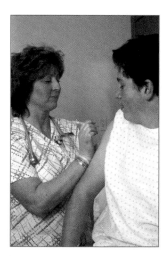

**Figure 19-6** *This person is getting a vaccine.*

The ability to resist a certain disease is called **immunity**. People are often born with certain immunities. These are called *natural immunities*. The other form of immunity is called *acquired immunity*. Acquired immunity is one that people develop or acquire over time.

One type of acquired immunity is called *passive acquired immunity*. Passive acquired immunity can occur when a mother passes antibodies for certain diseases to her developing baby. This method is called passive because your body did not actually produce the antibodies.

Another form of acquired immunity is called *active acquired immunity*. Sometimes if you are infected with a disease once, you will not get the disease again. This is a form of active acquired immunity. It is called active because your body produced antibodies to the disease. Getting a vaccine is another form of active acquired immunity. A **vaccine** is made from dead or weakened forms of pathogens. After receiving a vaccine, the body develops antibodies for the disease. This vaccination makes the body immune to that disease.

## Types of Disease

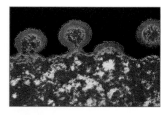

**Figure 19-7** *The virus HIV causes the disease AIDS.*

When pathogens enter and infect the body, a disease can occur. There are many different kinds of diseases. Some may only last a short time. Others can be life threatening. Diseases that can be passed from one person to another are **contagious**. Measles, colds, and chickenpox are contagious. Some contagious diseases can be life threatening. AIDS is a contagious disease.

Some diseases are not caused by a pathogen. These diseases are not contagious. Cancer and Alzheimer's disease are examples of diseases that cannot be spread from one person to another.

Write your answers in complete sentences.

**1.** What structures prevent pathogens from entering the body? the skin, mouth, and nose

**2.** How do white blood cells help to fight disease?
They kill pathogens by attacking them.

**3.** What are three examples of a contagious disease?
measles, mumps, and the cold

## Basic Nutrition

What do you think about when you bite into a hamburger? You probably do not think about the nutrients contained in the hamburger. Nutrients are the food molecules that each of your body cells needs. They are used for growth, repair, and energy. To be strong and healthy, you must get six kinds of nutrients: water, vitamins, minerals, carbohydrates, proteins, and fats.

Every cell in your body needs water to do its work. Water aids your body's chemical reactions. Water keeps your body at the right temperature. Water also helps your body get rid of wastes. You should drink from six to eight 8-ounce glasses of water a day.

### Safety Alert

Dieting can be dangerous to your health, especially fad diets. The smart way to stay in shape is to get plenty of exercise. At the same time, eat controlled amounts of healthy foods, but do not make yourself go hungry!

**Total Body Weight**

35% Other

65% Water

**Figure 19-8** *Water is necessary to stay alive and healthy. More than half of your body is water.*

Everyday Science

Most of the foods in your home should have a nutrition label on them. Read the labels on three different foods. Compare the numbers of Calories and nutrients in each.

### Vitamins and Minerals

**Vitamins** are a type of nutrient that your body must get from the foods you eat. Vitamins A, C, and D are vitamins your body needs. Vitamins aid growth. They also help control the chemical processes that go on in the body. Vitamins are called organic because they come from living things.

**Minerals** are inorganic substances. This means they are molecules that do not contain carbon. Zinc, iron, calcium, and phosphorus are all minerals needed by your body. These minerals help body tissues to function. For example, iron helps red blood cells carry oxygen. You get iron from foods such as red meats and spinach.

### Carbohydrates, Proteins, and Fats

Carbohydrates, proteins, and fats are the main types of nutrients found in foods. These nutrients are used by your cells to carry out life processes, such as cellular respiration. They are also needed to build and repair cells. Look at Figure 19-10 to learn about these nutrients.

## Nutrition Facts

Serving Size: 1 cup
Servings Per Container: About 9

| | |
|---|---|
| **Calories** 100 | |
| Calories from fat 0 | |

| | % Daily Value |
|---|---|
| **Total fat** 0g | 0% |
| Saturated Fat 0g | 0% |
| **Cholesterol** 0mg | 0% |
| **Sodium** 10mg | 0% |
| **Total Carbohydrate** 25g | 10% |
| Dietary Fiber 0g | |
| Sugars 25g | |

Figure 19-9 *Nutrition labels tell you the amount of Calories and certain nutrients in food.*

| Nutrients Found in Foods | | |
|---|---|---|
| **Nutrient** | **Role in the Body** | **Sources** |
| Carbohydrates | Used by the body for energy | Potatoes, pasta, grains, cereals, beans |
| Proteins | Used by the body for growth and healing; involved in most of the chemical reactions in the body | Beef, pork, chicken, fish, eggs, dairy products, beans |
| Fats | Used for energy and to help build cell parts | Butter, margarine, cooking oil, sausage, bacon, fried foods, potato chips |

Figure 19-10 *Carbohydrates, proteins, and fats are three basic nutrients found in foods.*

## Calories

The energy in carbohydrates, proteins, and fats is measured in *Calories*. One Calorie of food will give you a certain amount of energy. Most packaged foods have nutrition labels on them. The labels list the amount of Calories in the food. They also tell you the serving size and the amounts of different nutrients.

## The Food Guide Pyramid

How can you be sure you get all six nutrients in your diet every day? The chart in Figure 19-11 is called the **Food Guide Pyramid**. It has six groups of food. The chart shows you how many servings from each food group you should eat each day. If you follow the Food Guide Pyramid, you will get enough of the six nutrients.

### Science Fact

A calorie is a measure of heat energy. One calorie is the amount of energy needed to raise the temperature of 1 milliliter of water 1°C. The energy in food is actually measured in kilocalories. A kilocalorie is equal to 1,000 heat calories. Usually, kilocalories are written as Calories with a capital C. How many heat calories are in 100 Calories of food? 100,000

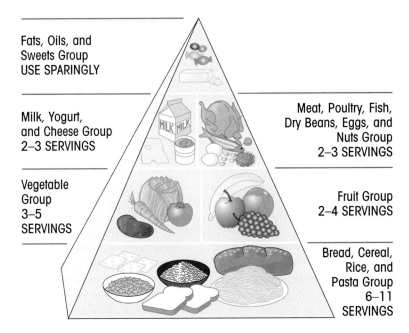

Fats, Oils, and Sweets Group
USE SPARINGLY

Milk, Yogurt, and Cheese Group
2–3 SERVINGS

Vegetable Group
3–5 SERVINGS

Meat, Poultry, Fish, Dry Beans, Eggs, and Nuts Group
2–3 SERVINGS

Fruit Group
2–4 SERVINGS

Bread, Cereal, Rice, and Pasta Group
6–11 SERVINGS

Figure 19-11 *The Food Guide Pyramid can be used daily to plan a healthy diet.*

Answers will vary. Remind students not to take any medication without speaking to their parent or guardian first.

### Everyday Science

Read the label of a non-prescription drug container. Write down the proper dosage amount. How many milligrams of the drug can an adult safely take per day? How does this amount compare if an adult were to take the extra-strength version of the same medicine?

## The Effects of Drugs and Alcohol

A *drug* is a chemical that causes a change in the body. Doctors prescribe drugs to treat illnesses. These drugs help a patient to get well. Antibiotics and cough medicine are both drugs.

Other drugs can be very harmful to the body. Cocaine, marijuana, and LSD are all harmful drugs. Caffeine, tobacco, and alcohol are also drugs. The improper use of a drug is called *drug abuse*. Using too much of a legal drug and using illegal drugs are both forms of drug abuse. Drug abuse can lead to serious illness or even death. Drugs are classified based on how they affect the body. Many different body functions are affected by drugs. Look at Figure 19-12 to learn more about drugs and their effects on the body.

| Effects of Drugs on the Body | | | |
|---|---|---|---|
| **Type of Drug** | **Examples** | **Short-Term Effects** | **Long-Term Effects** |
| Depressants | Alcohol, sleeping pills, tranquilizers, ketamine | Impaired judgement, impaired vision, confusion, loss of body control, loss of reflexes | Depression, liver damage, nerve and brain damage, respiratory failure, death |
| Narcotics | Heroin, morphine | Temporary happy feeling, impaired reflexes, impaired perception, sedation | Coma, respiratory failure, death |
| Stimulants | Amphetamines, caffeine, cocaine, crack | Temporary boost in energy, distorted thoughts, anxiety, higher blood pressure, increased heart rate | Irregular heartbeat, delusions, loss of coordination, permanent brain damage, respiratory failure, heart attack, death |
| Hallucinogens | Marijuana, ecstasy, PCP, LSD | Temporary dreamlike state, distorted thoughts, hallucinations, anxiety, slurred speech, violent behavior | Genetic damage, depression, loss of motivation, paranoia, brain damage, death |
| Inhalants | Fumes from glue, paint, and markers; aerosol cans | Disorientation, confusion, memory loss | Brain damage, hearing loss, spasms, liver and kidney damage, heart attack, respiratory failure, death |

**Figure 19-12** *Drugs have many harmful effects on the body.*

## ✓ Check Your Understanding

Write your answers on a separate sheet of paper.
Use complete sentences.

1. Your body is about two-thirds _____. water

2. _____ are the food molecules that each of your body cells need. nutrients

3. The energy contained in foods is often measured in _____. Calories

4. Caffeine, tobacco, and alcohol are all _____. drugs

5. CRITICAL THINKING Why do you need to drink enough water every day? Your body cells need water to carry out important functions.

## A Closer Look

### THE DANGERS OF DRIVING UNDER THE INFLUENCE

When alcohol enters the body, it is quickly absorbed into the bloodstream. Then, it travels to the brain. Alcohol has serious effects on how the brain functions. Alcohol greatly impairs a person's judgement. It also affects vision and other senses. Alcohol reduces the speed at which a person can react to things. All of these factors decrease a person's ability to drive safely.

Figure 19-13 *Accidents involving drunk driving are often fatal.*

Driving while under the influence of alcohol is dangerous to the driver and to others. It is also illegal. Every year, thousands of people are injured or killed in accidents related to drunk driving. Many people make the mistake of thinking they are able to drive, even after they have had a few drinks. This can lead to tragedy. In fact, alcohol is related to nearly 50 percent of all car accidents involving teenagers.

CRITICAL THINKING Suppose someone wants to drive after drinking alcohol. What should you do?
Do not get in the car and let them drive. Insist the person not drive.

# LAB ACTIVITY
## Testing Foods for Vitamin C

### BACKGROUND
Vitamin C helps promote healthy teeth, gums, and blood vessels. It is found in citrus fruits, tomatoes, and leafy vegetables.

### PURPOSE
In this activity, you will test foods to see if they contain vitamin C.

### MATERIALS
safety goggles, apron, protective gloves, milk, apple juice, orange juice, lemon juice, water, graduated cylinder, 6 test tubes, indophenol, test-tube rack, dropper

| Testing Foods for Vitamin C | | |
|---|---|---|
| Food | Number of Drops | Color Change |
| Milk | | |
| Apple juice | | |
| Orange juice | | |
| Lemon juice | | |
| Water | | |

Figure 19-14 *Copy the chart onto a sheet of paper.*

### WHAT TO DO
1. Make a hypothesis to predict which of the foods in the materials list will contain vitamin C. Write your hypothesis on a separate sheet of paper. Then, copy the chart in Figure 19-14 onto a sheet of paper. Put on your safety goggles, apron, and protective gloves.

2. Use a graduated cylinder to measure out 10 milliliters of indophenol. Then, pour the indophenol into a clean test tube.

3. Add milk drop by drop into the test tube of indophenol. Note how many drops were needed to cause a color change. Record your observation in your chart.

4. Repeat Steps 2 and 3 for apple juice, orange juice, lemon juice, and water. Make sure you rinse out the dropper after each food test.

### DRAW CONCLUSIONS
- Indophenol turns colorless when vitamin C is present. Which of the foods tested contain vitamin C? orange juice, lemon juice, apple juice

- The more vitamin C present, the less liquid it takes to turn indophenol colorless. Which of the liquids tested has the most vitamin C? orange juice and lemon juice

- Was your hypothesis correct? Explain why or why not. Answers will vary. Check to see that students gave reasons for their answer.

# ON-THE-JOB BIOLOGY
## Dietitian

Mark Williams is a dietitian (dy-uh-TIHSH-uhn). He plans meals for the patients in a hospital. Mark creates meals based on the special needs of the patients. People with high blood pressure, for example, need foods with less salt. People who are overweight need meals low in calories.

**Figure 19-15** *Mark Williams helps patients plan healthy meals.*

Mark makes sure the patients have a balanced diet. He uses the Food Guide Pyramid to plan meals for the day. Each day, patients are served plenty of grains, fruits, and vegetables. They also have foods made from milk and foods high in protein, such as meat, fish, poultry, and beans. Fats and sugars are used in small amounts. Mark knows the patients are getting the right amount of proteins, carbohydrates, and fats.

To prepare for his career, Mark took many classes in science, math, and health. He learned about nutrients and how the body uses energy. He also learned how to communicate well. Mark teaches patients how to prepare and eat a balanced diet when they return home.

**On a separate sheet of paper, list all of the foods you ate for the last two days. Then, answer the questions.**

1. Which of the foods that you listed above belong in the bread and cereal group?
   Possible answers: bagel, bran flakes, baked potato, pasta
2. Which of the foods belong in the fruit and vegetable group? Possible answers: string beans, carrots, orange, banana
3. Which of the foods might be high in fat?
   Possible answers: cheese, ice cream, French fries, candy

**Critical Thinking**
Possible answer: grilled fish, 1 cup broccoli, 1/2 cup rice, glass of milk

**Critical Thinking**
Plan a dinner that contains foods from at least four food groups.

## Summary

- A pathogen is a substance that causes disease. Viruses, bacteria, fungi, and protists are all types of pathogens.

- The skin, mouth, nose, and airways help prevent pathogens from entering the body.

- The immune system includes cells and special molecules that help to fight disease. White blood cells travel through the bloodstream and attack pathogens.

- Immunity is the resistance to a disease. Humans are born with or develop some immunities. Other immunities can come from a vaccine.

- Contagious diseases can be passed from one person to another. AIDS, measles, chickenpox, and the cold are contagious. Other diseases such as cancer and Alzheimer's disease are not contagious.

- There are six main nutrients: water, minerals, vitamins, carbohydrates, proteins, and fats. The best way to get all six nutrients daily is to eat from all six food groups in the Food Guide Pyramid.

- Drugs and alcohol can have serious effects on your health. They can lead to permanent damage to your body systems or even death.

| |
|---|
| antibody |
| contagious |
| immunity |
| pathogen |
| white blood cell |

## Vocabulary Review

**Complete each sentence with a term from the list.**

1. A molecule that attaches to a specific pathogen is called an _____.  antibody

2. Resistance to a specific disease is called _____.  immunity

3. A _____ is a type of blood cell that helps to protect the body against disease.  white blood cell

4. A _____ disease is one that can be spread from one person to another.  contagious

5. A _____ is any agent that causes disease.  pathogen

Research Project
Make sure students' plans include some form of exercise on at least three to four days of the week.
Encourage students to include forms of exercise that they enjoy, so they are more likely to follow the plan.

# Chapter Quiz

**Write your answers in complete sentences.**

1. What is a pathogen?  a substance that causes a disease

2. Which parts of your body help prevent pathogens from entering the body?  the skin, nose, and mouth

3. How does the immune system fight disease?
   by making cells and special molecules that attack pathogens

4. What is immunity?
   the resistance to a certain disease

5. What are three types of cells that help fight disease?  white blood cells, T cells, B cells

6. Name three contagious diseases. How are they different from noncontagious diseases?

7. What are the six nutrients needed by your body?
   water, vitamins, minerals, proteins, carbohydrates, and fats

8. How much water should you drink each day?
   six to eight 8-ounce glasses

**CRITICAL THINKING**

9. How do drugs and alcohol damage the body?

10. Suppose you have a friend who wants to lose weight. He decides to stop eating any foods that are rich in carbohydrates. Is this a good idea? Why or why not?

**Test Tip**
If you cannot think of an answer, try to remember a picture or chart related to that question. It may help you think of an answer.

6. measles, AIDS, chickenpox ; they can be spread from one person to another

**Critical Thinking**

9. They slow reflexes and coordination, damage important organs, and can lead to death.

10. Possible answer: It is not a good idea because carbohydrates are an important food group. They supply energy to the body. Also, he may eat too much fat, which can cause weight gain and health problems.

## SC*L*INKS

Go online to www.scilinks.org. Enter the code PMB136 to research **health benefits of sports.**

### Research Project

Exercising is an important factor when it comes to staying healthy. Playing sports is a good way to get exercise. Research how sports and other forms of exercise affect the body. Write a report about the benefits of exercise. Then, think about how much you exercise. Make a personal exercise schedule for one week. For each day, include the types of activity you might do.

See the *Classroom Resource Binder* for a scoring rubric for the Research Project.

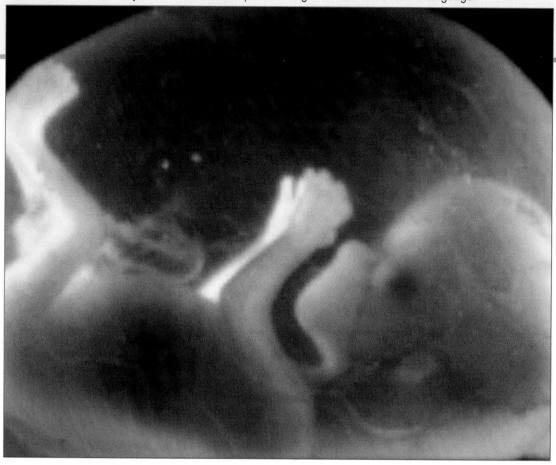

Figure 20-1 *At 14 weeks, the fetus shown above has developed all of the organs it will need to survive. However, it will continue to depend on its mother until it is ready to be born. What things does an unborn child get from its mother?*

Possible answer: The baby gets food and oxygen from its mother.

## Learning Objectives

- Identify the main organs of the male and female reproductive systems.

- Explain the menstrual cycle and its role in reproduction.

- Describe how human babies develop inside the mother.

- List the different stages of human development.

- **LAB ACTIVITY:** Graph and analyze fetal development.

- **BIOLOGY IN YOUR LIFE:** List some ways adolescents can reduce their risk of injury.

# Chapter 20 ▶ Reproduction and Development

## Words to Know

| | |
|---|---|
| **testes** | the male organs that make sperm cells and hormones |
| **penis** | the male organ that delivers sperm to the female reproductive system |
| **semen** | the mixture of fluids in which sperm leaves the body |
| **puberty** | time at which a person becomes sexually mature |
| **ovaries** | the female organs that make egg cells and hormones |
| **ovulation** | the monthly release of an egg cell from an ovary |
| **uterus** | the female organ in which a fertilized egg develops into a baby |
| **menstruation** | the monthly shedding of the lining of a woman's uterus |
| **embryo** | an organism that is developing from a fertilized egg |
| **placenta** | the organ through which nutrients, oxygen, and wastes pass between the mother and the embryo or fetus |
| **fetus** | term used to describe an embryo after 8 weeks of development in the uterus |

## Human Reproduction

You have already read about the different organs and systems that make up the human body. How does a new human being come about? In this chapter, you will learn how humans reproduce. You will also learn about how humans develop from one cell into a complete organism.

**Linking Prior Knowledge**
Students should recall that all living things reproduce and that sexual reproduction requires two parents (Chapter 2).

# The Male Reproductive System

**Remember**
Sperm cells are reproductive cells, or gametes. Like egg cells, they are produced during meiosis.

The male reproductive system has two functions. It must first produce sperm cells. It also delivers sperm to the female reproductive system.

In males, sperm cells are made in organs called **testes.** The testes are in a sac of skin. To stay alive, sperm must be kept cooler than other parts of the body. The sac of skin keeps the sperm cool.

Sperm travel from the testes through thin tubes called *sperm ducts.* These ducts carry the sperm to the urethra. In males, the *urethra* is a tube in the penis. The **penis** delivers the sperm to the female reproductive system. As the sperm move down the sperm ducts and urethra, three glands add fluids to it. The mixture of sperm and fluids is called **semen.**

The reproductive organs in a boy's body begin maturing between 13 and 16 years of age. This period of time is called **puberty.** Once puberty begins, the testes make sex hormones. The male sex hormone is called *testosterone.* Testosterone in boys and men makes thick facial and body hair, a deep voice, and a muscular body.

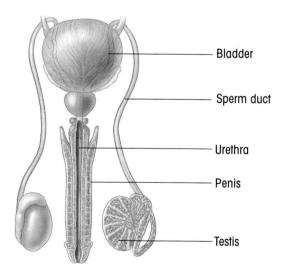

Figure 20-2 *The male reproductive system*

# The Female Reproductive System

Egg cells are made in the **ovaries** of females. The release of an egg is called **ovulation**. First, the egg moves into a tube called an *oviduct*. Tiny hairlike cilia push the egg along its path. Then, the egg eventually makes its way to the uterus.

The **uterus** is a muscular, pear-shaped organ. It is hollow and has thick walls. The uterus is where a baby develops if an egg cell is fertilized by a sperm cell. The opening to the uterus is called the *cervix*. The *vagina* is the canal that leads from a woman's uterus to the outside of her body.

Like the male testes, ovaries also make hormones. This begins during puberty. For most girls, puberty begins between 10 and 14 years of age. The female sex hormones produced in the ovaries are *estrogen* and *progesterone*. Estrogen gives women body hair and broad hips and causes their breasts to develop. Progesterone helps the female body prepare the uterus for a baby, if she becomes pregnant.

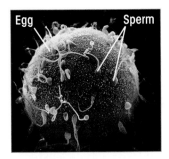

Figure 20-3 *The egg and sperm cells shown here have been magnified 345 times. Notice the difference in size between the egg cell and the sperm cells.*

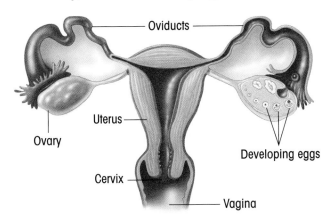

Figure 20-4 *The female reproductive system*

✓ **Check Your Understanding**

List and describe the parts of the male and female reproductive systems on a separate sheet of paper.

**Check Your Understanding**

Male reproductive system: testes produce sperm, sperm ducts carry sperm to the urethra, penis delivers sperm to the female; Female reproductive system: ovaries produce eggs, the oviduct carries the egg to the uterus, the uterus is where a baby develops, the vagina is the canal that leads from the uterus to the outside of the body.

# The Menstrual Cycle

In sexually mature women, an egg cell is released each month. The uterus also changes each month. The walls of the uterus thicken and become swollen with blood. In this state, the uterus is prepared to nourish a developing baby if the woman becomes pregnant.

Most of the time, however, the woman does not become pregnant. So, the extra lining of the uterus breaks down. The lining is made of mucus, blood, and dead cells. This material leaves a woman's body through the vagina in a process called **menstruation.**

Menstruation usually lasts from 3 to 6 days. Shortly after menstruation, another egg cell matures in the ovary. Ovulation usually occurs about 14 days after menstruation. Then, in another 14 days, menstruation occurs again. The menstrual cycle repeats itself about once a month. Look at Figure 20-5 to learn more about the menstrual cycle.

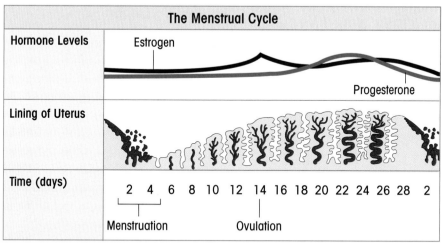

Figure 20-5 *One menstrual cycle usually takes about a month to complete.*

Most girls begin menstruating some time between 10 and 14 years of age. Then, some time between the ages of 45 and 55, a woman's menstrual cycle stops occurring each month. The permanent end of menstruation is called *menopause.*

# Fertilization and Development

Males release millions of sperm at a time. However, only one of these sperm can fertilize an egg. Once released into a female's vagina, sperm cells swim through the cervix and into the uterus. Then, they travel into the oviducts. Fertilization usually happens in the oviduct. Once the sperm and egg cells have united, the cell is called a zygote.

The zygote takes 4 to 5 days to travel to the uterus. As it travels, the zygote increases in the number of cells. By the time it reaches the uterus, it is made up of about 100 cells. This group of cells attaches itself to the wall of the uterus. It is now called an **embryo.**

**Think About It**

Once in a while, a woman releases two eggs at once. What do you think would be the result if both of these eggs were fertilized?

**Think About It**
Fraternal twins would develop.

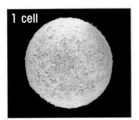
1 cell

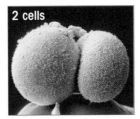
2 cells

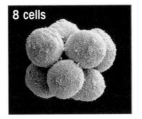
8 cells

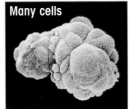
Many cells

Figure 20-6 *This zygote is increasing the number of cells it contains by going through cell division.*

An organ called the **placenta** forms in the wall of the uterus. Food, oxygen, and wastes pass between the embryo and the mother through this organ. However, the blood of the mother and of the embryo never mix. The capillaries from both mother and embryo lie very close together in the placenta. A ropelike structure called the *umbilical cord* connects the baby to the placenta.

**Remember**
Capillaries are tiny blood vessels.

Pregnancy usually lasts about 9 months. Pregnancy is often divided into three periods called *trimesters.* During each trimester, the embryo will grow in length and in weight. It will also develop organs that it will need to survive outside its mother's uterus. After about 8 weeks, the embryo is called a **fetus.**

The fetus grows in a pouchlike structure called the *amniotic sac*. Shortly before birth, this structure breaks. Strong muscle movements in the mother's uterus, called contractions, force the baby out of her body. The baby leaves the body through the vagina. After the baby is born, the placenta is also pushed out of the mother's body. Shortly after birth the mother's breasts begin to secrete milk. This is a result of the effects of hormones from the pregnancy.

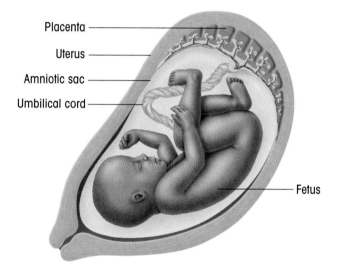

Placenta

Uterus

Amniotic sac

Umbilical cord

Fetus

Figure 20-7 *The fetus grows and develops inside the uterus for about 9 months.*

## ✓ Check Your Understanding

Write your answers on a separate sheet of paper. Use complete sentences.

**1.** In what structure does a fetus grow and develop?

**2.** About how long does a fetus develop inside its mother?

**3.** CRITICAL THINKING  People refer to the date a baby is expected to be born as the "due date." Why do you think due dates can only be thought of as estimates?

**Check Your Understanding**

1. the uterus

2. 9 months

3. A due date cannot be exact because each pregnancy is different. The date the baby will be born cannot be predicted.

# The Stages of Human Development

There are several stages of human development. The period before birth is called the prenatal stage. After birth, humans continue to grow and develop. There are five main stages of human development after birth. They are infancy, childhood, adolescence, adulthood, and the later years. The body changes in many ways during each of these stages.

## On the Cutting Edge

### PRENATAL CARE

Over the years, there have been many great advances in prenatal care. One of these is a procedure called ultrasound. An ultrasound machine uses sound waves to see inside the mother's uterus. An image can be seen on a monitor and can be printed out. Doctors use this to see how the baby is doing.

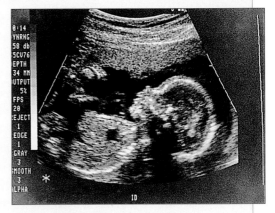

Figure 20-8 *The picture of this 20-week-old fetus was created using ultrasound.*

Two other prenatal tests are amniocentesis and chorionic villi sampling. In amniocentesis, the doctor inserts a long needle into the woman's abdomen. The doctor collects fluid that contains cells from the fetus. These cells can be tested for some diseases. Chorionic villi sampling, or CVS, is a technique that examines tissue from the placenta. This tissue contains cells from the fetus and can be tested for genetic diseases.

CRITICAL THINKING  Why do you think it is important that pregnant women receive prenatal care? Possible answer: It can help keep both the mother and the baby in good health during the pregnancy. It may also help doctors to detect diseases before the baby is born.

# LAB ACTIVITY
## Graphing Human Development

**BACKGROUND**

Many physical changes take place as an embryo grows and develops into a fetus. While inside the uterus, a fetus grows dramatically in size.

**PURPOSE**

In this activity, you will graph the changes that take place in a fetus during development.

**MATERIALS**

two sheets of graph paper, pencil

| Average Length and Mass During Fetal Development | | |
|---|---|---|
| Time (weeks) | Length (millimeters) | Mass (grams) |
| 4 | 2.5 | 1 |
| 8 | 30 | 5 |
| 12 | 87 | 45 |
| 16 | 140 | 200 |
| 20 | 190 | 460 |
| 24 | 230 | 820 |
| 28 | 270 | 1,300 |
| 32 | 300 | 2,100 |
| 36 | 340 | 2,900 |

Figure 20-9  *A fetus increases in both length and mass during pregnancy.*

**WHAT TO DO**

1. Look at the information in Figure 20-9.

2. Use the information to make a line graph showing the changes in the length of a fetus. Make your graph on a separate sheet of paper. Be sure to label the *x*- and *y*-axis and give your graph a title.

3. Now, make a separate line graph showing the changes in the mass of a fetus. Make your graph on another sheet of graph paper.

**DRAW CONCLUSIONS**

• What is the average length and mass of a 12-week-old fetus? 87 millimeters; 45 grams

• Between which weeks does the fetus show the greatest increase in length? During which trimester does the fetus show the fastest growth? between 8 and 12 weeks; first trimester

# BIOLOGY IN YOUR LIFE
## Keeping Healthy and Safe

The numbers are scary. About 15,000 adolescents die of injuries each year. That is one death each hour of each day. In fact, injuries kill more teens and preteens than all diseases combined. Car crashes cause the greatest number of teen deaths.

Why is this stage of human life so dangerous? It is hard to say. However, teens often avoid steps that can reduce the risk. They are far less likely to wear seat belts than any other age group. Even fewer wear bike helmets. Drinking alcohol was involved in 35% of teen driver deaths and about 40% of teen drownings. Boys seem to be at greater risk than girls are.

**Look at the figures below. Then, answer the questions that follow.**

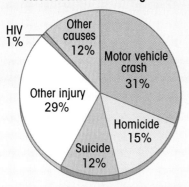

**Causes of Deaths in Adolescents and Young Adults**

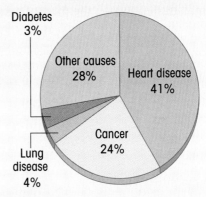

**Causes of Deaths in Adults**

Figure 20-10 *Injuries are common causes of death in adolescents.*

Figure 20-11 *Diseases are common causes of death in adults.*

1. What is the leading cause of death for adolescents and young adults?
   motor vehicle crashes and other injuries
2. What percent of deaths in the younger group is caused by motor vehicle crashes?
   31%
3. What is the leading cause of death in adults? heart disease

### Critical Thinking

List two actions that would decrease the risk of teen deaths.

Possible answers: wear seat belts, do not drink and drive

# Chapter

## 20 ▷ Review

## Summary

- The testes are male reproductive organs that make sperm cells and hormones. Sperm cells travel through sperm ducts to the urethra. The urethra is a tube in the penis.

- The ovaries are female reproductive organs that make egg cells and hormones. Egg cells are released during ovulation. They travel through oviducts to the uterus.

- Fertilization usually occurs in an oviduct. Once united, the egg and sperm cells are called a zygote. When the zygote attaches to the uterus, it becomes an embryo.

- If a fertilized egg does not enter the uterus, the lining of blood and mucus leaves the woman's body during a process called menstruation. Menstruation occurs about once each month.

- The placenta is an organ in the uterus through which oxygen, nutrients, and wastes pass between the mother and fetus.

- There are several stages of human development. These include the prenatal stage, infancy, childhood, adolescence, and adulthood.

fetus

ovaries

oviducts

placenta

sperm

zygote

## Vocabulary Review

**Complete each sentence with a term from the box.**

1. Males produce sex cells called _____. sperm

2. The term used to describe offspring after eight weeks of development is _____. fetus

3. Egg cells travel from the ovaries to the uterus by way of the _____. oviducts

4. A mother passes food and oxygen to her young through a tissue called the _____. placenta

5. The female organs that make egg cells and hormones are the _____. ovaries

6. A _____ is the cell that forms right after an egg cell has been fertilized. zygote

Research Project

Check students' charts for accuracy and completeness. Also explain that each pregnancy is different and that different women may experience changes at different times.

# Chapter Quiz

**Write your answers on a separate sheet of paper. Use complete sentences.**

1. In which organs are sperm cells made?  testes

2. In which organs are egg cells made?  ovaries

3. What is the purpose of the sac of skin holding the testes?  to keep the sperm cool

4. What is the male sex hormone called?  testosterone

5. Where are egg cells usually fertilized?  in the oviducts

6. What are the two female sex hormones called?
   progesterone and estrogen

7. What is puberty?
   the time at which a person becomes sexually mature

8. Why do blood and mucus leave a woman's body once a month? What is this process called?
   because the egg has not been fertilized; menstruation

9. Once an egg cell has been fertilized, what is it called?
   a zygote

10. How do human babies develop inside the mother?
    They grow in length and mass and they also develop organs such as the brain, spinal cord, and heart.

**CRITICAL THINKING**

11. How does reproduction in humans compare to reproduction in other animals, such as fish, birds and reptiles?

12. In what ways do hormones affect the body during puberty?

**Test Tip**

If you cannot think of an answer, try to remember a picture of something related to that topic. It may help you recall the answer.

**Critical Thinking**

11. Fish, birds, and reptiles usually lay eggs. The eggs develop outside the mother's body. The egg of a human develops inside the body of the mother.

12. In boys, testosterone causes the testes to produce sperm. The boy's voice may change, facial hair grows, and the body may become more muscular. In girls, progesterone and estrogen cause the ovaries to begin releasing eggs. This causes menstruation to begin. The hips and breasts usually enlarge also.

*SCLINKS*

Go online to www.scilinks.org. Enter the code PMB138 to research **pregnancy**.

## Research Project

There are many changes in the body of the mother and of the baby during pregnancy. Use the Internet or other references to find out what changes take place during each trimester of pregnancy. Then, make a chart that shows these changes and the approximate dates they occur.

See the *Classroom Resource Binder* for a scoring rubric for the Research Project.

# Unit 5 Review

**Standardized Test Preparation** This Unit Review follows the format of many standardized tests. A Scantron® sheet is provided in the *Classroom Resource Binder.*

Choose the letter for the correct answer to each question.

Use Figure U5-2 to answer questions 1–3.

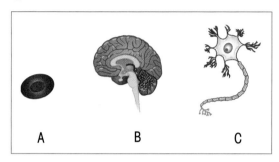

Figure U5-2 *Structures found in the body*

**1.** Which structure carries messages to the brain?  C (p. 268–269)

A. structure labeled A

B. structure labeled B

C. structure labeled C

D. both A and B

**2.** Which structure is part of the central nervous system? D (p. 268–272)

A. structure labeled A

B. structure labeled B

C. structure labeled C

D. both B and C

**3.** Which structure carries oxygen to the cells of the body?  A (p. 247)

A. structure labeled A

B. structure labeled B

C. structure labeled C

D. both A and B

**4.** Which of the following is an example of an involuntary muscle?  D (p. 232)

A. biceps

B. triceps

C. deltoid

D. intestine

**5.** Which organ removes excess wastes from the bloodstream?  A (p. 249)

A. kidney

B. stomach

C. spleen

D. lungs

**6.** In which structure are gases exchanged?  B (p. 247)

A. bronchi

B. alveoli

C. liver

D. kidneys

**7.** Which structure releases eggs and hormones?  C (p. 313)

A. uterus

B. testes

C. ovary

D. placenta

**Critical Thinking**

Check students' responses for accuracy and completeness. Make sure they list all of the components of the system and explain how they work together to form a system. (Chs. 14–20)

**Critical Thinking**

In the human body, organs and tissues work together to form systems. Choose one system of the human body and explain how its parts work together.

# Unit 6 ▶ Living Together on Earth

Chapter 21  **Ecology**

Chapter 22  **Animal Behavior**

Chapter 23  **Cycles in Nature**

Chapter 24  **Natural Resources**

Chapter 25  **Evolution**

Figure U6-1  *The red fox shown above is searching for food in a dry stream bed. Like all animals, the fox depends on its environment for survival.*

## Biology Journal

Plants, animals, soil, air, and water are all part of an ecosystem. In this unit, you will learn about ecosystems, and the living and nonliving things that make up ecosystems. You will also learn about how organisms change over time. In your journal, write a list of questions that you have about how living things relate to each other and their environment. As you read each chapter, go back and answer your questions.

**Biology Journal**  The journal activity can be an alternative assessment, a portfolio project, or an enrichment exercise. Have students write at least five questions. As they read the chapters, have them answer the questions.

**Figure 21-1** *The giraffe and zebras shown above are part of an ecosystem in Africa. What other things shown in the photo do you think are part of the ecosystem?*

Possible answers: the trees, soil, water, and air

## Learning Objectives

- Describe an ecosystem.

- Compare and contrast community, population, habitat, and niche.

- Explain how population growth curves show relationships.

- Describe the stages of succession.

- Give an example of how organisms work together in partnerships.

- Describe the characteristics of several biomes found on Earth.

- LAB ACTIVITY: Simulate a land biome.

- ON-THE-JOB BIOLOGY: Describe the activities of an ecologist.

## Words to Know

| | |
|---|---|
| **ecosystem** | the series of relationships between a community of organisms and the environment |
| **community** | all the living things in one ecosystem |
| **population** | a group of individual organisms of the same kind living in an ecosystem |
| **habitat** | the place in which an organism lives |
| **adaptation** | a characteristic of an organism that helps it to survive in its environment |
| **niche** | the job or function of an organism within its ecosystem |
| **symbiosis** | a relationship between two organisms in which at least one organism benefits from the other |
| **food web** | a diagram that shows many food chains interacting with each other |
| **producer** | an organism that can make its own food by using energy from the Sun or other inorganic sources |
| **consumer** | an organism that cannot make its own food and must eat other organisms |
| **decomposer** | an organism that breaks down and absorbs nutrients from dead organisms and wastes, and then returns the nutrients to the environment |
| **succession** | the gradual change of the populations in a given community over time |
| **biome** | (BY-ohm) a large region of Earth that has a characteristic climate and characteristic kinds of organisms |

**Linking Prior Knowledge** Students should recall the different kingdoms of living things and how each of these kinds of organisms obtains energy (Chapters 4 and 6–13).

Figure 21-2 *Sea stars are part of an ocean ecosystem.*

## Organisms and Their Environment

Many different organisms live in a given environment. For example, an ocean may contain many kinds of fish. It may also have a large variety of other animals, such as sea stars, oysters, and crabs. The ocean may contain other organisms such as seaweed. These organisms all relate to the other organisms in the ocean. They also relate to the many nonliving parts of their environment, such as the rocks, water, and air. The study of the relationship between living things and their environment is called ecology.

## Ecosystems

All the living and nonliving things together in an environment make an ecosystem. An **ecosystem** is the series of relationships between living things and their environment.

**Everyday Science**
Possible answers: living things—trees, grass, worms, insects, birds, squirrels; nonliving things—street, sidewalk, buildings, air, water

**Everyday Science**

Your own community is an ecosystem of its own. What are some living things in your town or city? What are some nonliving things? Make a chart with two columns, one labeled *Living Things* and the other labeled *Nonliving Things*.

Figure 21-3 *The open ocean and the rocky shoreline are both ecosystems. Different organisms live in each ecosystem.*

An ocean ecosystem includes fish, mammals, plants, and algae. It also includes the materials that make up the ocean floor, rocks, salt water, and much more. Organisms get the things they need from the ecosystem they live in. These needs include food, water, air, and shelter.

All together, the living members of an ecosystem are called a **community.** Communities are made up of populations. **Populations** are groups of individual organisms of the same species that live in a specific area. For example, a group of the same type of sea otter that all live in the same place is called a population of sea otters.

Figure 21-4 *This sea otter is part of one population that lives in the ocean.*

There are many different populations in a community. In a given ocean community, there might be many populations of fish. There may also be a population of sea otters, and a population of sea turtles all living together in the same ecosystem.

## A Place and Job for Every Organism

Ecosystems, when left alone, are usually balanced. Every organism has a place to live, called a **habitat.** The ocean is the habitat of sea turtles. Organisms adapt to their surroundings, or habitat, over many generations. **Adaptations** are characteristics of organisms that help them survive in their environment.

Figure 21-5 *Animals are adapted to their environment. How is the sea turtle adapted to living in water?* Possible answer: Its flippers are an adaptation to living in water.

Every organism also has a function within the ecosystem. This function, or job, is called a **niche.** For example, the job of green plants is to make the energy of the Sun available to other organisms. They do this through photosynthesis. A job of spiders is to eat insects. This helps the ecosystem by keeping the insect population down.

# Relationships Between Organisms

Organisms depend on each other for survival. Many organisms feed on other organisms. The organism that is eaten is called the *prey*. The organism that kills and eats the prey is called the *predator*. Predators are an important part of an ecosystem. Predators often kill the weak members of a population. They keep the population size of prey animals in balance.

Figure 21-6 *A barn owl is a predator. The mouse is its prey.*

Organisms have other relationships with each other that are not predator-prey based. Sometimes organisms live in such a close relationship that one organism is helped by another. Such a relationship is called a symbiotic relationship. **Symbiosis** means "living together." There are three types of symbiotic relationships—parasitism, commensalism, and mutualism.

### Parasitism

*Parasitism* is one kind of symbiotic relationship. A parasitic relationship is one in which one organism is helped and the other is harmed. The organism that a parasite lives on is called the *host*. Ticks, for example, can be parasites of dogs. The dog is the host. Ticks feed on the blood of the dog and may pass diseases to the dog. Although parasites harm their host, they usually do not kill them. This is because the parasite depends on the host for survival.

Figure 21-7 *Ticks are parasites. The tick on the right is larger because it is filled with blood from the host.*

### Commensalism

*Commensalism* is a relationship in which one organism benefits and the other organism is not harmed but does not benefit either. Spanish moss often grows on the limbs of trees. The moss benefits from the tree by having a place to live. The tree does not benefit from the moss, but it is also not harmed by the moss. This is an example of commensalism.

## Mutualism

*Mutualism* is a relationship that helps both organisms. For example, an acacia tree has thorns on its branches. The thorns prevent most animals from eating the leaves. However, there are some animals, such as the giraffe, that can eat around the thorns. When this happens, ants that live on the acacia tree attack the animal. The ants defend the tree from being eaten. At the same time, the ants feed on milky sap produced by the tree. So, the tree gets protection, and the ants get food.

Figure 21-8 *Acacia trees and certain species of ants have a mutualistic relationship.*

### ✔ Check Your Understanding

Answer the questions in complete sentences. Use a separate sheet of paper.

1. What is an ecosystem?

2. What is a community?

3. What is a niche?

4. What does *symbiosis* mean?

5. **CRITICAL THINKING** A bumble bee helps pollinate flowering plants while getting nectar from the flower. What type of relationship does this represent?

**Check Your Understanding**

1. the series of relationships between a community and its environment

2. all the living things in one ecosystem

3. the job or function of an organism

4. a relationship between two organisms in which at least one organism is benefited by the other

5. mutualism; the bee gets nectar and the plant gets pollinated.

# Food Chains and Food Webs

Organisms depend on each other for survival. Plants get energy from the Sun. The energy stored in plants is then passed along to the animals that eat the plants. Other animals eat the plant-eaters.

### Food Chains

Organisms get energy from the things they eat. This series of relationships is called a *food chain*. For example, in a prairie ecosystem, grasses use the energy from sunlight to make food. Animals such as meadow voles feed on the grasses. Other animals such as hawks feed on the voles. The relationship between the grasses, the vole, and the hawk is a food chain.

In a diagram of a food chain, arrows point from the energy source to the organism that gets the energy. Look at the food chain in Figure 21-9. Notice that energy flows from the grass to the vole and then to the hawk.

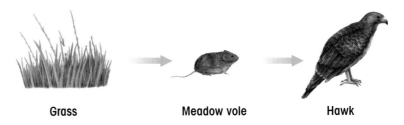

Grass          Meadow vole          Hawk

Figure 21-9 *A food chain shows how energy passes from one organism to another.*

### Food Webs

All organisms are part of a food chain. In fact, most organisms belong to more than one food chain. Usually food chains in an ecosystem interact with one another. Animals in one food chain often eat animals in other food chains. This network of related food chains is called a **food web**.

## Science Fact

Animals are usually part of more than one food chain. This is because animals do not eat the same thing all of the time. Their diets change depending on what food is available.

There can be many different organisms in a food web. For example, snakes often eat mice and rabbits. The snake is food for hawks, coyotes, and other animals. Besides the snake, a hawk might also eat voles or prairie chickens. The voles and prairie chickens probably eat some of the same insects. Look at the food web in Figure 21-10.

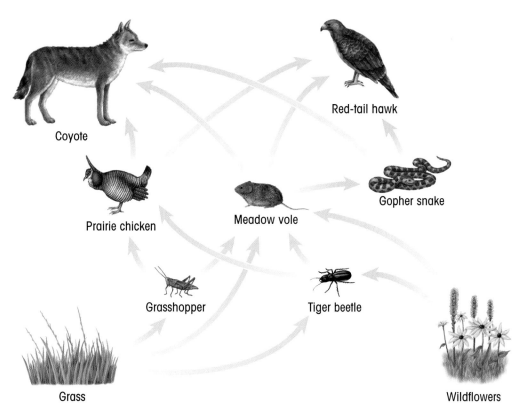

Figure 21-10 *In a food web, organisms form many food chains that indirectly affect each other.*

You are also part of food chains and food webs. Think about the foods you eat. They come from other living things. For example, the meat you eat may come from a cow. The cow probably ate grains, which come from plants.

## Everyday Science

Think about what foods you ate over the past two days. Draw a food web that includes these organisms and yourself.

## Energy Flows Through an Ecosystem

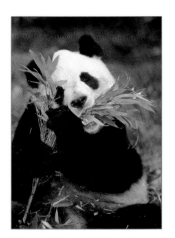

Figure 21-11 *The giant panda, a herbivore, eats bamboo.*

All ecosystems are powered by an energy source. For most ecosystems, this source is the Sun. In any ecosystem, some organisms can absorb energy and use it to produce food. The food is in the form of sugars. These organisms are called producers. Plants are **producers.** They use the Sun's energy to make food by photosynthesis.

Animals cannot make their own food. So they must eat plants and other animals to get energy. Organisms that cannot make their own food are called **consumers.** In every ecosystem, there are different types of consumer herbivores, carnivores, and omnivores. Herbivores eat plants. Carnivores eat animals. Omnivores eat both plants and animals. Look at Figure 21-12.

| Types of Consumers in an Ecosystem | | |
|---|---|---|
| **Type of Consumer** | **Examples** | **Diet** |
| Herbivores | Deer, rabbits, and some insects | Feed only on plants |
| Carnivores | Snakes, hawks, and sharks | Feed on other animals |
| Omnivores | Bears, raccoons, and opossums | Feed on both plants and animals |

Figure 21-12 *All ecosystems contain different types of consumers.*

**Think About It**

Most animals that eat other animals kill only the sick, weak, and aged in a population. How does this practice help keep an ecosystem in balance?

Possible answer: It can reduce the number of weaker organisms in a population. This makes the population stronger overall.

**Decomposers** are organisms that break down dead organisms and their waste products. In doing so, they return important nutrients to the soil. Bacteria, fungi, insects, and worms are examples of decomposers. Ecosystems contain some producers, some consumers, and some decomposers.

Energy flows one way through an ecosystem. The first level of consumers feed on the producers. The second level of consumers feed on the first level consumers. The energy that travels through each of these organisms can be represented in an energy pyramid. Look at the energy pyramid in Figure 21-13.

**Science Fact**

Notice that the amount of energy at each level decreases. This means that fewer organisms can survive at the higher levels. For example, there will be more meadow voles in a balanced ecosystem than hawks.

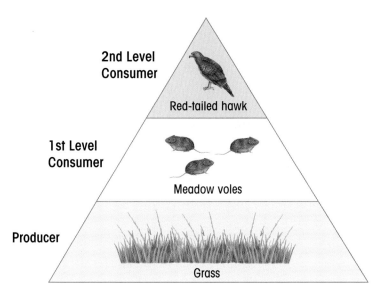

**Figure 21-13** *At each level of an energy pyramid, some energy is transferred to the next level. However, some of the energy is lost as heat.*

✓ **Check Your Understanding**

Answer the questions in complete sentences. Use a separate sheet of paper.

1. List the types of organisms that energy flows through in an ecosystem.

2. Why are decomposers important to an ecosystem?

3. **CRITICAL THINKING** Should humans be considered herbivores, carnivores, or omnivores? Explain your answer.

**Check Your Understanding**

1. producer, first level consumer, second level consumer

2. because they return nutrients to the soil

3. omnivores, because we eat both plants and animals

# Changes in Populations

The size of a population changes over time. The size of a population can depend on the amount of food available or the number of predators in a given area. These characteristics are called limiting factors. The *limiting factors* of an ecosystem prevent a population from growing beyond a certain size. This size is called the *carrying capacity.*

A good example of how populations depend on each other is the populations of lynxes and hares. A lynx is a member of the cat family. The lynx eats hares. As the population of hares rises and falls, the population of lynxes also rises and falls. A population graph can be used to analyze the relationship between two populations. Look at the graph in Figure 21-15 to learn more about the relationship between lynx and hare populations.

Figure 21-14 *Lynx depend on hare for survival.*

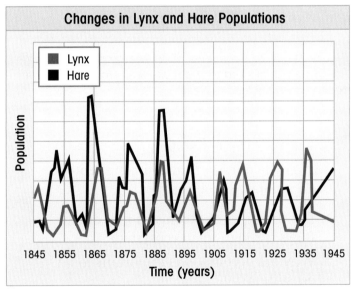

Figure 21-15 *Population graphs can be used to show how two populations are related to one another.*

# Succession

The populations of an ecosystem can change quite a bit over a long period of time. The gradual changes in the populations of a given area is called **succession**. Normally in succession, the plants are the populations that change first. Then, animal populations will change as well.

Succession often occurs after a major change in the ecosystem. For example, a volcano that erupts can destroy the plant and animal life in an area. Eventually, simple plant species and lichens will begin to grow in the area. These small plants and lichens are called *pioneer species*. They will add nutrients to the soil. After many years, the area may become populated with plants that are more complex, forming a meadow. Then, the meadow may have enough nutrients and soil to allow small shrubs to grow. Over time, the shrub land may become a forest. The first trees to grow in such an area are usually coniferous trees, such as pine trees or spruce trees. Later, hardwood trees will grow. Beech, maple, and oak are examples of hardwood trees.

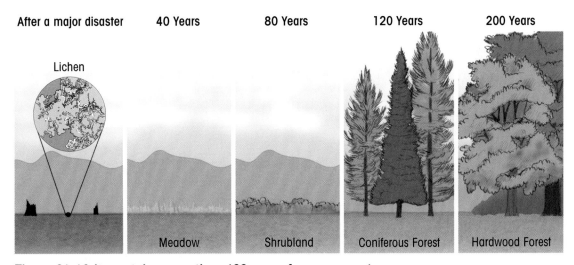

Figure 21-16 *It can take more than 100 years for an ecosystem to go through all of the stages of succession.*

Figure 21-17 *Human activity has affected the gray wolf population.*

# Humans Can Change Ecosystems

Humans have an effect on the ecosystems they live in. For many years, people in the western United States feared wolves. Gray wolves were hunted and killed until there were few left. People did not realize that the wolves were an important part of the ecosystem. The wolves kept the populations of other animals in balance by feeding on herbivores such as deer and rabbits. By reducing the number of wolves, humans changed the balance in the ecosystem.

## A Closer Look

### HUMAN POPULATION GROWTH

Populations generally can rise only to a certain level, called the carrying capacity. This occurs because there is only a certain amount of food and other resources available in an ecosystem to support the population.

Earth is home to more than 6 billion people. Many scientists predict that by the year 2050, there will be more than 10 billion people living on Earth. Some fear that Earth will not be able to provide enough food, water, and shelter for the growing human population.

**CRITICAL THINKING** Why do you think the human population has risen so dramatically?
The human population has risen because of improvements in farming, shelter, and medicine.

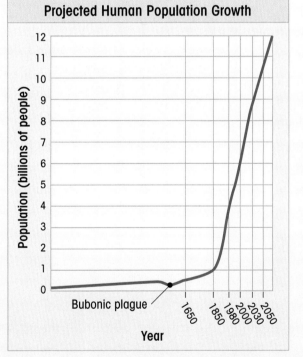

Figure 21-18 *The human population increased rapidly after the turn of the 19th century.*

## The Major Land Biomes

**Everyday Science** Possible answers: grassland, temperate forest, desert, chaparral, freshwater or saltwater biomes; the state bird and flower may exist in one or more biomes.

Different parts of the world have different temperature ranges, amounts of rainfall, and types of soil. These conditions only allow for certain plants and animals to survive. Therefore, different regions are able to support different kinds of plants and animals. A large region of Earth that has a characteristic climate and kinds of organisms is called a **biome**.

There are seven major land biomes found on Earth. They are called tundra, taiga, temperate forest, tropical rain forest, temperate grassland, savanna, and desert. Figure 21-19 shows the locations of these biomes. It also shows where large polar ice caps exist on Earth.

**Everyday Science**

There are several different biomes that cover the United States. Which biomes does your state cover? Use the Internet to find out your state plant, animal, and bird. Do they all live in the same area?

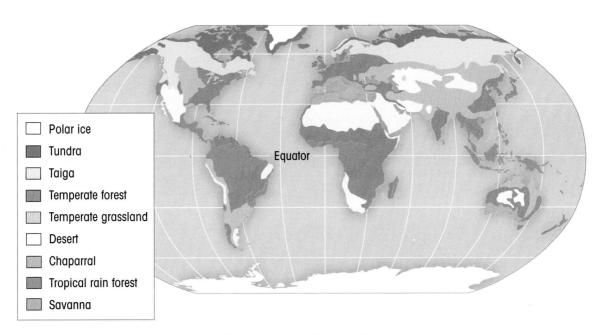

Polar ice
Tundra
Taiga
Temperate forest
Temperate grassland
Desert
Chaparral
Tropical rain forest
Savanna

Equator

Figure 21-19 *Earth can be divided into seven major land biomes.*

### Tundra

The *tundra* is a very cold, dry biome. Much of the ground is permanently frozen. This layer of permanently frozen soil is called *permafrost*. Only very small plants can grow in the tundra.

Figure 21-20 *Small plants cover the ground in the tundra. Animals such as caribou feed on these plants.*

### Taiga

The *taiga* has warmer temperatures and receives more rainfall than the tundra. Cone-bearing trees grow in the taiga.

### Temperate Forest

*Temperate forests* are warmer and wetter than the taiga. They differ in the types of trees they have. The leaves on most of the trees in this biome change colors in fall and drop off from the tree. This type of tree is called *deciduous*.

### Tropical Rain Forest

*Tropical rain forests* can be found near Earth's equator. This biome also has the greatest number of different organisms living together in one area. It is filled with dense forests and thousands of different kinds of organisms.

Figure 21-21 *This anteater lives in the tropical rain forests of South America.*

## Temperate Grassland

*Temperate grasslands* have a deep layer of soil that is rich in nutrients. Many tall and short grasses grow there. This biome is usually located in the interior of a continent.

## Savanna

*Savannas* are also covered with grasses, but differ from temperate grasslands in a few ways. The soil of the savanna is not as deep or rich. There are also shrubs and some kinds of trees scattered across the savanna. Animals such as lions, antelope, elephants, and zebras live in the savanna. Both temperate grasslands and savannas receive less rainfall than temperate forests and taigas.

Figure 21-22 *Lions live in the savanna.*

## Desert

*Deserts* are the driest biomes. The desert biome is characterized by having less than 25 centimeters (9.8 in.) of rain each year. However, this does not mean that plants and animals cannot live there. Desert plants are adapted to long periods of drought. They can store water for a long time. Animals such as mountain lions, birds, insects, and reptiles can be found in the desert. The climate of deserts varies greatly in temperature, depending on what time of day it is.

because the tundra is very cold and receives very little rainfall

**Think About It**

Tundras are often called "frozen deserts." Why do you think this term is used?

# Land Biomes and Climate

The climate plays a big part on the type of biomes that exist in a given area. Both temperature and average rainfall are part of the climate of an area. The table in Figure 21-23 summarizes the basic characteristics of the climates in the seven major biomes.

| Major Biomes Found on Earth | | | |
|---|---|---|---|
| Biome | Average Yearly Temperature | Average Yearly Rainfall | Plant and Animal Life |
| Tundra | −13° to 10.5°F (−20° to 12°C) | Less than 10 in. (less than 25 cm) | Grasses, mosses, and lichen; caribou, foxes, and lemmings |
| Taiga | 14° to 57°F (−10° to 14°C) | 14 to 30 in. (35 to 75 cm) | Fir and spruce trees; moose, wolves, grizzly bears, and birds |
| Temperate Forest | 43° to 82°F (6° to 28°C) | 30 to 50 in. (75 to 125 cm) | Maple, birch, hickory, and oak trees; squirrels, black bears, deer, raccoons, mice, frogs, and birds |
| Tropical Rain Forest | 68° to 93°F (20° to 34°C) | 80 to 160 in. (200 to 400 cm) | Very large trees, vines, ferns; many kinds of insects, amphibians, reptiles, and birds |
| Temperate Grassland | 32° to 77°F (0° to 25°C) | 10 to 30 in. (25 to 75 cm) | Tall or short grasses; buffalo, prairie dogs, snakes, and birds |
| Savanna | 60° to 93°F (16° to 34°C) | 30 to 60 in. (75 to 150 cm) | Tall grasses and some scattered trees; zebra, antelope, lions, and cheetahs |
| Desert | 44° to 100°F (7° to 38°C) | Less than 10 in. (less than 25 cm) | Cactus, shrubs, and some grasses; small mammals, reptiles, and insects |

Figure 21-23 *Temperature and amount of rainfall are important factors in categorizing biomes.*

# Other Land Biomes

Besides the seven major biomes, there are also a few minor biomes found on Earth. The chapparal is a biome that covers a small part of Earth. It has mild, rainy winters, and long dry summers. The polar ice biome surrounds the North and South Poles and has large sheets of ice.

# Water Biomes

Earth has many different water biomes as well. In fact, over 75 percent of Earth is covered with water. This water is divided into several different biomes.

Oceans are saltwater biomes. Different parts of the ocean receive different amounts of light. This is because light can only travel to a certain depth in the ocean. Organisms that require light, such as photosynthetic plankton, usually stay near the surface of the ocean. Other organisms are adapted to living at the bottom of the ocean.

Examples of freshwater biomes include lakes, ponds, rivers, and streams. Lakes and ponds are home to many kinds of fish as well as birds, mammals, and amphibians. Lakes and ponds also contain plants such as water lilies. The organisms that live in rivers and streams are adapted to constant movement of water. Many fish, such as salmon and brook trout, are able to swim upstream, against the flow of water.

An *estuary* is an area where freshwater streams or rivers flow into the ocean. This biome has almost as many different species of organisms as the tropical rain forest. Many organisms come to the estuaries to reproduce.

Figure 21-24 *This angler fish lives in the deepest parts of the ocean. It uses the extension on its head as a "fishing pole" to catch prey.*

Figure 21-25 *Many people are trying to protect estuaries from environmental changes such as pollution.*

## Science Fact

The water in the oceans contains about 3 percent salt. The water in freshwater biomes contains about 0.005 percent salt. Estuaries have a mix of salt water and fresh water, and the amount of salt changes all the time.

# LAB ACTIVITY
## Modeling a Land Biome

### BACKGROUND
A biome is a type of living community that's found a single ecological region, like a desert.

### PURPOSE
In this activity, you will create a community to model a land biome.

### MATERIALS
clean 2-liter plastic bottle, cup or spade, scissors, gravel, soil or sand, small plants, insects

**Figure 21-26** *Add gravel and soil to the plastic bottle.*

### WHAT TO DO
1. First, choose the biome you would like to simulate. You can simulate a desert, forest floor, or rain forest.

2. Cut the top off the bottle about 5 cm below the curved top. Put about 5 cm of gravel in the bottom of the bottle. Now, fill the bottle about half full with the appropriate soil for your ecosystem.

3. Add the appropriate plants. Be sure to water them according to the needs of the plants. Hint: Desert plants will need less water than rain forest plants.

4. Punch 8–12 holes in the top piece and attach it with clear tape. Place the planter in the right amount of light. Check on your bottle biome every day you are in class.

5. Once your plants are stable you may add insects. Ladybugs, crickets, spiders, and praying mantises are possible inhabitants for your bottle biomes. Make sure to only use insects that are appropriate for your biome.

### DRAW CONCLUSIONS
- How is your bottle biome similar to an actual biome? How is it different? It is similar because plants in a bottle need the same things as plants in the real world. It different because there is less variety of life and fewer of the complex interactions among members of the biotic communi
- What are the producers in your biome? What are some consumers? Answers will vary. Students should list a few plants for producers, and some insects for consumers.

# ON-THE-JOB BIOLOGY
## Ecologist

Bob Simon is an ecologist. He studies how living things relate to each other and their surroundings. Bob works for a nonprofit group. This group tries to protect wetlands. Marshes, bogs, and estuaries are examples of wetlands. The group's goal is to save plants and animals of the wetlands that are at risk. To do this, the group works to protect the land and waters these organisms need to survive.

Testing the water quality of the ecosystem is a big part of Bob's job. First, he collects a sample of water. Next, he analyzes the chemicals and living things found in the water. Then, he looks for trends. Sadly, he has seen some species decline as buildings, roads, and farms replace the natural wetlands environment. Pollution is also harming the wetlands. Bob and his group look for ways to stop the pollution and to help restore the ecosystem to its natural state.

**Figure 21-27** *Bob Simon checks for signs of pollution in the water.*

The decline of some species can harm other species, too. For example, if microscopic organisms in the water begin to die because of pollution, the fish that eat them may also die. Then, the birds and other organisms that eat the fish may be affected.

**Answer the questions on a separate sheet of paper. Use complete sentences.**

1. What is an ecologist? someone who studies how living things relate to each other and to their environment
2. What is one way an ecologist can monitor the health of a wetlands ecosystem? by testing the water for signs of pollution
3. What may cause the populations of some wetlands' and organisms to decline? loss of habitat and pollution

**Critical Thinking**

How does the decline of one species harm other species?

**Critical Thinking** Other species may be harmed because they depend on the first species as a source of food.

## Summary

- All the living and nonliving things together in an environment make an ecosystem. The group of living things in an ecosystem is called a community. The members of one kind of organism in an ecosystem make up a population. The place where an organism lives is called its habitat. The job an organism has within the community is called its niche.

- Energy moves through food chains. Most organisms are a part of several food chains, called a food web. Food webs contain producers, consumers, and decomposers.

- In a symbiotic relationship, at least one organism is helped. There are different kinds of symbiotic relationships. In parasitism, one organism is helped and one is harmed. In mutualism, both organisms are helped. In commensalism, one organism benefits while the other is unaffected.

- A biome is a region of Earth that has a characteristic climate and kinds of organisms. The seven major land biomes are tundra, taiga, temperate forest, tropical rain forest, temperate grassland, savanna, and desert. Other biomes include the chaparral, polar ice biome, and several different water biomes.

| |
|---|
| biome |
| community |
| ecosystem |
| food web |
| population |
| symbiosis |

## Vocabulary Review

**Write a word from the list that matches each definition.**

1. the series of relationships between a community of organisms and the environment   ecosystem

2. groups of individual organisms of the same kind living in an ecosystem   population

3. a relationship between two organisms in which at least one organism is benefited by the other   symbiosis

4. many food chains that cross over one another   food web

5. large region of Earth that has a characteristic climate and kinds of organisms   biome

6. all the living things in one ecosystem   community

## Chapter Quiz

**Write your answers on a separate sheet of paper. Use complete sentences.**

1. What is an ecosystem?
a series of relationships between a community of organisms and the environment

2. What are five living and five nonliving things in your community? trees, grass, birds, dogs, people; houses, schools, streets, air, water

3. How is a community related to a population?
Different populations make up a community.

4. How does a population growth curve show the relationship between predator and prey? It shows how the rise and fall of one population affects the rise and fall of another.

5. What might be the habitat of a frog? What is its niche? a pond or stream; it catches insects and provides food for other animals

6. What are three types of symbiotic relationships? parasitism, mutualism, and commensalism

7. As you move up an energy pyramid, what happens to the number of organisms? The number of organisms decreases.

8. Compare the characteristics of the temperate forest biome and tropical rain forest biome.

**8.** The temperate forest has a cooler and drier climate than the tropical rain forest. Also, the temperate forest has deciduous trees and the tropical rain forest has trees that keep their leaves all year. There are more species of organisms in the tropical rain forest.

**Critical Thinking**
**9.** Answers will vary but should include at least one producer and several consumers. Check to make sure arrows are pointing in the correct direction (with the flow of energy).

**10.** newly formed land with few plants and animals, meadow, shrub land, forest of pines and spruces, hardwood forest

### CRITICAL THINKING

9. On a separate sheet of paper, draw a food chain. Show and label at least one producer, one consumer, and one decomposer.

10. Describe the stages of succession.

*SCi*LINKS

Go online to www.scilinks.org. Enter the code PMB140 to research **threats to rain forests.**

### Research Project

Many of the rain forests in South America are being cut down. Use the Internet and other resources to research this problem. Find out how the loss of trees affects the ecosystem. Also, find out why people are doing it. Decide if you are for or against the cutting of these trees. Write a report that supports your decision.

See the *Classroom Resource Binder* for a scoring rubric for the Research Project.

Figure 22-1 *Monarch butterflies migrate in the winter from the northern United States and Canada to as far south as Mexico. Why do you think these insects migrate so far?*

Possible answer: to get to a warmer climate where food and water are available

## Learning Objectives

- Describe how behaviors help animals to survive.
- Define *innate behavior*. Give some examples.
- Define *instinct*. Give some examples.
- Define *learned behavior*. Give some examples.
- **LAB ACTIVITY:** Solve a puzzle.
- **ON-THE-JOB BIOLOGY:** Find out how an animal trainer trains seals.

## Words to Know

| | |
|---|---|
| **behavior** | the way an organism responds to its environment |
| **innate behavior** | a behavior that an animal has at birth |
| **reflex** | a simple inherited behavior that an animal automatically performs in response to a stimulus |
| **instinct** | an inherited pattern of behavior that an animal can control |
| **competition** | the struggle among organisms for resources |
| **hibernation** | a resting state during which an animal's body temperature drops and its heart rate and breathing slow down |
| **pheromone** | a chemical that animals use to affect the behavior of other animals of the same species |
| **learned behavior** | a behavior that develops as a result of experience |

## Behaviors Help Animals Survive

Some animals are born knowing how to swim or how to hunt for food. These behaviors help the animals to survive. **Behavior** is the way an animal responds to its environment.

There are some behaviors that an animal must perform in order to survive. For example, a newborn dolphin knows that it must rise to the surface to get air. If it does not, it will drown.

There are some behaviors that an animal learns. These behaviors help the animal to adjust to a changing environment. For example, certain plant-eating animals learn to eat different types of plants. Therefore, when one type of plant is not available, the animals will not starve. They can eat another type of plant.

**Linking Prior Knowledge**
Students should recall that a stimulus is a change that causes a response (Chapter 2).

Scientists believe that behavior is partly controlled by genes. A behavior is a response to a stimulus. The behavior involves an animal's nervous system. The way an animal's nervous system functions depends partly on the genes that it inherits from its parents.

In general, there are two main types of behavior: innate behavior and learned behavior.

## Innate Behavior

A newborn mammal knows to drink its mother's milk. Usually, the baby is not taught this behavior. Behaviors that an animal is born with are called **innate behaviors.** These behaviors are usually performed correctly the first time the animal acts.

A **reflex** is an innate behavior. It is a simple inherited behavior that an organism automatically performs. It is a quick movement in response to a stimulus. Most reflexes cannot be controlled. Blinking when an object comes close to your face is a reflex. Animals, such as the one shown in Figure 22-2, also have reflexes.

Figure 22-2 *When a chameleon senses a moving insect, it flips out its tongue to capture it. This action is a reflex.*

**Think About It** Because the sea turtles survived after being hatched, the place where they were born may be a safe place to lay their own eggs.

An **instinct** is another type of innate behavior. It is an inherited pattern of behavior that an organism can control. A spider spinning a web is an example of instinctive behavior. When the spider needs to make a web, it knows how to make it. Sea turtles have nesting instincts. They return to the place where they were hatched to lay their own eggs.

**Think About It**

Why do you think sea turtles return to the place where they were hatched to lay their eggs?

✓ **Check Your Understanding**

Write your answers in complete sentences.

**1.** What is a behavior?

**2.** How are a reflex and an instinct similar? How are they different?

**3.** CRITICAL THINKING  Is swallowing a reflex or an instinct? Explain your answer.

**Check Your Understanding**

**1.** the way an organism responds to its environment

**2.** They are both innate behaviors. However, a reflex is just a quick movement in response to a stimulus. An organism has control over its instincts.

**3.** It is a reflex because it is a quick response to the stimulus of food.

## Modern Leaders in Biology

**EDWARD OSBORNE WILSON, Biologist**

Dr. Edward Osborne Wilson is a biologist. He is well known for his research on social insects. These insects include bees, wasps, ants, and termites.

Dr. Wilson studied ant colonies. He found that ants and other insects that live in groups are divided into "social" classes. These classes include the queen, drones, and worker, or soldier, ants. Dr. Wilson observed that every ant within a social class has a certain job. The queen's job is to lay eggs. The drones are male ants that mate with the queen. The worker, or soldier, ants take care of the young and feed and protect the colony.

Dr. Wilson believes that the way in which ants behave is due to genetics. He also believes that certain human behavior is a result of genetics.

Figure 22-3 *Dr. Edward Osborne Wilson believes that ant behavior is controlled by genetics.*

**CRITICAL THINKING**  How are ant behavior and human behavior similar?

Possible answer: Ants live in colonies. Humans live in groups. Ants have specific jobs, and humans have specific jobs, too.

## Types of Instinctive Behavior

Instinctive behavior can be seen in the ways that some animals mate or compete for space. They can also be seen in the ways that certain animals respond to changes in the environment.

### Mating

The behavior of some animals to attract other animals is instinctive. This behavior, called *courtship,* is performed before two animals of the same species mate. This behavior helps animals to identify healthy mates. Courtship leads to mating only if the animal performing the behavior is not rejected by the other animal. Peacocks show off their feathers to attract mates. Whales sing a special song. Rabbits may run and jump around each other.

Mating instincts help animal species to continue through reproduction. They also help animals to recognize other animals of the same species. The song that a certain type of bird sings may attract only birds of the same species.

Figure 22-4 *A peacock shows off his feathers to attract a mate.*

### Competition

Some animals compete with each other for resources. This struggle among organisms is called **competition.** One resource that animals compete for is space. This space is called a *territory.* Within this space, the animal may find its food, its nesting site, and its mate. It may also reproduce there. An animal will defend its space from another animal. The instinct to defend its territory helps an animal and its offspring to survive. A territory may also be claimed and protected by a group of animals.

Food is another resource for which organisms compete. For example, within a pride of lions, some individuals will compete to see who gets to fill its belly first.

**Everyday Science**

Have you ever heard a bird sing? Why do you think the bird was singing? The next time that you hear a bird singing, listen for a reply to the song from another bird.

When competing for resources, an animal may show threatening behavior. This threatening behavior, or *aggression,* can include growling, snarling, or clawing at another animal. Aggression helps an animal gain control over another animal.

Figure 22-5 *This snow leopard is showing aggression.*

## Migration

You have read that some birds migrate when seasons change. Instincts make other animals migrate, too. Migration helps animals to survive cold seasons. It also helps them to find food and places to breed or to mate.

Some animals travel to warm areas because they cannot survive cold weather. For example, monarch butterflies and some moths migrate south in the winter. Some species of fish, such as mullets and green sturgeons, migrate to warm waters.

Some animals migrate looking for food. Caribou and elk travel from place to place in search of food. The wolves that feed on caribou may migrate with them. Other animals migrate for breeding purposes. Gray whales migrate south so that they can give birth to their young in warm waters. Some species of crabs move to shallow water to breed. Then, they return to deeper ocean waters.

**Remember**
Migration is the traveling over long distances of some animals from season to season for feeding, nesting, and warmth.

**Remember**
Hormones are chemical messengers produced in endocrine glands.

## Science Fact

Bears are not true hibernators. They sleep through the winter and their heart rate and breathing rate drop, but their body temperatures remain about the same. They can also be awakened from their sleep.

### Hibernation

Not all animals migrate. Shorter days and colder temperatures cause the release of certain hormones in many animals. These hormones cause them to hibernate. **Hibernation** is a resting state. During this sleep-like state, the animal is not active. The animal's metabolism slows so that it does not need much energy. Its body temperature drops. Its heart rate and breathing slow down. Before it goes into hibernation, the animal eats a lot of food. The food is stored as fat in the animal's body. During hibernation, the animal lives off the stored fat.

Animals that hibernate include chipmunks, squirrels, bats, lizards, turtles, swifts, and nighthawks. Hibernators do not usually sleep through the entire winter. They may awaken to eat or to get rid of wastes. They may also awaken if their body temperatures fall too low.

Hibernation is an instinct. This behavior helps animals to survive cold weather. It also helps them to survive when food may not be available.

Figure 22-6 *A chipmunk hibernates to survive cold weather.*

## Communication

Some animals make sounds, give off scents, perform "dances," touch each other, give off light signals, or change colors. These behaviors are instincts. They are also forms of communication. Communication is the sharing of information. Most animals communicate with members of their own species.

Some animals use chemicals to send information to other members of the same species. These chemicals are called **pheromones**. They have an odor, although humans may not be able to smell them. They affect the behavior of other animals within the same species. Ants use these chemicals to lead other ants to food sources. Some animals use them to mark their territories. Many insects use them to attract mates.

Animals need to pass information to each other in order to survive. The information may lead other animals to food, to shelter, or away from harm. Animals also use forms of communication when finding a mate. Mating to produce offspring is needed for species to survive.

Figure 22-7 *Dolphins communicate using sound. They can make these sounds underwater.*

**Think About It**

Humans can communicate by using language. Language uses symbols to represent ideas. Why are humans the only animals that can use language?

Humans are the only animals with a well-developed brain and memory.

✓ **Check Your Understanding**

Write your answers on a separate sheet of paper. Use complete sentences.

**1.** Name some types of instinctive behavior.

**2.** Why does an animal defend its territory?

**3.** What happens during hibernation?

**4.** CRITICAL THINKING  Why is it important for animals to communicate with each other?

## A Closer Look

### PHEROMONES

Researchers have found a way of using pheromones to benefit humans. They have studied the chemistry of the pheromones made by certain insects. They can make these pheromones in a lab. The chemicals are then used in traps. These traps attract the insects. They also contain poisons. The insects are lured to the traps and then killed inside them. These traps have been effective in controlling the population size of the Mediterranean fruit fly, the melon fly, and the Oriental fruit fly.

Figure 22-8 *Using pheromones has helped control the sizes of populations of this Mediterranean fruit fly.*

The use of pheromones has also helped farmers control the populations of certain insects. The pheromones of certain female insects are made in a lab. They are then placed in different areas of a field. The male insects of the species are attracted to the chemicals. Because the males do not meet any females, mating does not occur. Offspring are not produced. The population of that species of insect is controlled. This method has been successful in controlling the populations of the pink bollworm, the Oriental fruit moth, and the European grape moth.

CRITICAL THINKING  How do you think pheromones can be used to control a species of insect that is spreading a disease?

Pheromones made in a lab can be used in traps that attract the disease-causing insects.

# Learned Behavior

The other main type of behavior is learned behavior. **Learned behavior** is behavior that is developed through experience. In humans, reading, writing, playing sports, and playing a musical instrument are learned behaviors.

For animals, learned behavior is important for survival. Behavior has to change in response to changes in the environment. An animal must learn to adapt to change. If an animal does not learn certain new behaviors, it can die.

Imprinting, classical conditioning, operant conditioning, and insight learning are different ways that animals learn.

### Imprinting

*Imprinting* is learning that occurs in the early stages of development of an animal. It occurs when an animal observes an object and forms an attachment to the object. Newborn ducks identify their mothers through imprinting. The first animal that a duckling sees is usually its mother. The duckling will follow its mother. This behavior will help the duckling to survive. It will receive food and protection.

Figure 22-9 *Playing a musical instrument is a learned behavior. It usually takes practice in order to play well.*

Figure 22-10 *Ducklings identify their mother through imprinting.*

## Classical Conditioning

In *classical conditioning,* a response to a certain stimulus becomes connected to another stimulus. An example of this type of learning is the work of the Russian biologist Ivan Pavlov. He tested how certain stimuli could make a dog salivate. Pavlov's experiment is shown in Figure 22-11.

**STEP 1**
**Control**

The dog is not conditioned.

**STEP 2**
**Before conditioning**

Pavlov noticed that the dog salivated when it saw food.

**STEP 3**
**Conditioning Phase**

Pavlov rang a bell whenever he fed the dog. The dog learned to connect the ringing of the bell with food.

**STEP 4**
**After conditioning**

The dog would salivate when it heard the bell, even when there was no food present.

Figure 22-11 *An example of classical conditioning is Pavlov's dog.*

## Operant Conditioning

*Operant conditioning* is also known as trial-and-error learning. This type of learning is based on rewards or punishments. For example, a dog may learn a certain behavior because it expects to receive a reward. This behavior is learned through *positive reinforcement.*

Animals may also learn a behavior in order to avoid a punishment. This type of learning occurs, for example, when an animal eats an insect that causes it to get sick. The animal learns never to eat that insect again. This behavior is learned through *negative reinforcement.*

## Insight Learning

*Insight learning* occurs when an animal uses something that it has already learned in a new situation. There is no trial-and-error learning involved. For example, a chimpanzee learns to stack boxes. It then stacks boxes to reach for bananas hanging high above it. A student applies insight learning when he uses addition and subtraction to solve multiplication or division problems.

**Figure 22-12** *A student applies insight learning when he uses addition to solve a multiplication problem.*

## ✓ Check Your Understanding

Write your answers in complete sentences.

1. Give three examples of learned behavior in humans.

2. What is imprinting?

3. What is another name for operant conditioning?

4. **CRITICAL THINKING** Every time a goldfish is fed, a light is turned on. What do you think will happen every time the light is turned on? What kind of learning has occurred in the goldfish?

**Check Your Understanding**

1. Possible answer: reading, writing, and playing a musical instrument

2. learning that occurs in the early stages of development of an animal

3. trial-and-error learning

4. The fish will behave the same way as when it is getting food; classical conditioning

# LAB ACTIVITY
## Learning to Solve a Puzzle

**BACKGROUND**
Learned behavior is behavior that is changed through experience.

**PURPOSE**
In this activity, you will learn how to solve a puzzle.

**MATERIALS**
puzzle pieces, timer or watch with a second hand, pencil, paper You can provide students with a commercial puzzle or one made from an enlarged copy of Figure 22-13.

**WHAT TO DO**
1. Copy the chart in Figure 22-14 onto a separate sheet of paper.

2. Work with a partner. Decide which one of you will put the puzzle pieces together. The other partner will record how long it takes for the puzzle to be pieced together.

3. Put the pieces together and record the time it takes to put the puzzle together in the chart next to Trial 1.

4. Take the puzzle apart. Mix up the pieces.

5. Repeat Steps 3 and 4 four more times. Record the times next to the correct trials.

**DRAW CONCLUSIONS**
- How did the time change from Trial 1 to Trial 5? The times decreased.

- What kind of learning has taken place? trial-and-error learning or operant conditioning
- What caused the times to change? The puzzle became easier to complete due to practice.

Figure 22-13 *Record the time it takes to put a puzzle like this together.*

| Putting a Puzzle Together | |
|---|---|
| Trial | Time (seconds) |
| 1 | |
| 2 | |
| 3 | |
| 4 | |
| 5 | |

Figure 22-14 *Copy this chart onto a separate sheet of paper.*

# ON-THE-JOB BIOLOGY
## Animal Trainer

Pam Keene is an animal trainer. She teaches seals to perform at a zoo. Pam trains the animals to jump high and to follow moving targets. She uses rewards, such as food and toys, to get the seals to behave correctly. Her job requires knowledge of how the seals learn.

Pam trains seals, but many of the methods that she uses work with other animals, such as dolphins, sea lions, horses, and dogs. Training takes time and lots of patience.

**Figure 22-15** *An animal trainer uses rewards to train animals.*

**Copy the chart in Figure 22-16 onto a separate sheet of paper. For each animal listed, write a behavior that you would like to train it to do. Also, write the reward that you would give each animal for performing the behavior.**

| Training Animals | | |
|---|---|---|
| Animal | Behavior | Reward |
| 1. Dog | | |
| 2. Cat | | |
| 3. Bird | | |
| 4. Dolphin | | |

**Figure 22-16** *Copy and complete the chart.*

Possible answers: Dog: laying down, dog biscuit; Cat: sitting up, cat treat; Bird: singing, bird seed; Dolphin: jumping out of the water, fish

**Critical Thinking**

Humans learn in much the same way as other animals. How might parents use a training method to teach a child?

**Critical Thinking**

Possible answers: Parents can reward their children with hugs, smiles, praise, and treats.

## Summary

- Behavior is the way an animal responds to its environment. In general, there are two main types of behavior: innate behavior and learned behavior.

- Innate behavior is behavior that an animal is born knowing. A reflex is an innate behavior. It is a simple inherited behavior that an animal automatically performs in response to a stimulus. An instinct is another type of innate behavior. It is a an inherited pattern of behavior that an animal can control.

- Mating helps animal species to continue. Competition is the struggle among organisms for resources. Migration and hibernation help animals to survive cold weather and the lack of food. Communication helps animals to share information with other animals that can help them to survive.

- Learned behavior is behavior that is developed through experience. Imprinting, classical conditioning, operant conditioning, and insight learning are different ways that animals learn.

---

behavior

hibernation

innate behavior

instinct

learned behavior

pheromones

---

# Vocabulary Review

**Complete each sentence with a term from the list.**

1. Chemicals that animals use to affect the behavior of other animals of the same species are called _____. pheromones

2. Behavior that an animal is born knowing is called _____. innate behavior

3. The way an organism responds to its environment is called _____. behavior

4. Behavior that is changed as a result of experience is called _____. learned behavior

5. A resting state during which an animal's body temperature drops and its heart rate and breathing slow down is called _____. hibernation

6. An inherited pattern of behavior that an animal can control is called an _____. instinct

# Chapter Quiz

Write your answers on a separate sheet of paper.
Use complete sentences.

1. What are the two main types of behavior?
   innate behavior and learned behavior
2. What is a reflex?
   a simple inherited behavior that an animal automatically performs in response to a stimulus
3. Why is mating an important instinct?
   It helps animals species to continue.
4. What are some ways that an animal defends
   its territory? Accept any of the following : it marks its territory with a special substance; it may growl; it may fight to defend its territory.
5. What are three reasons that some animals migrate?
   to survive cold weather, to find food, and to breed or nest
6. What is learned behavior?

7. Name three ways that animals communicate.

8. When does imprinting occur?
   in the early stages of development of an animal
9. What kind of learning occurred in Pavlov's experiment? classical conditioning

10. What kind of learning uses positive reinforcement and negative reinforcement? operant conditioning

6. a behavior that develops as a result of experience

7. Accept any three of the following: they give off scents, they perform dances, they touch each other, they give off light signals, or they change color.

11. Possible answer: Migrating to find food helps an animal to survive.

12. A reflex is a quick response to a stimulus that an animal does not control. An instinct is a behavior that an animal can control.

## CRITICAL THINKING

11. Describe how a certain behavior helps an animal to survive.

12. What is the difference between a reflex and an instinct?

SCiLINKS

Go online to www.scilinks.org. Enter the code **PMB142** to research **animal behavior**.

## Research Project

Walruses live in large herds. They use their tusks to communicate in different ways. Use the Internet and other sources to write a paragraph on how walruses use their tusks to communicate.

See the *Classroom Resource Binder* for a scoring rubric for the Research Project.

**ESL Note** Use photographs from magazines and newspapers to show the parts of the water cycle. Have students cut out pictures that represent the parts of the water cycle and place them in the correct order.

Figure 23-1 *Water covers more than 70% of Earth. This water is continuously cycled between air, land, and the ocean. In what forms is water shown in this photo?*

liquid water in the ocean and the rain, water vapor in the clouds

## Learning Objectives

- Explain the water cycle.
- Explain the carbon cycle.
- Describe how human activity can affect the carbon cycle.
- Explain the nitrogen cycle.
- **LAB ACTIVITY:** Determine whether acid rain occurs in your community.
- **BIOLOGY IN YOUR LIFE:** Explain the ways droughts affect daily life.

# Chapter 23 ▷ Cycles in Nature

## Words to Know

| | |
|---|---|
| **evaporate** | to change from a liquid to a gas |
| **condense** | to change from a gas to a liquid |
| **transpiration** | the process by which plants lose water through the stomata of their leaves |
| **precipitation** | condensed water vapor that falls as rain, hail, sleet, or snow |
| **greenhouse effect** | a natural process by which carbon dioxide and other gases trap heat in the atmosphere |
| **nitrate** | a form of nitrogen used by plants |
| **nitrogen fixation** | process that changes nitrogen gas into a form that living things can use |

## Nature Recycles Itself

Think about the last time you had a glass of water. Now, think about where that water had been before you drank it. It is possible that it could have been part of a swamp where alligators swam. It could have fallen as rain in Italy. It could have been given off by plants in Africa as water vapor.

The point is that all water is recycled. Recycled means it is used over and over again. The same is true for oxygen, carbon, and nitrogen. These important life substances must be shared—used and reused—by all life on Earth. In this chapter, you will learn how water, carbon, and nitrogen are recycled.

**Linking Prior Knowledge**
Students should recall the processes of photosynthesis and cellular respiration (Chapters 4 and 9). They should also recall the action of nitrogen-fixing bacteria (Chapter 6).

# The Water Cycle

**Think About It**

Think about the water cycle. Why do you think it is important to keep the water in lakes, rivers, and seas clean?

Possible answer: Because water is recycled throughout the environment, pollution in one place can affect many other places.

The recycling of water between the air, land, and organisms is called the water cycle. Water does not stay as a liquid because it can evaporate. To **evaporate** means to turn into a gas. Water in its gas form is called water vapor. The water vapor rises into the sky.

When water vapor rises, it cools, or **condenses**. That means the gas turns back into a liquid. The water vapor condenses into tiny droplets that make clouds. In the clouds, the tiny droplets combine into larger drops of water. When the drops are heavy enough, the water falls as rain.

Some of the water falls onto soil. Plants use the water that falls onto soil. Later, the plants will give off some of the water as water vapor. This process is called **transpiration.** Water that is not used by plants seeps down through the soil. The water collects underground. This water is called *groundwater.* Some of the water falls into lakes, rivers, and oceans. Small rivers run into larger rivers. Eventually larger rivers dump their water into the ocean. From there, the cycle starts all over again. Figure 23-2 shows the parts of the water cycle.

## Science Fact

Raindrops are often shown as a teardrop shaped object. This is actually not the real shape of raindrops. Small droplets or rain are round. Larger droplets are oval-shaped and look something like a hamburger bun!

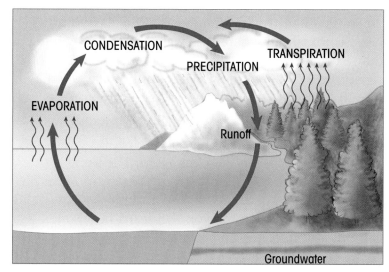

Figure 23-2 *The water cycle*

Water does not always fall to Earth in the form of rain. If the temperature is very cold, water can turn to snow, sleet, or hail as it falls to Earth. These are solid forms of water. These different forms of water are all called **precipitation.**

Figure 23-3 *Snow is a solid form of water. It forms when temperatures are at or below freezing.*

✓ **Check Your Understanding**

On a separate sheet of paper, write the term that best goes in each blank.

**1.** When liquid water turns into a gas, it _____.
evaporates
**2.** When water vapor cools, it _____.
condenses
**3.** Plants release water vapor during _____.
transpiration
**4.** Rain, hail, sleet, and snow are all forms of

_____. precipitation

## A Closer Look

**ACID RAIN**

Human activity can affect the cycles in nature. What we do to one part of the environment can affect many other parts.

Acid rain is formed by air pollution. Sulfur and nitrogenous-oxide waste gas rises from autos and factories. These gases can join with water vapor in the air. This makes the water acidic. Later, the water falls as acid rain. Acid rain pollutes water and kills organisms. It also affects stone buildings and sculptures. Many people fear that acid rain may permanently damage ecosystems unless we find ways to prevent it.

**CRITICAL THINKING** What do you think is one way that acid rain can be reduced?
by reducing the amount of air pollution

Figure 23-4 *Acid rain has damaged this statue.*

# The Carbon Cycle

Carbon is a key element to all life. In the air and water, it is part of carbon dioxide. It is also found in the tissues of living things. Carbon is cycled between the air, water, and organisms. Figure 23-5 shows the parts of the carbon cycle.

In photosynthesis, plants change carbon dioxide and water into sugar molecules. Sugar molecules contain carbon. Oxygen, a byproduct, is released into the air.

Both plants and animals use the food from photosynthesis for cellular respiration. In cellular respiration, oxygen and food are used to release energy. The byproducts are carbon dioxide and water. Carbon dioxide and water are released into the air. Some of this carbon dioxide dissolves in water. Some carbon dioxide is used during photosynthesis. The cycle starts all over again.

**Remember**
Decomposers are organisms that break down and absorb nutrients from dead organisms and wastes.

Decomposers also break down wastes and dead organisms and return carbon to the air. Carbon is also added to the air when fossil fuels, like coal and oil, are burned.

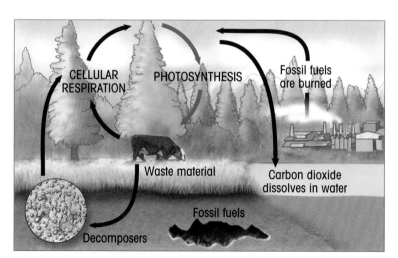

Figure 23-5 *The red arrows in this diagram of the carbon cycle show the movement of carbon molecules. The blue arrows show the movement of oxygen.*

# Human Impact on the Carbon Cycle

People use oil and coal to heat homes. They are also used to produce electricity. These fuels contain carbon. As the fuels are burned, carbon is released into the atmosphere. The exhaust from cars and buses also produces a large amount of carbon. This is because gasoline is a product of oil, so it also contains carbon.

At the same time, more of Earth's forests are being cut down. Like all plants, trees take in carbon dioxide during photosynthesis. As the trees are cut down, less carbon dioxide is absorbed by plants. As a result, more carbon dioxide is left in the atmosphere.

Carbon dioxide builds up in the air surrounding Earth. This blanket of carbon dioxide traps heat on Earth. As light energy from the Sun hits Earth, it is turned into heat energy. Some of the heat energy is trapped by the atmosphere. This is similar to the way the glass of a greenhouse traps heat. Scientists call this the **greenhouse effect.**

Figure 23-6 *Factories release carbon into the atmosphere.*

**Think About It**

The amount of carbon in the atmosphere began rising significantly about 200 years ago. Why do you think this occurred? Hint: Remember that carbon is released by factories and cars.

because the Industrial Revolution occurred about 200 years ago and this increased the amount of fuel being burned

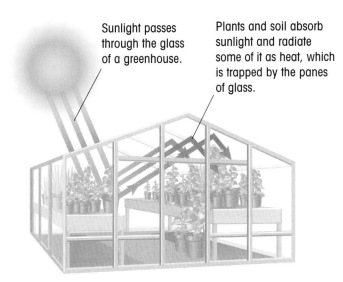

Sunlight passes through the glass of a greenhouse.

Plants and soil absorb sunlight and radiate some of it as heat, which is trapped by the panes of glass.

Figure 23-7 *In a greenhouse, heat energy from the Sun is trapped by the glass. This is similar to the way Earth's atmosphere traps heat.*

## On the Cutting Edge

### MONITORING GLOBAL WARMING

Scientists all over the world want to learn more about climate changes. Is Earth getting warmer or colder? At what rate are temperatures changing? The tools they use to answer these questions range from basic ground thermometers to satellite monitoring systems.

One way to get average global temperatures is by using satellites to monitor objects that are sensitive to heat. Researchers have found that glaciers are good indicators of climate change.

Satellite images are also taken of Earth as a whole. These pictures show the temperatures of different areas. Taken over a number of years, scientists can compare the images. Once enough information is gathered from around the world, researchers can determine whether Earth is getting warmer or colder.

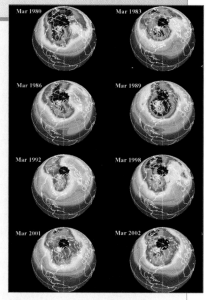

Figure 23-8 *This satellite image shows changes in Earth's temperature over time.*

**CRITICAL THINKING** Why would smaller glaciers be better to use than larger glaciers when monitoring climate changes?

The larger the glacier, the harder it would be to accurately monitor changes in its size.

## The Nitrogen Cycle

All living things need nitrogen. Nitrogen is one of the most important nutrients that helps to form cell proteins. Almost 80 percent of the air is nitrogen. However, plants and animals cannot use nitrogen from the air. So the nitrogen must be changed into a form that plants and animals can use. This process of change is called *nitrogen-fixing*. To do this, special kinds of bacteria called nitrogen-fixing bacteria change nitrogen gas into another form called ammonia. Then, other bacteria change the ammonia into compounds called nitrates. **Nitrogen fixation** allows organisms to get the nitrogen they need.

### Science Fact

Earth's atmosphere is 78% nitrogen and 21% oxygen. The remaining 1% is made up of various gases such as argon and carbon dioxide.

Nitrogen-fixing bacteria live in the soil and on the roots of some plants. Once the ammonia has been changed into nitrogen compounds, they can be used by plants. Plants use nitrates to make proteins. Animals eat the plants to make their own proteins. In this way, nitrogen compounds are transferred from plants to animals.

Another form of nitrogen is returned to the soil in animal waste. Also, when plants and animals die, they decay. Decomposers such as bacteria and fungi change waste and decayed matter back into nitrogen in the air or nitrates in the soil. This completes the nitrogen cycle. Figure 23-9 shows the parts of the nitrogen cycle.

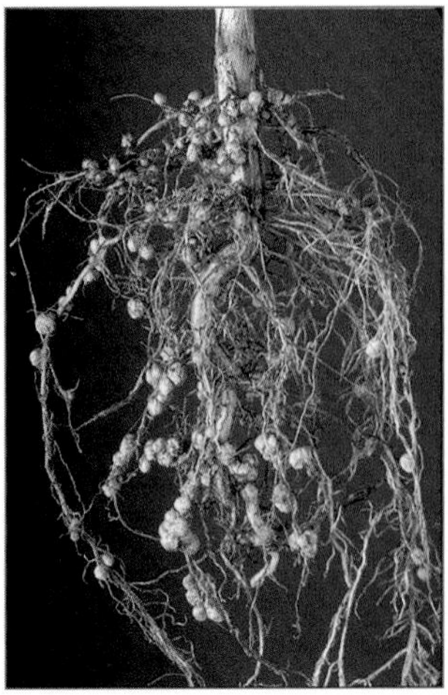

Figure 23-10 *Nodules on the roots of some plants contain nitrogen-fixing bacteria.*

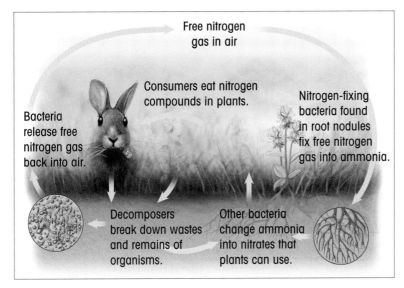

Free nitrogen gas in air

Consumers eat nitrogen compounds in plants.

Bacteria release free nitrogen gas back into air.

Nitrogen-fixing bacteria found in root nodules fix free nitrogen gas into ammonia.

Decomposers break down wastes and remains of organisms.

Other bacteria change ammonia into nitrates that plants can use.

Figure 23-9 *The nitrogen cycle*

✓ Check Your Understanding

Answer the questions in complete sentences.

1. Which cell processes are involved in the carbon cycle?

2. What is the greenhouse effect?

3. **CRITICAL THINKING** Why do you think scientists want to find more information on global warming?

Check Your Understanding

1. photosynthesis and cellular respiration

2. the way heat energy from the Sun is trapped by the blanket of gases on Earth

3. because global warming may cause ice sheets and glaciers to melt, causing flooding and disrupting whole ecosystems

# LAB ACTIVITY
## Testing the Acidity of Rainwater

### BACKGROUND
Whether or not something is acidic depends on its pH level. The pH of a substance is measured using a scale of 1 to 14. Substances with a pH of 1 to 6.9 are acidic. Substances with a pH greater than 7 are called basic.

### PURPOSE
In this activity, you will determine the acidity level of the rain in your community.

### MATERIALS
safety goggles, plastic bottle (1- or 2-liter), scissors, 2 test tubes, dropper, tap water, pH test kit

Figure 23-11 *You can use a plastic bottle to collect rainwater.*

### WHAT TO DO
1. Remove the cap of a plastic bottle. Cut off the top of the bottle. **Caution: Be very careful using scissors.** Turn the top upside down and place it on the soda bottle, as shown in Figure 23-11.

2. Place the soda bottle outside and allow it to collect at least 1 inch of rainwater.

3. Use a dropper to take a few drops of the rainwater and place it in a clean test tube. Use the test kit to determine the pH of the sample. Write the pH of your sample on a separate sheet of paper.

4. Fill the second test tube with a few drops of tap water. Use the test kit to determine the pH of the tap water. Write your observations on a separate sheet of paper.

### DRAW CONCLUSIONS
- Does your community have acid rain?
  Answers will vary. Any sample with a pH of less than 6 should be classified as acid rain.
- How did the acidity of the rainwater you collected compare to that of tap water?
  The rainwater sample should have had a lower pH than the tap water sample.

# BIOLOGY IN YOUR LIFE
## Dealing With Drought

Droughts can be a serious problem. With too little rain, the rivers, lakes, and wells that supply our water can become very low. If we do not limit our use of water, a drought can lead to a costly disaster. A 3-year drought in the late 1980s cost 39 billion dollars!

Farmers are usually hardest hit. Crops dry up and die. Cattle cannot graze on dried up grass. Even if you do not live in a rural area, you can still feel the effects. Most likely, your family will pay more for food.

Droughts can have other effects. The risk of forest fires is very high. At times, the fires have come so close to towns that people have had to leave their homes. Governments often restrict water use during a drought. They place limits on when lawns can be watered, cars washed, and pools filled.

There are a few ways you can conserve water. You can take shorter showers and turn off the water while you brush your teeth. You can also make sure that your faucets do not leak. When it comes to a drought, every drop of water counts.

Figure 23-12 *This reservoir is completely drained of water because of a drought.*

**Critical Thinking** Possible answer: It is important to restrict water for these uses during a drought because people need water for drinking, cooking, and bathing. Most people feel these other uses are more important than washing cars or watering lawns.

**Critical Thinking**

Many people complain when water use for lawn care or for washing cars is restricted. Why do you think it is important to restrict water use during a drought?

**Answer the questions in complete sentences.**

**1.** How do droughts affect farmers?
Droughts kill their crops, costing farmers a great deal of money.

**2.** What are some other effects of droughts?
Droughts increase the risk of forest fires and cause water shortages.

**3.** What are some ways you can deal with droughts? You can conserve water by taking shorter showers and fixing faucets that leak.

## Chapter Summary

- All water on Earth is recycled within the water cycle. Water evaporates and goes into the air. There it condenses into tiny droplets of water and forms clouds. Clouds release water as precipitation. The water falls to Earth. It is used by plants, and water vapor is released during transpiration. Some of the water runs into rivers and lakes. Rivers carry the water to the sea.

- Oxygen and carbon are recycled by plants and animals. Plants give off oxygen as a byproduct of photosynthesis. Most organisms use the oxygen in cellular respiration. They give off carbon dioxide as a byproduct. Plants use this carbon dioxide in photosynthesis. This exchange is at the center of the carbon cycle.

- All living things need nitrogen. Plants and animals cannot acquire nitrogen through respiration. Bacteria change this nitrogen gas into nitrates in the soil. This process is called nitrogen-fixation. Plants use the nitrates in the soil. Animals get a form of nitrogen from eating the plants. When plants and animals die, other bacteria break down the decayed matter and put nitrates back into the soil. Still other bacteria change nitrogenous compounds back into nitrogen gas. This nitrogen gas goes back into the air. This is called the nitrogen cycle.

---

condense

evaporate

nitrate

nitrogen fixation

precipitation

transpiration

---

## Vocabulary Review

**Match each term to its definition.**

1. process by which plants lose water through the stomata of their leaves  transpiration

2. to change from a liquid to a gas  evaporate

3. condensed water vapor that falls as rain, hail, sleet, or snow  precipitation

4. to change from a gas to a liquid  condense

5. a form of nitrogen used by plants  nitrate

6. the process that changes nitrogen gas into a form living things can use  nitrogen fixation

# Chapter Quiz

**Write your answers on a separate sheet of paper.**
**Use complete sentences.**

1.  What are the three main steps in the water cycle?
    evaporation, condensation, and precipitation
2.  How does water get into the air?
    It evaporates into a gas.
3.  What are four forms of precipitation?
    rain, sleet, hail, and snow
4.  How does oxygen get into the air?  Plants and other
    organisms that carry out photosynthesis release it into the air.
5.  What byproduct of respiration do plants use in
    photosynthesis?  carbon dioxide

6.  What form must nitrogen take before plants and
    animals can use it?  It has to be made into nitrates.

7.  How do bacteria help plants to get nitrates?

8.  How do animals get a form of nitrogen?

**CRITICAL THINKING**

9.  What could happen if the greenhouse effect
    continues to increase?

10. Scientists are trying to genetically engineer plants
    and soil bacteria to fix nitrogen at a faster rate.
    How do you think this may affect farming?

7. They turn nitrogen from the atmosphere into nitrates that plants can use.

8. They eat plants, or they eat other animals that have eaten plants.

**Critical Thinking**

9. The average temperature of Earth could rise at a faster rate, causing climate change and effects on ecosystems.

10. It may increase the growth of crops and decrease the need for artificial nitrogen fertilizers.

*SC*$i$*INKS*

Go online to www.scilinks.org.
Enter the code PMB144 to
research **pollution**.

## Research Project

You have studied three cycles: the water cycle,
the carbon cycle, and the nitrogen cycle. All these
substances must be used over and over again by
living things on Earth. Use the Internet to research the effects of air and
water pollution. What does pollution have to do with the three cycles
you have studied? What do you think should be done about pollution?
Summarize your findings in a written report or oral presentation.

See the *Classroom Resource Binder* for a scoring rubric for the Research Project.

Figure 24-1 *Wind farms, like the one in the photograph, use the energy in wind to produce electricity. Energy from the wind does not pollute air, water, or soil. Where do you think wind farms can be found?*

in windy areas

## Learning Objectives

- List examples of natural resources.

- Compare renewable resources and nonrenewable resources.

- Explain the importance of conservation.

- Describe the different types of pollution.

- Explain why some animals are extinct, endangered, or threatened.

- List alternative energy resources.

- **LAB ACTIVITY:** Model one way in which an oil spill is cleaned up.

- **ON-THE-JOB BIOLOGY:** Compare a hybrid electric car to a gasoline-powered car.

## Words to Know

| | |
|---|---|
| **natural resource** | a material that comes from nature and is used by living things |
| **renewable resource** | a natural resource that can be replaced at about the same rate as it is being used |
| **weathering** | the process of wearing away rock |
| **erosion** | the removal of materials such as rock and soil by wearing them away |
| **nonrenewable resource** | a natural resource that is being used faster than it can be replaced |
| **conservation** | the wise use of natural resources to prevent resources from being used up; to maintain the balance of ecosystems |
| **recycle** | to reprocess a waste material for use in a new product |
| **pollution** | the release of materials into the environment, usually causing harm |
| **biodiversity** | the total variety of living things in an environment; the healthy ecosystem that different groups of species create |
| **extinct species** | a once-living species that no longer exists |
| **endangered species** | organisms in danger of becoming extinct |
| **threatened species** | organisms that are likely to become endangered in the near future |

## Supplies of Materials

In this chapter, you will learn about the supplies of materials on Earth that people need to live. You will find out that some materials can be replaced. Other materials cannot be replaced. You will also read about the need to use these materials wisely.

**Linking Prior Knowledge**

Students should recall facts about plants and animals that they have learned (Chapters 6 through 13), how organisms live together in ecosystems (Chapter 21), how animals behave (Chapter 22), and the different cycles in nature (Chapter 23).

# Natural Resources

The materials that come from nature and are used by living things are called **natural resources.** Without many of these materials, life could not exist. Nonliving natural resources include air, water, minerals, fossil fuels, and soil. Living natural resources include plants and animals.

# Renewable Resources

Many materials can be used again and again. They can be replaced at about the same rate as they are being used. These materials are called **renewable resources.** Air, water, soil, and some living things are examples of renewable resources. Currently, these materials are renewable. However, overuse of them can cause these materials to become limited.

### The Air We Breathe

The air that we breathe is a mixture of gases. These gases include nitrogen, oxygen, and carbon dioxide. The nitrogen cycle, photosynthesis, and the carbon cycle help renew the gases that make up air.

**The Air We Breathe**

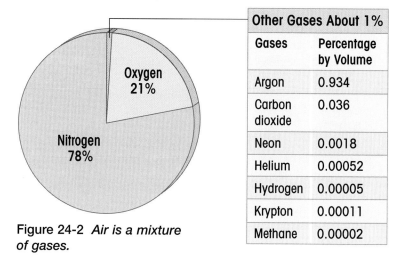

| Other Gases About 1% | |
| --- | --- |
| Gases | Percentage by Volume |
| Argon | 0.934 |
| Carbon dioxide | 0.036 |
| Neon | 0.0018 |
| Helium | 0.00052 |
| Hydrogen | 0.00005 |
| Krypton | 0.00011 |
| Methane | 0.00002 |

Figure 24-2 *Air is a mixture of gases.*

## Water

Water is one of the most important resources for all living things. People need clean water to drink, to cook food, and to bathe. Farmers need water to grow crops. Industries and businesses also need water. The water cycle constantly replaces water on Earth.

### Water on Earth

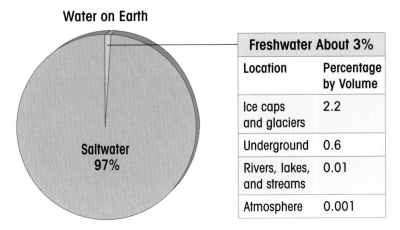

| Freshwater About 3% | |
| --- | --- |
| Location | Percentage by Volume |
| Ice caps and glaciers | 2.2 |
| Underground | 0.6 |
| Rivers, lakes, and streams | 0.01 |
| Atmosphere | 0.001 |

Figure 24-3 *Freshwater is the main source of drinking water for many people.*

## Soil

Soil is a renewable resource. Soil is needed to grow food. The breaking down of rocks helps form soil.

People used to believe that the statues they carved from rock would last forever. Today, we know that all rock wears away over time. The process of wearing away rock is called **weathering.** Weathering causes rock to break down into smaller and smaller pieces, forming soil. This process can take a very long time.

Weathering can also cause chemical changes in a rock. These chemical changes can break down rocks to form soil.

The process by which weathered material is removed from one place and carried to another place is called **erosion.** The erosion of rock can form soil. It also adds nutrients to the soil.

Figure 24-4 *These rock formations have been weathered over time.*

## Everyday Science

Trees are a very important natural resource. Take a look at the objects around you. Make a list of the products that come from trees. Share your list with the class.

Possible answers: paper, furniture, maple syrup, fruits, rubber

**Check Your Understanding**

**1.** a material that comes from nature and is used by living things

**2.** Accept three of the following: water, air, soil, plants, and animals

**Figure 24-5** *Iron is a mineral that people use to make products. A certain form of iron can also be taken to maintain healthy body functions.*

### Plants, Animals, and Other Organisms

Plants are very important renewable resources. People eat plants. Plants put oxygen into the air and in water. Plants also provide people with medicines, building materials, and clothing.

Animals are renewable resources. Animals are a source of food. Some animals can provide transportation for people. Some animals are trained to help people with disabilities. Animals are also pets. Reproduction renews the populations of animals.

Some types of fungi, bacteria, and protists are also important renewable resources. Many medicines are made from fungi and bacteria. Certain protists that live in oceans put oxygen into the water.

✓ **Check Your Understanding**

Write your answers on a separate sheet of paper. Use complete sentences.

**1.** What is a natural resource?

**2.** Name three renewable resources.

## Nonrenewable Resources

Some natural resources are being used faster than they can be replaced. These resources are called **nonrenewable resources.** These resources include minerals, and fossil fuels, such as coal, oil, and natural gas.

### Minerals

Minerals help make soil fertile. Plants need minerals to grow healthy. Animals need them to keep healthy body functions. Minerals are also used to produce many different products important to people. Products that are made with minerals include cookware, automobile parts, computer parts, and jewelry.

### Fossil Fuels

Coal, oil, and natural gas are called fossil fuels. Fossil fuels formed from the buried remains of organisms, mostly plant matter, that lived long ago. The remains of these organisms became buried under layers of sand, rock, and mud. Over millions of years, heat and pressure built up and changed the remains into fossil fuels. People depend on fossil fuels as sources of energy. All three fossil fuels are found underground in Earth's crust.

Oil is used to heat homes. Oil is found in the Earth's crust. It can even be found in the ocean floor. Oil rigs are used to drill deep into the Earth's crust to get oil.

Natural gas is also used to heat homes. It is less dense than oil. It rises above oil deposits. Pipelines are used to transport natural gas from their sources to the places natural gas is used.

**Figure 24-6** *Oil rigs are used to drill deep into the ocean floor to get oil.*

Coal is burned in power plants to produce electricity. Coal has to be removed from Earth's crust, or mined. Today, machines, robots, and drills make it easier and safer to mine for coal.

# Conservation

The number of people living on Earth is increasing. By the year 2050, there may be as many as 10 billion people living on Earth. However, the supplies of natural resources are not increasing. As the number of people living on Earth grows, the supplies of materials that people use decrease.

People must learn how to make natural resources last longer. We must also make sure that the use of natural resources does not harm ecosystems. The wise use of natural resources to prevent resources from being used up is called **conservation.** Conservation also maintains the balance of ecosystems. Conservation of all natural resources, renewable and nonrenewable, is important.

## Composting

*Composting* is one way to conserve natural resources. Composting recycles food waste. It can turn food waste into fertilizer for soil. It can also reduce the amount of solid waste that has to be put into landfills. In a compost pile, bacteria and other organisms break down plant material. This process makes the soil more fertile for growing plants.

## Reusing

Many items that you might throw away can be reused. These items can be used for other purposes. Reusing an item helps conserve resources. Old clothes can be used as rags or given away. Empty bottles and containers can be used to hold other items. Old newspapers can be used as packing material.

Products that can be used over and over again also help conserve resources. For example, if you use cloth napkins instead of paper napkins, you save paper. Cloth napkins can be washed and used again. You can also use lunch containers that can be washed and used again.

**Think About It**

Conservation practices include taking showers instead of baths. Make a list of other conservation practices. Share your list with the rest of the class.

Possible answer: Not running water when brushing teeth and keeping the thermostat at a low temperature

### Recycling

Products made from paper, glass, aluminum, or iron can often be recycled. To **recycle** means to reprocess a waste material for use in a new product. Glass, aluminum, or iron waste products can be melted and used to make new products. Recycling waste paper into new paper products saves trees from being cut down. Recycling also saves energy. Less energy is needed to recycle materials than to find new resources. Many cities have recycling programs.

Figure 24-7 *This symbol is printed on products that can be recycled.*

### Total Household Waste Materials

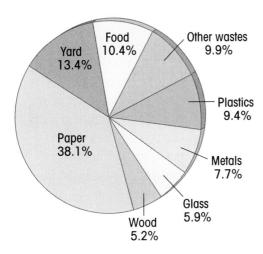

Figure 24-8 *Many types of household wastes can be recycled.*

✔ **Check Your Understanding**

Write your answers in complete sentences.

1. What are some nonrenewable resources?
2. What happens to the supplies of materials on Earth as the human population increases?
3. What are two ways to conserve natural resources?
4. **CRITICAL THINKING** Why do you think coal, oil, and natural gas are called fossil fuels?

**Check Your Understanding**

1. minerals, coal, oil, and natural gas.
2. The supplies decrease.
3. Possible answers: recycling and composting
4. Fossils are the remains of living things that lived long ago. Coal, oil, and natural gas come from living things that lived long ago.

# Pollution

To protect the supplies of renewable resources, people must not pollute the air, water, or soil. **Pollution** is the release of materials into the environment. These materials are harmful to the environment.

Pollution can be caused by natural events. Erupting volcanoes can put harmful particles into the air, water, and soil. Water from floods carry materials that pollute soil and bodies of water. Forest fires caused by lightning can pollute the air, water, and soil with soot and ash. However, most pollution is caused by human activities.

Figure 24-9 *Smog often covers Mexico City.*

### Air Pollution

The burning of fossil fuels can pollute the air. When people burn coal, oil, or natural gas, harmful particles and gases are put into the air. Smog occurs when harmful particles are put in the air. Smog is a mixture of smoke and fog. It pollutes the air and makes it very difficult for people to breathe. Carbon monoxide is a deadly gas. It is produced by cars, trucks, factories, home furnaces, and burning cigarettes.

### Acid Rain

Pollutants in the air can also combine chemically with water. Sulfur dioxide is a pollutant that is put in the air by power plants. When it combines chemically with rain, acid rain forms. *Acid rain* is harmful to nonliving things and living things. It speeds the process of weathering. It damages buildings, statues, and other structures. It can also harm trees, grasses, and other organisms. When it falls into lakes and streams, it can kill fish and other organisms living in the water.

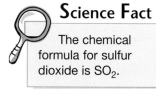

### Science Fact

The chemical formula for sulfur dioxide is $SO_2$.

### Water Pollution

Wastes from homes, factories, boats, cities, and towns are sources of water pollution. Water pollution reduces the supply of clean water. It also harms the organisms living in the water.

Many chemicals pollute water sources. Some of these chemicals include pesticides and fertilizers used by farmers and gardeners. These chemicals are washed away into rivers and lakes. They also leak into groundwater. Groundwater is a major source of drinking water for many people.

**Think About It**

Most cities have water treatment plants. Water treatment plants remove harmful substances from water. What kinds of substances do you think are removed by water treatment plants?

Possible answers: sewage, dirt, sediment, microscopic organisms

| Freshwater Pollutants | | |
|---|---|---|
| **Kinds of Pollutants** | **Examples** | **Sources** |
| Disease-causing organisms | Protozoans, bacteria | Human wastes, runoff from livestock pens |
| Pesticides and fertilizers | DDT, nitrates, phosphates | Runoff from farm fields, golf courses |
| Industrial chemicals | PCBs, carbon tetrachloride, dioxin | Factories, industrial waste, disposal sites |
| Metals | Lead, mercury, copper | Factories, waste disposal sites |
| Radioactive wastes | Uranium, carbon-14 | Medical and scientific disposal sites, nuclear power plants |
| Petroleum products | Oil, gasoline | Road runoff, leaking underground storage tanks |

Figure 24-10 *These sources are just some of the different types of freshwater pollutants.*

## Soil Conservation

Soil can take a very long time to form. Soil can also be eroded, or worn away, and carried to new places. Conservation practices can be used to prevent soil from eroding.

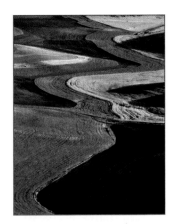

Figure 24-11 *Contour plowing helps prevent soil from eroding down hills.*

In contour plowing, crops are planted across the slope of a hill. This planting method helps prevent soil from being carried away by water running down the hill. Planting rows of trees as windbreaks also prevents soil from being blown away. The trees block the wind.

Write your answers on a separate sheet of paper. Use complete sentences.

1. What is pollution? the release of harmful materials into the environment

2. What happens when coal, oil, or natural gas is burned? Harmful particles and gases are put into the air.

3. What are two ways to conserve soil? using contour plowing and planting windbreaks

4. **CRITICAL THINKING** Why does burning fossil fuels cause pollution? When people burn coal, oil, or natural gas, harmful particles and gases are put into the air.

## Great Moments in Biology

### RACHEL CARSON'S *SILENT SPRING*

Rachel Carson was a biologist and a writer. One of her books, *Silent Spring,* was published in 1962. This book played a large part in creating environmental awareness in the United States.

In *Silent Spring,* Rachel Carson made the public aware of the dangers of insect sprays. The book describes how these chemicals, called pesticides, can harm other living things. Plants and animals could be affected by the chemicals. People who eat those plants and animals could become affected, too.

Figure 24-12 *Rachel Carson made the public aware of the dangers of using pesticides.*

Rachel Carson's book caused the U.S. government and the public to examine the use of pesticides. Some of the chemicals outlined in the book have been banned from use in the United States. The response also resulted in the modern environmental movement. In 1970, a special committee was formed to deal with environmental concerns. This committee is called the Environmental Protection Agency (EPA).

**CRITICAL THINKING** Why did *Silent Spring* have such a strong influence on the public? It increased the awareness of the dangers of using pesticides.

## Maintaining Biodiversity

Plants, animals, and other organisms make up the biodiversity in a particular area. **Biodiversity** is the total variety of living things in an environment. It also is the healthy ecosystem that the different groups of species create. Usually, the more variety there is in an ecosystem, the stronger, the healthier, and the more balanced the ecosystem. Healthier ecosystems provide more resources for people. Some of these resources include medicines, such as painkillers, anticancer drugs, and antibiotics.

Figure 24-13 *Healthy ecosystems provide many resources, including medicines, for people. The strychnos vine is used to make a medicine that treats rabies.*

### The Disappearance of Some Species

Wildlife is another resource that must be conserved. Many species of plants and animals that once lived on Earth have disappeared. These species have become **extinct species.** Many other plants and animals are members of endangered species. **Endangered species** are organisms in danger of becoming extinct. There are very few members of these species left. Some species are not close to extinction, but they may become extinct in the near future. These species are called **threatened species.**

Figure 24-14 *The black rhinoceros is an endangered species.*

### The Loss of Habitat

The main threat to wildlife is the loss of habitat as more land is used for human activities. *Deforestation* is the cutting and clearing of forests. It is often done to create farmland or to build areas for human use. However, it destroys the homes and the hunting and mating grounds of many organisms. Hunting, trapping and commercial fishing can also endanger or threaten animal species.

### Protecting Wildlife

To protect plants and animals, many governments have passed laws that forbid the hunting of endangered species. There are also many laws that regulate hunting and fishing. Land has also been set aside to provide homes for endangered animals. Wildlife refuges are large areas of land where many different species of animals and plants can be protected from hunting, human activity, and loss of habitat. Biologists and zoos have also set up special breeding programs. These programs help endangered species to increase in number.

**Figure 24-15** *A breeding program helped baby chicks of California condors to hatch safely. This program has helped the species from becoming extinct.*

## Alternative Energy Resources

Another way to conserve natural resources is to use alternative energy resources.

- *Solar energy* is energy from the Sun. Special materials can absorb energy from the Sun. This energy can be converted into heat energy. The heat energy can be used to produce electricity and heat water.

- *Wind energy* is energy from wind. Windmills and wind farms use the energy in wind to generate electricity.

- *Geothermal energy* is energy stored as heat within Earth's crust. If the heat is concentrated near Earth's surface, it can be used to heat homes and to supply energy for electric generators.

- *Nuclear energy* is energy obtained from the nuclei of certain atoms. There is a great amount of energy stored inside the nuclei of atoms. When this energy is released, it can be used to produce electricity.

- *Hydroelectric energy* is energy obtained from falling water. It can be used to produce electricity. Dams are built on rivers. These dams convert the energy of falling water into electricity.

**Science Fact**

There are two types of nuclear energy: nuclear fission and nuclear fusion. Fission is the splitting of atoms. Current nuclear power plants use fission to release energy. Fusion is the joining of atoms. It takes place in the Sun.

Figure 24-16 *Dams produce hydroelectric energy.*

These alternative energy resources are already used in some areas. They are renewable and nonpolluting. However, not all of them are practical for everyone to use. For example, wind farms need a steady source of wind. Hydroelectric dams are expensive and can harm wildlife. The process used to obtain nuclear energy can be dangerous.

✓ **Check Your Understanding**

Write your answers in complete sentences.

1. What is biodiversity?

2. Name three alternative energy resources.

3. **CRITICAL THINKING** What are some ways that pollution can be stopped?

**Check Your Understanding**

**1.** the total variety of living things in an environment; the healthy ecosystem that different groups of species create

**2.** Accept any three of the following: solar energy, hydroelectric energy, energy in wind, nuclear energy, and geothermal energy.

**3.** Possible answer: Laws can be passed to fine people and businesses caught polluting water and air.

# LAB ACTIVITY
## Cleaning Up an Oil Spill

### BACKGROUND

An oil spill usually occurs when an oil tanker or pipeline is damaged. The oil pours out of the damaged ship and into the sea. Oil and water do not mix. The oil usually floats on the surface of the water. Eventually, the oil may float to shore. The thick oil can kill ocean life, including birds, mammals, fish, and mollusks.

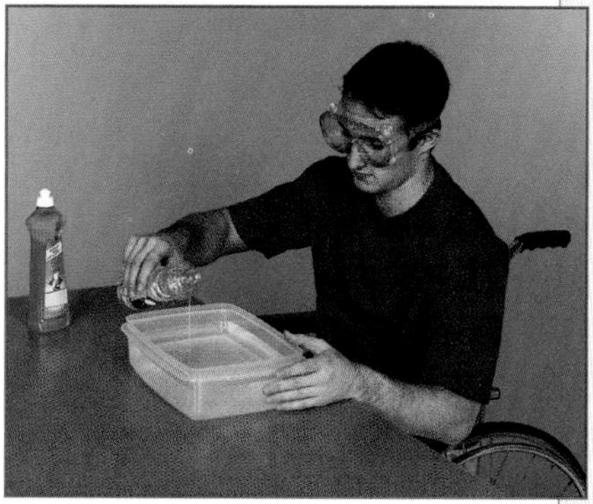

### PURPOSE

In this activity, you will model an oil spill and a way to clean it up.

### MATERIALS

safety goggles, liquid dishwashing detergent, plastic container, cooking oil, water

**Figure 24-17** *Predict what will happen if you pour oil into water.*

### WHAT TO DO

1. Put on safety goggles. Do not remove them until you have finished the activity.

2. Fill a plastic container with water. Predict what will happen if you pour oil into the water. Write your prediction on a separate sheet of paper.

3. Pour a small amount of cooking oil into the water. Observe what happens to the water.

4. Add some liquid detergent to the middle of the "oil spill." Observe what happens to the oil spill.

### DRAW CONCLUSIONS

- Was your prediction correct? Did the oil mix with the water in the container? Describe how it looked. The oil did not mix with the water. It floated on top of the water.

- A certain type of bacteria acts the way the detergent did when you added it to the oil spill. Describe what happened when you added the detergent. The detergent broke up the oil.

- How are oil spills harmful to the environment? They can kill ocean life.

# BIOLOGY IN YOUR LIFE
## Driving Hybrid Electric Cars

Every gallon of gas that a car burns puts about 20 pounds (9 kilograms) of carbon dioxide into the air. Cars that burn less gas create less pollution. Laws now require car makers to build cars that are less polluting.

Electric cars add the least amount of pollution to the air. However, they are only good for short trips. A new type of car combines gas and electric power. It is called a hybrid electric vehicle (HEV). HEVs use far less gas than standard cars do, and they can travel longer distances.

Figure 24-18 *An electric car is less polluting than other types of cars.*

An HEV can travel 20 or 30 more miles on one gallon of gas than a standard car can travel. Over time, this better mileage can lead to cleaner air.

**Look at the chart in Figure 24-19. It shows how far three different kinds of cars can travel on 1 gallon of gas. Use the chart to answer the questions that follow.**

| Comparing Different Types of Cars | |
|---|---|
| **Car** | **Miles per Gallon** |
| Hybrid electric car | 50 |
| Standard compact car | 25 |
| Midsize car | 20 |

Figure 24-19 *Different types of cars get different gas mileage.*

**1.** How many miles can an HEV travel on 1 gallon of gas?   50

**2.** Which car burns the least fuel?   HEV

**3.** Which car burns the most fuel?   midsize

**Critical Thinking**
How much gas would each car use on a 100-mile trip?

HEV, 2 gallons; standard compact, 4 gallons; midsize, 5 gallons

## Summary

- The materials that come from nature and are used by living things are called natural resources. Air, water, and soil are renewable resources. Minerals, coal, oil, and natural gas are nonrenewable resources.

- Pollution is the release of materials into the environment, usually causing harm. The burning of fossil fuels and the dumping of wastes cause pollution.

- Biodiversity is the total variety of living things in an environment. It also is the healthy ecosystem that the different groups of species create. Conservation is the wise use of natural resources to prevent resources from being used up. It maintains the balance of ecosystems.

- The loss of habitat, commercial fishing, hunting, and other human activities have caused many species to become extinct, endangered, or threatened. To protect wildlife, laws have been passed that regulate hunting and fishing. Wildlife refuges also protect plants and animals.

- Alternative energy resources include solar energy, wind energy, hydroelectric energy, nuclear energy, and geothermal energy. They are renewable and nonpolluting. However, not all of them are practical for everyone to use.

---

biodiversity

natural resource

nonrenewable resource

pollution

recycle

---

1. natural resource
2. biodiversity
3. recycle
4. pollution
5. nonrenewable resource

## Vocabulary Review

**Complete each sentence with a term from the list.**

1. A material that comes from nature and is used by living things is called a _____.

2. The total variety of living things in an environment is called _____.

3. To _____ is to reprocess a waste material for use in a new product.

4. The release of materials into the environment, usually causing harm, is called _____.

5. A natural resource that is being used faster than it can be replaced is called a _____.

# Chapter Quiz

**Write your answers in complete sentences.**

1. Name six natural resources.
   Accept any six of the following: air, water, soil, coal, oil, natural gas, plants, and animals.
2. How does weathering form soil?
   Weathering causes rock to break down into smaller and smaller pieces forming soil.
3. Why is an ecosystem with great variety
   important? Possible answer: because it can provide people with many resources

4. Where do fossil fuels come from? from the buried remains of organisms

5. How is coal used as a source of energy?
   It is used to produce electricity in factories or to heat buildings.
6. Why is conservation important?
   Possible answer: because people must learn to make resources last longer
7. Name two types of pollution. What is the main
   cause of pollution? Possible answer: deforestation,
   air pollution and water pollution, human activity hunting, commercial fishing,
8. What kinds of human activity threaten wildlife? and trapping

9. Name three alternative energy resources. Accept any three of the following:
   solar energy, wind energy, hydroelectric energy, nuclear energy, and geothermal energy.
10. What is a disadvantage of wind energy?
    It needs a constant source of wind.

**CRITICAL THINKING**

11. How do natural resources affect life?
    Without natural resources life could not exist.
12. What is the difference between renewable
    resources and nonrenewable resources?

12. Renewable resources can be replaced at about the same rate as they are being used. Nonrenewable resources are being used faster than they can be replaced.

**Research Project:** Students' reports may include an indirect system that uses a heat exchanger to heat a fluid that circulates in tubes through a water storage tank.

---

## Research Project

Solar hot-water heaters are sometimes used to supply homes with hot water and heat. Use the Internet and other resources to find out more about these heating systems. Write a short paragraph that describes how one of these systems works.

---

$SC\overset{i}{L}INKS$

Go online to www.scilinks.org.
Enter the code **PMB146** to
research **solar heated homes**.

See the *Classroom Resource Binder* for a scoring rubric for the Research Project.

**Figure 25-1** *The imprint of the dragonfly above is a fossil. It was formed millions of years ago. Scientists learn much about organisms that lived long ago by studying fossils. What kind of information do you think can be found by studying a fossil?*

Possible answers: How large the organism was and what body structures it had

## Learning Objectives

• Define *evolution*.

• Explain Darwin's theory of natural selection.

• Describe the relationship between genetics and natural selection.

• Identify evidence of evolution.

• **LAB ACTIVITY:** Model natural selection.

• **ON-THE-JOB BIOLOGY:** Find out about the work of a paleontologist.

# Chapter 25

# Evolution

## Words to Know

| | |
|---|---|
| **evolution** | the process by which a species gradually changes over time |
| **fossil** | the remains of an organism that lived a long time ago |
| **natural selection** | the survival of offspring that have favorable traits |
| **variation** | a difference in a trait among individuals in a species |
| **adaptation** | a trait that helps an organism to survive in its environment |
| **gene pool** | all the genes in a population |
| **genetic drift** | a random change in a gene pool caused by chance events |
| **relative dating** | a way to estimate the age of a fossil by comparing its position in a rock layer |
| **radioactive dating** | a way to determine the age of a fossil by measuring the amount of radioactive elements in it |
| **geologic time scale** | the record of Earth's history based on the types of organisms that lived at different times |

## Changes Over Time

Most species change over time. This slow, gradual change is called **evolution**. In the case of evolution, "slow" usually means thousands to millions of years. The first living things were single-celled organisms with no nucleus. Evidence shows that those organisms lived in the oceans about 3 billion years ago. Many theories state that all of the organisms that have lived on Earth since then evolved from those single-celled organisms. The process of evolution goes on even today.

### Linking Prior Knowledge

Remind students that offspring do not always look like their parents and that genetics determines which traits are passed to offspring (Chapter 5).

Many species have become extinct. Extinct species were once living but no longer exist. New species have taken their place. Scientists believe new species develop from older species as a result of evolution.

## An Early Theory of Evolution

Until the early 1800s, most scientists believed that organisms were exactly the same as they were when they first appeared on Earth. Around the same time, scientists were discovering and studying new **fossils.** A fossil is the remains of an organism that lived a long time ago. The fossils showed that living things had changed over time. Some scientists began to think about evolution.

Jean Baptiste Lamarck proposed one of the first theories of evolution. He said organisms develop traits by using or not using a part of their body. He used the giraffe as an example. According to Lamarck, the earliest giraffes ate grass. When the grass died out, giraffes needed to find new food sources. They stretched their necks to reach the leaves on trees. As they stretched, their necks grew longer. Traits that develop during an organism's lifetime are called acquired traits.

Lamarck thought that acquired traits could be passed to offspring. He said that the giraffes with stretched necks would have offspring with long necks. Lamarck did not have much evidence for his theory. He was eventually proven wrong.

Figure 25-2 *Lamarck believed that the long necks of giraffes are passed to their offspring through acquired traits.*

**Think About It**

Use what you know about DNA to explain why Lamarck's theory of evolution is wrong.

**Think About It**

Acquired traits are not genetic. Only genetic traits can be inherited.

## Darwin and Natural Selection

In 1831, an Englishman named Charles Darwin took a job on a ship called the HMS *Beagle*. The *Beagle's* voyage was to the South Pacific and South America. The purpose of the voyage was to map the coast of South America.

## The Voyage of the *Beagle*

One of the places that the *Beagle* visited was the Galápagos Islands. The Galápagos Islands are located off the west coast of South America. Darwin noticed that although the islands were close together, they had very different climates. He also noticed that the different islands had different species of animals. For example, Darwin observed that each of the Galápagos Islands had a different species of tortoise.

Darwin also discovered 13 different species of finches. Most of the finches' traits were the same. However, each species of finch had a differently shaped beak. Each beak was adapted for eating a different type of food. Because of the similarities between the finch species, Darwin inferred that the finches shared a common ancestor.

**Figure 25-3**
*Charles Darwin studied different species on the Galápagos Islands.*

Figure 25-4 *Darwin drew pictures of the different species of finches that he saw.*

## Natural Selection

Darwin spent the next 20 years thinking about what he had seen on the Galápagos Islands. He used the information that he had gathered about the tortoises and the finches as the basis for his theory of evolution.

Darwin used the term **natural selection** to describe his theory of evolution. Natural selection is the way organisms that are best suited to an environment survive. These organisms can pass their helpful traits to their offspring.

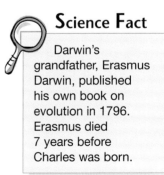

**Science Fact**

Darwin's grandfather, Erasmus Darwin, published his own book on evolution in 1796. Erasmus died 7 years before Charles was born.

**Remember**
Competition is the struggle among organisms for resources.

There are four main points to Darwin's theory of natural selection: Overproduction, competition, variation, and survival of the fittest.

1) Overproduction occurs when organisms give birth to many offspring. Not all offspring can survive because there are not enough resources for all of them.

Figure 25-5 *Many species overproduce. Certain scorpion species can have up to 30 offspring, which they carry on their backs.*

2) There is competition between all offspring for available resources. Only those offspring that can get the resources will survive to reproduce. The rest will die or not reproduce.

3) There are **variations**, or differences, among the members of a species. Some members have **adaptations** that make them better suited to their environment. For example, some individuals in a species may be faster or stronger than others.

4) The term *survival of the fittest* describes how some variations make organisms better able to survive in their environment. These organisms have a better chance of surviving and reproducing. The organisms that do not have these variations may not survive.

The organisms that survive will pass their traits to their offspring. Organisms with harmful traits are less likely to survive and reproduce. So, harmful traits are likely to disappear over generations. Eventually, these changes can result in the appearance of a new species.

## Genetic Variations

Darwin knew that traits are passed on, but he did not know how it occurred. Today, we know about DNA. With each new generation of a species, there is a new combination of genes. A **gene pool** is the combined genetic information of all the members of a population. By studying the gene pool of one population, scientists can see how variations in traits are passed to offspring.

Evolution can only occur when there is a change in the frequency of genes for certain traits within a gene pool. Several events can cause these changes.

1) Natural selection can affect a gene pool. Organisms with helpful traits will survive and reproduce. The genes for their helpful traits will remain in the gene pool.

2) A mutation to a gene can affect a gene pool. If the mutation is harmful, the organisms might die before passing on the gene. If the mutation is helpful, however, the organisms live and the gene may be passed to future generations. In this way, a new gene becomes part of the gene pool.

3) Chance events can also affect a gene pool. This random change is called **genetic drift**. An example of this change can occur in a small population. If some members with a certain gene have more offspring than other members, that gene will appear more often in the gene pool. In addition, a gene pool can change if individuals leave the population or new individuals join the population.

Figure 25-6 *This white kangaroo is an albino. It has no color in its fur and eyes. Albinism is a mutation that can be harmful if it makes organisms more likely to be seen by predators.*

### Evolution of New Species

How do variations in the gene pool of a population lead to the evolution of a new species? New species evolve when members of similar populations are no longer able to reproduce with each other. This separation is known as reproductive isolation. Reproductive isolation can occur in three ways.

1) Populations may not be able to reproduce with each other because they are in different geographic locations. Physical barriers, such as rivers and mountains, can keep them separated.

2) Populations may not be able to reproduce with each other because they have different courtship behaviors.

3) Populations may not be able to reproduce with each other because they reproduce at different times.

## Different Types of Evolution

There are different types of evolution. In *divergent evolution,* members of the same species develop into different species. The new species are very different from one another. They develop different traits to adapt to different environments. The finches Darwin discovered on the Galápagos Islands are examples of divergent evolution. Each finch species had different adaptations for different environments.

Another type of evolution is *convergent evolution.* In convergent evolution, unrelated species sharing the same environment begin to develop similar traits. An example of convergent evolution is a comparison of dolphins and fish. They are not related, but they have similar body structures that are adapted to life underwater.

*Coevolution* occurs when two organisms evolve in response to changes in each other. Coevolution often occurs between plants and their pollinators as shown in Figure 25-7.

Figure 25-7 *The yucca moth and the yucca plant have coevolved. The moth is the only pollinator of that plant. The plant is the only food that the yucca moth larva eat.*

✓ **Check Your Understanding**

Write your answers in complete sentences.

1. What is evolution?

2. How can a gene pool change?

3. How does reproductive isolation occur?

4. **CRITICAL THINKING** Compare and contrast Lamarck's and Darwin's theories of evolution.

**Check Your Understanding**

**1.** the process by which a species gradually changes

**2.** through natural selection, mutations, or chance events

**3.** when two populations of the same species can no longer reproduce together

**4.** Both men thought that traits were passed along, however, Lamarck thought that acquired traits could be passed along and that a new species developed quickly. Darwin thought that helpful adaptations were passed along, and that a new species developed slowly over time.

## Modern Leaders in Biology

**STEPHEN JAY GOULD, Evolutionary Biologist**

Stephen Jay Gould was a professor of geology, biology, and the history of science at Harvard University. In addition, he was an author of popular science books and essays.

Gould had a strong interest in the theory of evolution. Gould and his coworker, Nile Eldredge, had studied fossils. They noted that some organisms changed form little by little. Yet, there were many examples of organisms going through a rapid change in form. When their study was completed, they proposed a theory. Their theory stated that the evolution of a species comes from rapid changes to a specific population. The changes occur after a natural event, such as a major climate change. After these quick changes, a species would have a period of steady development. Gould's theory caused debate because it was different from the belief that species evolved slowly over time.

Gould died in 2002. His books and essays will continue to inform general audiences about science.

**CRITICAL THINKING** Why would a major climate change cause a rapid change in a species' form?

Figure 25-8 *Stephen Jay Gould proposed that the evolution of a species comes from rapid changes to a specific population.*

**Critical Thinking**

A climate change would either be from hot to cold or from cold to hot. If this type of change happened quickly, an organism would have to make adjustments in order to adapt to the new environment.

# The Fossil Record

How do we know that life has changed over time? What were organisms like millions of years ago? To find evidence of evolution, scientists use fossils to study the past.

Most living things never become fossils. Their tissues and organs decay quickly after they die. To become a fossil, an organism must be preserved. There are several ways that this preservation can occur.

One way is in tree sap. Many insects have been preserved when tree sap hardened around them. The hard sap is called amber. Scientists can study these insects.

The mammoth was an early ancestor of the elephant. It has been extinct for almost 10,000 years. In March 1999, scientists found a fossil of a mammoth frozen in ice. The ice preserved parts of the body. Even some flesh and hair remained on the mammoth.

**Figure 25-9** *By defrosting the mammoth fossil, scientists can study the extinct animal.*

Most fossils, however, are the remains of hard parts of organisms, such as teeth and bones. Teeth and bones are made up mostly of minerals. They do not decay as easily as cell tissue and organs.

**Figure 25-10** *This insect was fossilized in hardened tree sap.*

### Fossils Found in Sedimentary Rocks

Most fossils are found in sedimentary rocks. Sedimentary rocks are made up of layers of sediments, or bits of clay, soil, sand, and other materials. Sedimentary rocks usually form in water.

Sometimes when an organism dies, its soft parts decay and its hard parts become buried in sediments. The sediments are pressed together by great pressure. Eventually, the sediments and bones form rocks. The outlines of the organisms are left in those rocks.

**Figure 25-11** *Scientists use the positions of fossils in sedimentary rock to determine their relative ages.*

### Determining the Age of Fossils

Earth's crust is made up of layers and layers of rock. The bottom layers are the oldest. The upper layers are the youngest. Scientists can learn how organisms change over time by studying fossils from different layers of rock. In **relative dating**, scientists can determine the age of a fossil by comparing its placement in a layer of rock with fossils in other layers of the same rock. For example, in Figure 25-11 the scorpian fossil is older than the human fossil.

Scientists use index fossils in relative dating. An index fossil is an easily recognized species that lived for a short time, but in many different places. Although index fossils can only be found in a few layers of rock, they can be found in many different locations.

Relative dating allows scientists to estimate how old a fossil is. It does not allow scientists to determine the exact age of a fossil. Scientists can measure the actual age of a fossil through **radioactive dating**.

In radioactive dating, scientists measure the number of radioactive elements in a fossil sample. The unit used to measure the rate of decay of a radioactive element is a *half-life*. A half-life is the time it takes for one-half of the atoms of a radioactive element to decay. By measuring the amount of radioactive elements remaining in a fossil, scientists can determine how old the fossil is.

**Think About It**

The half-life of carbon-14 is 5,370 years. If you had an 80-gram sample of carbon-14 today, how much would remain in 5,370 years?

40 grams

## Information From Fossils

The fossil record shows that changes have taken place over Earth's history. Scientists that study fossils are called paleontologists. By studying fossils, these scientists have learned that Earth has undergone great changes in climate. For example, fossils of palm trees and other tropical plants have been found in Greenland, an island near the North Pole.

Fossils help scientists to study how organisms from millions of years ago have evolved into organisms today. Many scientists believe that the most complete record of evolutionary change is that of the horse. According to this theory, modern horses appear to have evolved from an animal about the size of a fox. Over time, the horse slowly developed longer legs. Earlier horses had four toes. The modern horse has only one toe, the hoof. The hoof carries the great weight of the modern horse better. This trait helps the modern horse to run faster. Figure 25-12 shows how horses may have evolved over millions of years.

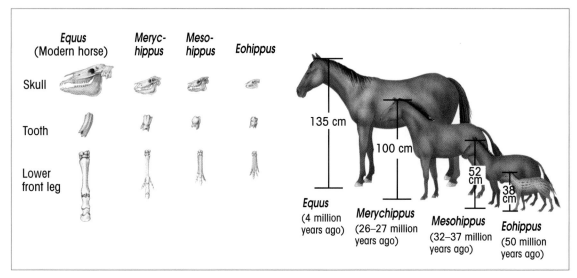

Figure 25-12 *One theory proposes that the horse has evolved from an animal about the size of a fox over 50 million years.*

Scientists also can learn about organisms that are extinct by studying fossils. One of the best-known groups of extinct animals is the dinosaurs. Fossils of more than 400 different dinosaur species have been found.

### Science Fact

Fossils of many different kinds of dinosaurs have been discovered. The largest complete dinosaur found is called the brachiosaurus. It was about 82 feet in length and about 36 feet in height. Bones from dinosaurs that may have been much larger have also been found.

### The Geologic Time Scale

Scientists have used the information they have gathered from fossils to create a geologic time scale. The **geologic time scale** is a record of Earth's history. It is based on the types of organisms that lived during different times. It shows that living things have become more complex over time. The geologic time scale is divided into four eras. Some eras are divided into periods. The chart in Figure 25-13 shows some information from the geologic time scale.

| The Geologic Time Scale | | | |
|---|---|---|---|
| **Era** | **Period** | **Start Date (Millions of Years Ago)** | **Some Organisms that Developed** |
| Precambrian | | 4,600 | Bacteria began appearing about 3 billion years ago. |
| Paleozoic | Cambrian | 600 | Invertebrates |
| | Ordovician | 480 | Algae, fungi |
| | Silurian | 435 | Fish |
| | Devonian | 405 | Insects, amphibians, ferns |
| | Carboniferous | 345 | Reptiles |
| | Permian | 275 | Seed plants |
| Mesozoic | Triassic | 225 | Conifers |
| | Jurassic | 180 | Dinosaurs, birds |
| | Cretaceous | 130 | Flowering plants |
| Cenozoic | Tertiary | 65 | Early horses, primitive apes, large carnivores |
| | Quaternary | 1.75 | Mammoths, humans |

Figure 25-13 *The geologic time scale shows a history of Earth's organisms.*

Lion foreleg

Bat wing

Dolphin flipper

**Figure 25-14** *Similar body structures are clues that some organisms may have evolved from the same ancestor.*

## Other Evidence of Evolution

Fossils are not the only evidence of evolution. Modern scientists also study living organisms to find evidence of evolution.

Some animals have similar body structures. Look at Figure 25-14. Notice how the lion's foreleg, the bat's wing, and the dolphin's flipper are similar in bone structure. These similarities are clues that these organisms may have evolved from the same ancestor.

In addition, scientists look for similarities in the early stages of animal development. Many animal embryos look similar as they develop. These similarities suggest that these organisms may have a common ancestor.

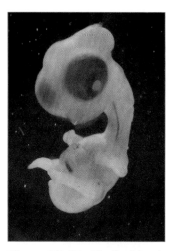 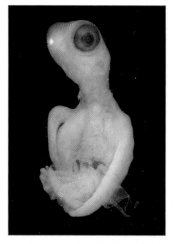

**Figure 25-15** *Similarities in the embryos of a chicken (left) and a turtle (right) may provide evidence of a common ancestor.*

Scientists can also learn more about evolution by studying an organism's DNA. Organisms that are closely related have similar DNA. Would it surprise you to learn that dogs are the closest relatives to bears? Dogs and bears have very similar DNA. They may have evolved from the same ancestor.

# Humans Change Over Time

The fossil record of human evolution is not complete. Scientists are still trying to find evidence of human evolution. In 1974, a fossil of a skeleton was discovered by Donald Johanson. Johanson called his find Lucy. Lucy is about 3.5 million years old. Her skeleton indicates that she walked upright on two feet. Some scientists believe that Lucy may be an ancestor to humans.

All modern humans are members of the species *Homo sapiens.* The name means "wise human." Fossils of several early types of humanlike species have been found. One type is the Neanderthal. Another is the Cro-Magnon. Both types are now extinct.

Neanderthals lived from 130,000 to 35,000 years ago. They were shorter than modern humans. They also had larger skulls. Some evidence suggests that they made tools and lived in groups.

Cro-Magnons looked more like modern humans. They lived from 35,000 to 10,000 years ago. They were tall and had large brain cases and skulls. Evidence shows that Cro-Magnons made more complex tools than the Neanderthals did. Cro-Magnons also lived in groups. They had ceremonies and painted on cave walls. Some scientists believe that modern humans are the descendants of the Cro-Magnons. Compare the skulls of a Neanderthal and a Cro-Magnon in Figures 25-16 and 25-17.

**Figure 25-16**
*This Neanderthal skull is larger than the skull of a modern human.*

**Figure 25-17**
*This Cro-Magnon skull is more similar to the skull of a modern human than the Neanderthal skull.*

✓ **Check Your Understanding**

Write your answers in complete sentences.

1. Where are most fossils found?
2. What is the geologic time scale?
3. **CRITICAL THINKING** How can living organisms provide evidence about the past?

**Check Your Understanding**
1. in sedimentary rock
2. a record of how Earth has changed over time
3. Living organisms can provide evidence of possible ancestor organisms through their DNA, body structures, or embryonic development.

# LAB ACTIVITY
## Modeling Natural Selection

### BACKGROUND
Organisms that have helpful adaptations are more likely to survive and reproduce.

### PURPOSE
In this activity, you will model natural selection.

### MATERIALS
squares of green paper and red paper, green fabric, stopwatch

| Modeling Natural Selection | | |
|---|---|---|
| Generation | Green Squares | Red Squares |
| 1 | 20 | 20 |
| 2 | | |
| 3 | | |

**Figure 25-18** *Copy this chart onto a separate sheet of paper.*

### WHAT TO DO
1. Copy the chart in Figure 25-18 onto a separate sheet of paper.

2. With a partner, place the green fabric on the floor. The fabric represents the environment. Scatter 20 green squares and 20 red squares on the fabric. Each square represents an insect in the first generation.

3. You will model a predator. Pick up as many of the squares as you can, one at a time, as your partner times you for 15 seconds.

4. The squares left on the fabric represent the insects that survived. Count the number of green squares and the number of red squares. Multiply each number by 2. Each doubled number represents the offspring insects in the second generation. Record these numbers in your chart.

5. Show the number of offspring insects in the second generation by scattering more squares of each color on the fabric. Repeat Steps 3 and 4 to get the number of offspring in the third generation.

### DRAW CONCLUSIONS
- In a forest, which color insect can be seen by predators?  red
- According to natural selection, what happens to helpful traits?
  They get passed to offspring.

# ON-THE-JOB BIOLOGY
## Paleontologist

Pat Worth is a paleontologist. She studies fossils. Her work helps her learn how life has changed over time.

Pat studies fossils in layers of rock. Fossils in the bottom layers are the oldest. Fossils in the upper layers are the most recent. Pat enjoys making displays of the fossils that she and others find.

**Look at Figure 25-20. Use the rock layers and fossils to answer the questions.**

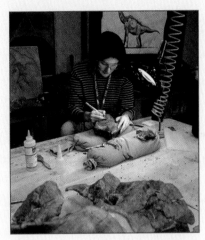

**Figure 25-19** *Paleontologists are scientists that study fossils.*

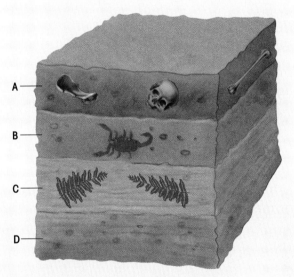

A
B
C
D

**Figure 25-20** *Fossils in layers of rock*

**1.** Which is the oldest layer? D

**2.** Which fossils are the most recent?
the human fossils

**3.** List the rock layers in the order that they formed. D, C, B, A

**Critical Thinking**

Each rock layer is older than the layer above it, so fossils found in the lowest layer are likely from organisms that lived earlier in time.

**Critical Thinking**

Why do you think the oldest organisms are found in the lowest rock layers?

## Summary

- Evolution is the process by which a species gradually changes over time.

- Darwin's theory of natural selection states that more offspring are born in each generation than can survive. These offspring compete for resources. The organisms best fitted for survival pass their traits to their offspring. Over time, a new species can develop.

- Modern scientists study a population's gene pool to study evolution. Natural selection and mutations can affect a population's gene pool.

- Scientists have gathered evidence of evolution by studying fossils. They also compare the embryos, the body structures, and the DNA of animals.

adaptation

evolution

fossil

gene pool

genetic drift

natural selection

radioactive dating

relative dating

## Vocabulary Review

### Match each term to its definition.

1. the remains of an organism that lived a long time ago   fossil

2. the survival of offspring that have favorable traits   natural selection

3. the process by which a species gradually changes over time   evolution

4. a trait that helps an organism to survive in its environment   adaptation

5. a random change in a gene pool caused by chance events   genetic drift

6. all the genes in a population   gene pool

7. a way to determine the age of a fossil by measuring the amount of radioactive elements in it   radioactive dating

8. a way to estimate the age of a fossil by comparing its position in a rock layer   relative dating

# Chapter Quiz

**Write your answers on a separate sheet of paper. Use complete sentences.**

1. What happens to a species as it evolves?
   It gradually changes over time.
2. What can biologists learn from the different layers of rock on Earth? They can learn how organisms have changed over time by studying the fossils in the rock.
3. What can similar body structures in different species tell scientists?
   that these organisms may have had a common ancestor
4. What is Darwin's theory of natural selection?
   that organisms best suited for the environment will survive to reproduce
5. What formed the basis for Darwin's theory of evolution? his voyage to the Galápagos Islands and what he discovered about the tortoise and finch species there
6. How can a mutation change a population's gene pool? A mutation is a sudden change in a gene that can be added to the gene pool.
7. What is coevolution?
   when two organisms evolve in response to changes in each other
8. What is genetic drift?
   random changes to a gene pool caused by chance events

## CRITICAL THINKING

9. Why might it be easier for a scientist to study the gene pool in a small population rather than in a large population?

10. Compare divergent evolution and convergent evolution.

## Research Project

Students should find that permineralized fossil bones have their original pore space filled in with minerals. Petrified fossils have all of their original material replaced and filled in with minerals.

9. The small population may have a smaller gene pool making it easier for scientists to study variations in the population.

10. Possible answer: In divergent evolution, members of the same species develop different traits from one another. In convergent evolution, different species develop similar traits.

*SciLINKS*
Go online to www.scilinks.org.
Enter the code PMB148 to research **fossils**.

## Research Project

Permineralization and petrification are two ways in which bones can be preserved with minerals to form fossils. Use the Internet and other resources to compare these two types of mineral preservation. Write your findings in a short report.

See the *Classroom Resource Binder* for a scoring rubric for the Research Project.

# Unit 6 **Review**

**Choose the letter of the correct answer to each question.**

**Use Figure U6-2 to answer Questions 1 to 3.**

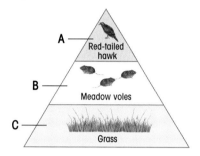

**Figure U6-2** *An energy pyramid*

**1.** Which level represents the producer?  D (p. 332)
   A.  the very top
   B.  level A
   C.  level B
   D.  level C

**2.** Which level represents the least amount of energy?  A (p. 333)
   A.  level A
   B.  level B
   C.  level C
   D.  each level has the same amount of energy

**3.** Which organisms are consumers?
   A.  the grass and the hawk  B (p. 332)
   B.  the vole and the hawk
   C.  the grass and the vole
   D.  the hawk only

**4.** Which type of behavior is hibernation?  B (p. 352)
   A.  classical conditioning
   B.  an instinct
   C.  a reflex
   D.  insight learning

**5.** What happens to water vapor as it cools?  D (p. 364)
   A.  It evaporates.
   B.  It goes through transpiration.
   C.  It forms precipitation.
   D.  It condenses.

**6.** Coal, oil, and natural gas are  D (p. 379)
   A.  renewable resources.
   B.  alternate energy resources.
   C.  pollutants.
   D.  fossil fuels.

**7.** In which type of evolution do unrelated species sharing the same environment begin to develop similar traits?  A (p. 398)
   A.  convergent evolution
   B.  divergent evolution
   C.  human evolution
   D.  coevolution

**Critical Thinking**

How is Earth's atmosphere similar to a greenhouse?

Earth's atmosphere is like the glass of a greenhouse. It traps heat and keeps Earth warm. (p. 367)

# Appendix A: Science Terms

The following list of prefixes and suffixes provides clues to the meaning of many science terms. Each word part usually comes from the Latin or Greek language.

| Word Part | Meaning | Example |
|---|---|---|
| a- | not, without | asexual |
| aero- | air | aerobic |
| anti- | against | antibody |
| bi- | two | biceps, bivalves |
| bio- | life | biology, biodiversity |
| carn- | meat, flesh | carnivore |
| chemo- | of, with, or by chemicals | chemosynthesis |
| chloro- | green | chloroplasts |
| cyt- | cell | cytoplasm |
| -derm | skin, covering | echinoderm, epidermis |
| di- | twice, double | dicot |
| eco- | environment, habitat | ecosystem, ecology |
| ecto- | outer | ectoderm |
| endo- | inside | endocrine |
| epi- | on, on the outside | epidermis |
| exo- | outside | exoskeleton |
| -gen | produce, generate | pathogen |
| geo- | Earth | geologic, geographic |
| hemo- | blood | hemoglobin |
| hydro- | water | hydroponics, hydroelectric |
| -itis | disease of | appendicitis, dermititis |
| leuko- | white | leukocyte |
| -logy | study of, science of | biology, zoology |
| mono- | one | monocot |
| -ose | carbohydrate | glucose, cellulose |
| photo- | light | photosynthesis, phototropism |
| -phyll | leaf | chlorophyll, mesophyll |
| -phyte | a plant | bryophyte |
| -scope | instrument for viewing | microscope |
| syn- | to put together, with | photosynthesis |
| trans- | across, through | transpiration |
| trop- | turn, respond to | tropism |
| uni- | one | unicellular |

# Appendix B: The Microscope

## Using a Microscope

To look at something under a microscope, take a sample of the thing you are studying. The sample is placed on a piece of glass or plastic called a slide. Another piece of glass or plastic is then placed on top of the sample to hold it in place. The sample is then slipped onto the stage of the microscope. Two stage clips hold the slide in place.

Always begin using low power. You can tell if you are using low power by making sure that the shortest objective lens is facing the stage of the microscope. To see the sample, you look through the eyepiece. There are two knobs for focusing a microscope. One knob is for major adjustments. The other is for fine tuning. Turn these knobs until the image becomes clear.

## Power of Magnification

When you view something through a microscope, you will see a magnified image of that object. How much larger will the image be? The size of the image depends on the strength, or magnifying power, of the microscope. It partly depends on how many lenses the microscope has. It also depends on the size and shape of the lenses.

Each lens has its own magnification power. If a lens can magnify an object ten times its actual size, we say the lens has a power of 10×. A typical classroom microscope has a magnifying power of 100× to 1,000×.

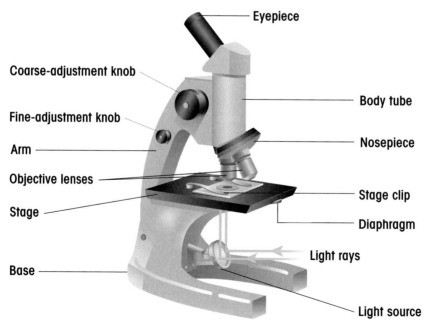

*The parts of a microscope*

# The Microscope *(continued)*

## Preparing a Slide

### BACKGROUND

You can look at small organisms or tissue samples under a microscope. To do so, you must first prepare a slide. The sample material you examine is called a specimen. Sometimes specimens are treated with a stain. The stain colors certain parts of the specimen, making them easier to see.

### MATERIALS

safety goggles, slide, specimen, dropper, stain (optional), cover slip, tissue or paper towel, microscope, lens paper

### WHAT TO DO

1. Put on safety goggles. You can take off the goggles when you look through the microscope.

2. Place the specimen on a clean slide.

3. Add one to two drops of water or stain to the specimen. Follow your teacher's instructions.

4. Carefully place one edge of the cover slip onto the slide next to the specimen. Gently lower the cover slip onto your slide. Make sure that there are no bubbles in the liquid covering the specimen.

5. If you need to clean up any liquid around the cover slip, place a small piece of tissue or paper towel next to an edge of the cover slip.

6. Carefully place your slide on the stage of the microscope. Make sure that the specimen is centered over the opening in the stage. Use the stage clips to hold the slide in place.

7. Look at the slide from the side. Position the low-power lens over your slide. Turn the coarse-adjustment knob so that the lens almost touches the slide.

8. Look through the eyepiece. You may have to move the slide to find your specimen.

9. When you can see the specimen, focus using the fine-adjustment knob.

10. Make sure that there is enough light entering the opening in the stage. Turn the diaphragm to adjust the amount of light.

11. To look at the specimen under a higher power lens, turn the nosepiece until that lens clicks into place over the slide. Look through the eyepiece and focus using the fine-adjustment knob only.

12. After every use, remove the slide and clean the stage and the lenses using lens paper. Do not use any other types of paper to clean the lenses. They can scratch the lenses.

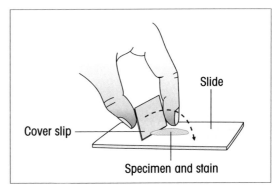

*Place one edge of the cover slip next to the specimen.*

# Appendix C: The Six Kingdoms

This appendix lists classification groups based on a six-kingdom system. Other classification systems have also been suggested. Classification continuously changes as new information about organisms is discovered. Latin names are given in parentheses next to the common names. Because of the large number of organisms, not all groups are listed.

**\*Note:** The terms *phylum, division, class,* and *order* are used to further classify organisms within a kingdom.

**ARCHAEBACTERIA KINGDOM (Kingdom Archaea)**
Single-celled organisms that have no cell nucleus; these organisms are able to live in extreme environments.

**BACTERIA KINGDOM (Kingdom Eubacteria)**
Single-celled organisms that have no cell nucleus

**PROTIST KINGDOM (Kingdom Protista)**
Mostly single-celled organisms that have cell walls

**Animal-like Protists**
Amoebas (phylum Rhizopoda)
Paramecia (phylum Ciliata)

**Plantlike Protists**
Red algae (division Rhodophyta)
Green algae (division Chlorophyta)
Brown algae (division Phaeophyta)
Euglena (phylum Euglenophyta)

**Funguslike Protists**
Slime molds (division Myxomycota)

**FUNGUS KINGDOM (Kingdom Fungi)**
Organisms that cannot move, have no chlorophyll, but do have cell walls

Molds (division Zygomycota)
Yeasts (division Ascomycota)
Mushrooms (division Basidiomycota)

**PLANT KINGDOM (Kingdom Plantae)**
Multicellular organisms that have a cell wall and chloroplasts; these organisms go through photosynthesis.

Mosses and liverworts (division Bryophyta)
Ferns (division Pterophyta)
Horsetails (division Sphenophyta)
Cone-bearing plants (division Coniferophyta)
Flowering plants (division Anthophyta)

*Diatoms are plantlike protists.*

*Mushrooms are fungi.*

*Liverworts are plants.*

# The Six Kingdoms *(continued)*

**ANIMAL KINGDOM (Kingdom Animalia)**

Many-celled organisms; most have specialized tissues and can move freely.

Sponges (phylum Porifera)

Cnidarians (phylum Cnidaria)

Flatworms (phylum Platyhelminthes)

Roundworms (phylum Nematoda)

Segmented worms (phylum Annelida)

Mollusks (phylum Mollusca)

Echinoderms (phylum Echinodermata)

Arthropods (phylum Arthropoda)
    Horseshoe crabs (class Merostomata)
    Sea spiders (class Pycnogonida)
    Arachnids (class Arachnida)
    Centipedes (class Chilopoda)
    Millipedes (class Diplopoda)
    Crustaceans (class Crustacea)
    Insects (class Insecta)

Chordates (phylum Chordata)
    Jawless fish (class Agnatha)
    Cartilagenous fish (class Chondrichthyes)
    Bony fish (class Osteichthyes)
    Amphibians (class Amphibia)
    Reptiles (class Reptilia)
    Birds (class Aves)
    Mammals (class Mammalia)
        Monotremes (order Monotremata)
        Marsupials (order Marsupialia)
        Rodents (order Rodentia)
        Flying mammals (order Chiroptera)
        Meat-eating mammals (order Carnivora)
        Trunk-nosed mammals (order Proboscidea)
        Insect-eating mammals (order Insectivora)
        Rabbit-like mammals (order Lagomorpha)
        Toothless mammals (order Edentata)
        Even-toed hoofed mammals (order Artiodactyla)
        Odd-toed hoofed mammals (order Perissodactyla)
        Water mammals (order Cetacea)
        Primates (order Primata)

*A centipede is an arthropod.*

*A frog is an amphibian.*

*A koala is a marsupial mammal.*

# Appendix D: Some Important Vitamins and Minerals

| Vitamin | Source | Body Function |
|---|---|---|
| A | Leafy green and yellow vegetables, egg yolk, milk, liver, butter, margarine | Good eyesight; healthy skin and hair; growth |
| B$_1$ | Whole grains, yeast, milk, green vegetables, egg yolk, liver, fish, soybeans, peas | Strong heart, nerves, and muscles; growth; respiration |
| B$_2$ | Lean meat, wheat germ, yeast, milk, cheese, eggs, liver, bread, leafy green vegetables | Healthy skin; growth; good eyesight; reproduction |
| B$_{12}$ | Liver, lean meat, milk, fresh fish, egg yolk, shellfish | Helps make blood; helps nervous system |
| C | Oranges, grapefruit, lemons, limes, berries, vegetables, tomatoes | Healthy bones; strong blood vessels; helps prevent sickness; helps heal wounds |
| D | Egg yolk, milk, fresh fish | Strong teeth and bones; growth |
| E | Leafy green vegetables, wheat germ, oils | Antioxidant |
| K | Vegetables, soybeans | Helps blood clot |

| Mineral | Source | Body Function |
|---|---|---|
| Calcium | Milk, vegetables, meats, dried fruits, whole-grain cereals | Healthy bones and teeth; helps blood clotting; prevention of muscle spasms |
| Iodine | Saltwater fish, shellfish, iodized salt | Functioning of thyroid gland; regulation of use of energy in cells |
| Iron | Liver, meats, eggs, nuts, dried fruits, leafy green vegetables | Formation of red blood cells |
| Magnesium | Milk, meats, whole grain cereals, peas, beans, nuts, vegetables | Normal muscle and nerve action; regulation of body temperature; helps build strong bones |
| Phosphorus | Milk, meat, fish, poultry, nuts, vegetables, whole-grain cereals | Bone and teeth formation; nerve and muscle function; energy production |
| Potassium and sodium | Most foods, table salt (sodium) | Blood and cell functions; helps maintain balance of fluids in tissues |

# Glossary

## A

**abdomen** the rear segment of an arthropod's body

**acid rain** a form of air pollution produced when sulfur dioxide or nitrous oxide and rain chemically combine to form acid compounds that fall to Earth as precipitation

**acquired immunity** the type of immunity that people develop or acquire over time

**active acquired immunity** the type of immunity acquired from a vaccine

**active transport** the movement of materials through a cell membrane using energy

**adaptation** a characteristic of an organism that helps it to survive in its environment

**adrenaline** a hormone that increases heart rate and blood pressure

**adult** the final stage of an organism's development

**aggression** a threatening behavior, such as growling or snarling

**algae** plantlike protists

**alternation of generations** reproduction that occurs in two stages—one asexual and one sexual

**alveoli** the air sacs in the lungs in which gas exchange occurs

**amniotic sac** the fluid-filled pouchlike structure that a bird, reptilian, or mammalian fetus grows in

**amoebocyte** a cell in a sponge that carries food to other cells

**amphibian** a coldblooded vertebrate that lives part of its life in water and part on land

**anaphase** the third stage of mitosis in which the chromosomes separate

**anatomy** the study of the parts of the body

**angiosperm** a plant that produces covered seeds and flowers

**antennae** body parts used to sense smell and touch

**anther** the tip of the stamen

**antibody** a molecule that attaches to a specific pathogen

**appendage** a movable part that extends out from the body, such as arms, legs, wings, and claws

**archaebacteria** bacteria that live in extreme environments

**artery** a blood vessel that carries blood away from the heart

**arthropod** an invertebrate with an exoskeleton and jointed appendages

**asymmetry** a type of body plan that does not form mirror images along a line

**atom** the smallest part of an element that can be identified as that element

**ATP** a form of chemical energy that can be used by a cell to carry out life processes

**atrium** the smaller chamber of a heart

**auditory nerve** the bundle of nerves that carries information from the ear to the brain

**axon** the fiber on neurons that carries messages away from the cell

## B

**B cell** a type of lymphocyte that produces antibodies

**bacteria** simple unicellular organisms that do not have nuclei

**ball-and-socket joint** a joint that allows bones to move in almost any direction

**behavior** the way that an organism responds to its environment

**bilateral symmetry** a type of body plan that forms two mirror images along a line

**bile** a green liquid produced by the liver that helps digest fats

**biodiversity** the total variety of living things in an environment; the healthy ecosystem that different groups of species create

**biology** the study of living things

**biome** a large region of Earth that has a characteristic climate and characteristic kinds of organisms

**bivalve** a type of mollusk that has two shells, such as clams and oysters

**blade** the broad, flat part of a leaf

**bony fish** a group of fish that has a skeleton made of bone

**botany** the study of plants

**Bowman's capsule** a cuplike structure located on one end of a nephron

**brainstem** the part of the brain that connects the brain to the spinal cord

**bronchus** one of two small tubes that leads from the trachea to the lungs

**budding** a form of asexual reproduction in which a small section of an organism develops and then pinches off

**bulb** a type of underground stem

# C

**Calorie** a measure of food energy, also known as kilocalorie; equal to 1,000 calories

**cambium** a special layer of growth tissue that produces new layers of xylem and phloem cells in woody stems

**canine** the type of tooth used for tearing food

**capillary** a tiny blood vessel that connects arteries to veins

**carbohydrate** a compound made up of carbon, hydrogen, and oxygen that is needed by living things for energy

**cardiac muscle** the muscle tissue that makes up the heart

**carnivore** a type of consumer that eats animals

**carrying capacity** the largest population size an environment can support

**cartilage** a strong, flexible connective tissue

**cartilagenous fish** a group of fish that has a skeleton made of cartilage

**cell** the basic unit of structure and function in all living things

**cell membrane** a thin, protective covering around a cell

**cell theory** a theory that states that all living things are made of one or more cells, that cells carry out all life functions, and that new cells come only from other living cells

**cell wall** a thick, stiff covering around some cells, such as plant cells

**cellular respiration** the process by which cells release energy from food

**central nervous system** the division of the nervous system made up of the brain and the spinal cord

**cephalopod** the most highly developed type of mollusk, such as a squid and octopus

**cephalothorax** the body part of a crustacean where the head and thorax are tightly joined

**cerebellum** the part of the brain that controls balance and coordination

**cerebrum** the part of the brain that contains voluntary muscle movements, thinking, learning, memory, speech, and the senses

**cervix** the opening to the uterus

**chitin** the hard material making up the exoskeleton of arthropods

**chlorophyll** a substance in chloroplasts that absorbs light energy

**chloroplast** a part in a plant cell that uses sunlight to make food

**chordate** an organism that has a notochord at some stage of development

**chromosome** a thick strand of nuclear material that passes on traits from the parent cell to the daughter cells during cell reproduction

**chyme** partially digested food material that leaves the stomach

**cilia** tiny hairs used for movement

**circulatory system** the organ system that moves blood throughout the body

**classical conditioning** a type of learning response in which a certain stimulus becomes connected to another stimulus

**classification** a system of grouping things according to similarities

**clone** an organism that has the same genetic information as another organism

**cnidarian** an invertebrate with a hollow body and stinging cells

**cochlea** a snail-shaped organ in the ear that receives sound waves

**codominance** a pattern of heredity that can occur when two genes are present for a certain trait but neither gene is dominant nor recessive

**coevolution** a process in which two different species evolve in response to changes in each other

**coldblooded** having a body temperature that changes with the temperature of the surroundings

**collar cell** a specialized cell in a sponge

**commensalism** a relationship between two organisms in which one organism benefits while the other organism is unaffected

**community** all the living things in one ecosystem

**compact bone** the hardest part of a bone

**competition** the struggle among organisms for resources

**complete metamorphosis** the four stages of development of some organisms: egg, larva, pupa, and adult

**composting** the conservation method that recycles food wastes

**compound** a combination of two or more elements

**condense** to change from a gas to a liquid

**conservation** the wise use of natural resources to prevent resources from being used up; to maintain the balance of ecosystems

**consumer** an organism that cannot make its own food and must eat other organisms

**contagious** can be spread from one person to another

**control** the part of an experiment in which conditions are kept constant; something used to make comparisons to

**convergent evolution** a process in which unrelated species living in similar environments begin to develop similar traits

**cornea** the clear, curved covering on the outer surface of the eye

**cortex** the second layer of a root made of soft, loose tissue

**cotyledon** a seed leaf

**courtship** the behavior performed before two animals of the same species mate

**cross-pollination** the movement of pollen from the stamen of one flower to the pistil of another flower of the same or different species

**cud** the food that a hoofed animal can bring back up to its mouth during digestion

**cutting** a plant piece that can be used to grow new plants

**cytoplasm** a jellylike substance that fills a cell

# D

**decomposer** an organism that breaks down and absorbs nutrients from dead organisms and wastes, and then returns the nutrients to the environment

**decomposition** the process of breaking down dead matter into smaller molecules

**deforestation** the cutting and clearing away of forests

**dendrite** the fiber on neurons that carries messages into the cell

**deoxyribonucleic acid** (see DNA)

**dermis** a thick layer of skin under the epidermis

**desert** a biome that receives less than 10 inches (25 centimeters) of rain each year

**diaphragm** a sheet of muscle below the lungs that assists in breathing

**dicot** a flowering plant with two seed leaves in its seeds

**diffusion** the movement of a substance from an area that is crowded to an area that is less crowded

**digestion** the process by which the body breaks down food into nutrients that can be absorbed by cells

**divergent evolution** a process in which members of closely related species develop different traits due to different environments

**DNA** the large molecules in a cell that store genetic information

**DNA cloning** the production of many copies of DNA

**dominant** describes a gene or trait that will always show itself

**drug** a chemical that causes a change in the body

**drug abuse** the improper use of a drug

# E

**eardrum** the tightly stretched layer of tissue inside the ear that vibrates when hit with sound waves

**echinoderm** an invertebrate that has an inner skeleton and rays that extend from a central point

**ecology** the study of how all living things relate to one another and their world

**ecosystem** the series of relationships between a community of organisms and the environment

**egg** the female reproductive cell

**element** a pure substance made of the same kinds of atoms

**embryo** an organism that develops from a fertilized egg

**enamel** a hard material covering a tooth

**endangered species** organisms in danger of becoming extinct

**endocrine gland** an organ that produces chemical substances called hormones that are released into the bloodstream

**endoplasmic reticulum** a network of tubes inside a cell that helps in protein synthesis and transport

**energy** the ability to do work and cause change

**environment** all the things around you

**enzyme** a substance that helps to change chemical reaction rates in the body

**epidermis** a thin outer layer of cells; the outermost layer of skin

**epiglottis** a flap of tissue that prevents food and liquids from getting into the windpipe

**erosion** the removal of materials such as rock and soil by wearing them away

**esophagus** a tube behind the windpipe that carries food from the mouth to the stomach

**estrogen** a female sex hormone produced mainly in the ovaries that stimulates and maintains female secondary sex characteristics

**estuary** an area where freshwater streams or rivers flow into the ocean

**evaporate** to change from a liquid to a gas

**evolution** the process by which a species gradually changes over time

**excretory system** the organ system that removes wastes from the body

**exocrine glands** glands that secrete a substance onto a surface either directly or by using ducts

**exoskeleton** a tough, stiff covering around the body of an organism

**experiment** a special kind of test to get information

**extensor** a muscle that extends a limb

**extinct species** a once-living species that no longer exists

# F

**Fallopian tube** another term for the oviduct

**fertilization** the joining of a male reproductive cell and a female reproductive cell

**fetus** an embryo after 8 weeks of development in the uterus

**fibrous roots** many small, thin roots

**filament** a thin stalk that holds up the anther in a flower

**fission** a form of asexual reproduction in which an organism splits in two to form two complete daughter cells

**flagella** long, thin, threadlike structures that are used for movement

**flexor** a muscle that bends a limb at the joint

**food chain** a diagram that shows how energy is transferred from one organism to another

**Food Guide Pyramid** a diagram that shows the kinds of food and the numbers of servings of each that you should eat each day

**food web** a diagram that shows many food chains interacting with each other

**fossil** the remains of an organism that lived a long time ago

**fracture** a break or crack in a bone

**frond** the leaf of a fern

**fungi** organisms that may be unicellular or multicellular, have cells with a nucleus, and do not have chlorophyll

# G

**gallbladder** an organ that stores bile

**gamete** a sex cell

**gastropod** a type of mollusk that has one shell or no shell at all, such as snails and slugs

**gene** a factor that controls inherited traits

**gene pool** all the genes in a population

**gene splicing** the process in which new DNA that contains a certain gene can be inserted into the plasmid of a bacterial cell

**genetic drift** a random change in a gene pool caused by chance events

**genetic engineering** the methods used to change the genetic information in the DNA of cells or organisms

**genetics** the study of how the characteristics of a living thing are passed along from generation to generation; the study of heredity

**genome** all of the genes of an organism

**genotype** the gene combination for a given trait in an organism

**geologic time scale** the record of Earth's history based on the types of organisms that lived at different times

**geothermal energy** the form of energy stored as heat within Earth's crust

**germination** the process by which a plant embryo develops and breaks out of the seed

**gill cover** a flap of skin and bones that covers and protects the gills of a fish

**gills** organs used for getting oxygen from water

**gland** an organ that makes a chemical substance that is used or released by the body

**gliding joint** a type of joint that allows bones to glide over one another

**Golgi bodies** flattened, folded sacs that package and move materials within a cell

**grafting** the production of a new plant by joining the stems of two different plants

**gravitropism** a plant's growth in a certain direction in response to the pull of gravity

**greenhouse effect** a natural process by which carbon dioxide and other gases trap heat in the atmosphere

**groundwater** the water contained in underground rock

**guard cell** a cell that controls the size of a stoma

**gymnosperm** a plant that produces uncovered seeds

# H

**habitat** the place in which an organism lives

**half-life** the time it takes for one-half of the atoms of a radioactive element to decay

**hemoglobin** a protein in red blood cells that contains iron and binds to oxygen

**hemophilia** a sex-linked blood disorder

**herbaceous stem** a smooth, soft, and green stem

**herbivore** a type of consumer that eats plants

**heredity** the passing of traits from parents to offspring

**heterozygous** having two unlike genes for the same trait

**hibernation** a resting state during which an animal's body temperature drops and its heart rate and breathing slow down

**hinge joint** a type of joint that allows back-and-forth motion

**homeostasis** the process by which organisms keep their internal conditions relatively stable

*Homo sapiens* the species in which modern humans are classified

**homozygous** having two like genes for the same trait

**hormone** a type of chemical messenger in the body that is produced by endocrine glands

**host** an organism that a parasite lives on or in

**humus** the organic material in soil

**hybridization** the breeding of two organisms of different species

**hydroelectric energy** a form of energy obtained from falling or running water

**hydroponics** growing plants without soil

**hydrotropism** a plant's growth in a certain direction in response to the location of a water source

**hyphae** chains of cells that form fibers in fungi

**hypothesis** a possible answer to a scientific question

# I

**immovable joint** a type of joint that does not move

**immune system** the body system that defends against foreign and disease-causing substances

**immunity** resistance to a specific disease

**imprinting** a type of learning that is acquired in the early stages of animal development based on what is seen, heard, touched, or smelled

**inbreeding** the crossing of closely related organisms

**incisor** the type of tooth used to cut food

**incomplete dominance** a pattern of heredity in which one gene for a trait is not completely dominant over the other gene

**incomplete metamorphosis** the type of development in which the larval stage looks similar to the adult

**index fossil** a fossil used to compare the relative ages of other fossils

**innate behavior** a behavior that an animal has at birth

**insight learning** a type of behavior in which an animal applies something that it has already learned to a new situation

**instinct**   an inherited pattern of behavior than an animal can control

**insulin**   a chemical that controls the level of glucose in the body

**interneuron**   a neuron that processes impulses from the sensory neurons and sends the response impulses to the motor neurons

**invertebrate**   an animal without a backbone

**involuntary muscle**   a muscle that moves without your control

**iris**   the colored part of the eye that controls the size of the pupil

## J

**jawless fish**   a group of fish that does not have jaws

**joint**   a place in the body where two or more bones meet

## K

**karyotype**   an organized display of pairs of chromosomes

**kingdom**   the most general of the seven levels of classification

## L

**larva**   the immature stage in the development of some organisms that looks different from the adult

**larynx**   the organ located at the top of the trachea that contains the vocal cords

**law**   a theory that has passed many tests

**learned behavior**   a behavior that develops as a result of experience

**lens**   the part of the eye that focuses light on the retina

**ligament**   a type of tissue that holds bones together at a joint

**limiting factor**   a condition in an environment that prevents a population from growing beyond a certain size

**lipid**   a fat or an oil

**liver**   a large organ that produces bile

**liverwort**   a type of nonvascular plant

**loam**   soil that has a well-balanced mixture of sand, silt, and clay

**loop of Henle**   the area of a nephron where urine is concentrated

**lymph**   a clear fluid that circulates throughout the body and contains special white blood cells

**lymph node**   a type of tissue that filters lymph

**lymphatic system**   the organ system that returns fluid lost by blood to the circulatory system

**lymphocyte**   a white blood cell that fights disease

**lysosome**   a cell part that contains chemicals that break down substances and old cell parts that the cell does not need

## M

**mammal**   a warmblooded vertebrate that is highly developed, fed milk from its mother, and covered with fur or hair

**mammary gland**   a gland that produces milk

**mandible**   the mouthpart of an insect adapted for biting and grinding food

**mantle**   the thin membrane in all mollusks that surrounds the internal organs and makes the shell if one is present

**marrow**   the soft tissue inside bones that makes blood cells

**marsupials**   a group of mammals that has pouches

**medulla**   the part of the brain that controls involuntary functions

**medusa**   an umbrella-shaped body form of some cnidarians

**meiosis**   the type of cellular reproduction that produces sex cells

**menopause**   the stage when the frequency of menstruation slows and eventually stops

**menstruation**   the monthly shedding of the lining of a woman's uterus

**metamorphosis**   the process of change and development that an insect goes through as it becomes an adult insect

**metaphase**   the second stage of mitosis in which the chromosomes line up in the center of the cell

**microbiology**   the study of living things too small to be seen with the naked eye

**migration**   the traveling over long distances of some animals from season to season for feeding, nesting, and warmth

**minerals**   inorganic substances in foods that the body needs to function properly and to remain healthy

**mitochondria**   structures in a cell that release most of a cell's energy

**mitosis**   the division of the nucleus of a cell

**molar**   the type of tooth used to crush and grind food

**molecule**   the smallest part of a compound

**mollusk**   an invertebrate with a soft body that may or may not have a shell

**molting**   the process by which an arthropod sheds its exoskeleton and produces a new one

**monocot**   a flowering plant with one seed leaf in its seeds

**monotremes**   a group of mammals that lays eggs

**moss**   a type of nonvascular plant

**motor neuron**   a neuron that carries messages from the central nervous system to muscles and glands

**multicellular**   having more than one cell

**multiple alleles**   when one gene has more than two possible variations

**muscular dystrophy**   an inherited disease that affects the skeletal muscles of the body

**mutation**   a change in a gene

**mutualism**   a relationship between two organisms of different species in which both organisms are helped

**mycology**   the study of fungi

# N

**nastic response**   a nongrowth plant response to a nondirectional stimulus

**natural immunity**   the natural ability to resist disease

**natural resource**   a material that comes from nature and is used by living things

**natural selection**   the survival of offspring that have favorable traits

**negative feedback**   a type of regulation in which one factor will inhibit the production or release of another factor in the body

**negative reinforcement**   an action that causes an animal to learn a certain behavior in order to avoid a punishment

**nephron**   a network of tubes and blood vessels in a kidney that filters wastes from blood

**neuron**   a nerve cell

**neutral buoyancy**   the ability an object has when it can neither sink nor float in a fluid

**niche**   the job or function of an organism within its ecosystem

**nitrate**   a form of nitrogen used by plants

**nitrogen-fixation**   the process that changes nitrogen gas into a form that living things can use

**nonrenewable resource**   a natural resource that is being used faster than it can be replaced

**nonvascular plant**   a plant that does not have well-developed structures for transporting water

**notochord**   a strong but flexible support rod found below the nerve cord

**nuclear energy**   the form of energy obtained from splitting or fusing the nuclei of certain atoms

**nuclear membrane**   the membrane surrounding the nucleus in a cell

**nucleic acid**   the molecules that form nucleotide chains in cells and contain the genetic information of an organism

**nucleus**   the part of a cell that controls all the other parts

**nymph**   the stage of development of some organisms in which the young looks like the adult but organs are not fully developed

# O

**offspring**   a new organism produced by a living thing

**omnivore**   a type of consumer that eats both plants and animals

**operant conditioning**   a type of learning that occurs based on rewards or punishments

**optic nerve**   the bundle of nerves that carries information from the eye to the brain

**organ**   a group of tissues that work together to do a certain job

**organ system**   a group of organs and tissues that work together to perform a certain job

**organelle**   a part in a cell that floats in the cytoplasm and helps the cell to carry out life functions

**organic**   having to do with living things or things that were once living; contains carbon

**organism**   any living thing

**osmosis**   the movement of water through a cell membrane

**ovary**   in plants, the bottom part of a pistil where female reproductive cells are formed; in animals, the female organ that makes egg cells and hormones

**overproduction**   the reproduction of many more offspring than are expected to survive

**oviduct**   a tube through which the egg cell travels from an ovary, in mammals, to the uterus

**ovulation**   the monthly release of an egg cell from an ovary

**ovule**   the part of the ovary that contains the female reproductive cells

# P

**paleobotany**   the study of prehistoric plants

**palisade layer**   the layer of cells under the epidermis of a leaf that contains many chloroplasts

**pancreas**   an organ that produces digestive enzymes and hormones

**parasite**   an organism that lives on or inside another organism, usually causing it harm

**parasitism**   a relationship between two organisms in which one is helped while the other is harmed

**passive acquired immunity**   the immunity acquired when a mother passes antibodies to her developing fetus

**passive transport**   the movement of materials through a cell membrane without the use of energy

**pathogen**   any substance that causes disease

**pedigree**   a chart that shows how a certain trait is passed from generation to generation

**penis**   the male organ that delivers sperm to the female reproductive system

**periosteum**   a strong layer of connective tissue covering bone

**peripheral nervous system**   the part of the nervous system that lies outside the brain and the spinal cord

**permafrost**   a layer of permanently frozen soil

**petal**   a part of a flower that is often brightly colored

**petiole** a stalk that connects the leaf to the plant

**pharynx** a tube that air and food pass through; in vertebrates, the throat

**phenotype** the appearance of a trait in an organism

**pheromone** a chemical that animals use to affect the behavior of other animals of the same species

**phloem** a type of tissue in roots, stems, and leaves that carries food from the leaves to other plant parts

**photosynthesis** the process by which light energy is used to make food

**phototropism** a plant's growth in a certain direction in response to light

**phylum** a level of classification below kingdom

**physiology** the study of how the body functions

**pioneer species** the first species to grow in a place

**pistil** the female part of a flower

**pivotal joint** the type of joint that allows side-to-side and up-and-down motion

**placenta** the organ through which nutrients, oxygen, and wastes pass between the mother and the embryo or fetus

**placentals** a group of mammals that is fully developed at birth

**plant embryo** the early, undeveloped stage of a new plant

**plaque** a sticky film of bacteria that can form on the teeth and gums

**plasma** the liquid part of blood that carries nutrients and wastes throughout the circulatory system

**plasmid** a small, circle-shaped DNA molecule found in some organisms

**platelet** a piece of a cell found in blood that helps blood cells clump together

**pollen** a light and powdery substance that contains the male reproductive cells of a plant

**pollination** the movement of pollen from a stamen to a pistil

**pollution** the release of materials into the environment, usually causing harm

**polygenic trait** a trait controlled by two or more genes interacting with one another

**polyp** a tubelike body form of some cnidarians

**population** a group of individual organisms of the same kind living in an ecosystem

**positive reinforcement** an action that causes an animal to learn a certain behavior because it expects to receive a reward

**precipitation** condensed water vapor that falls as rain, hail, sleet, or snow

**predator** an organism that hunts and kills another organism for food

**premolar** the type of tooth found between canines and molars that are used to crush and grind food

**prey** an organism that is hunted and killed by another organism for food

**primates** a group of mammals that includes apes, monkeys, and humans

**probability** the chance of an event occurring

**producer** an organism that can make its own food by using energy from the Sun or other inorganic sources

**progesterone** a female sex hormone that stimulates the uterus to prepare for a fertilized embryo and, later, to continue pregnancy

**prophase** the first stage of mitosis in which chromosomes can be seen under a microscope

**protein** a compound containing carbon, nitrogen, hydrogen, and oxygen that helps build and repair living cells and makes up enzymes

**protist**   a simple organism that is neither plant nor animal but that often has characteristics of both

**protozoan**   an animal-like protist

**pseudopod**   an extension of cytoplasm used for movement

**puberty**   the time at which a person becomes sexually mature

**Punnett square**   a chart that shows possible gene combinations

**pupa**   the stage of development in some organisms in which the larva develops into an adult

**pupil**   the opening in the eye that lets light in

# R

**radial**   a type of body plan that forms more than two mirror images along a line

**radioactive dating**   a way to determine the age of a fossil by measuring the amount of radioactive elements in it

**recessive**   describes a gene or trait that will be hidden when the dominant gene or trait is present

**recombinant DNA**   DNA extracted and joined together from at least two different sources to form a single fragment of genetic information

**recycle**   to reprocess a waste material for use in a new product

**red blood cell**   a blood cell that delivers oxygen to other body cells

**reflex**   an automatic and involuntary response to an outside stimulus; a simple, inherited behavior that an animal automatically performs in response to a stimulus

**reflex arc**   the nerve pathway that triggers a reflex action

**regeneration**   the ability to regrow lost body parts

**relative dating**   a way to estimate the age of a fossil by comparing its position in a rock layer

**renewable resource**   a natural resource that can be replaced at about the same rate as it is being used

**replication**   the process by which DNA molecules are duplicated to make an exact copy

**reproduction**   the way organisms make more of their own kind

**reproductive isolation**   the separation of populations that prevents them from mating and producing offspring

**reptile**   a coldblooded vertebrate that breathes air with lungs and has scaly skin

**respiratory system**   the organ system that moves oxygen into the body and carbon dioxide out of the body

**restriction enzyme**   a chemical substance that cuts a strand of DNA at a certain location; used to remove genes

**retina**   the layer of sensory neurons at the back of the eye that detects light

**rhizome**   the underground stem of a fern

**ribosome**   an organelle that makes proteins for the cell

**RNA**   a molecule in a cell used in the making of proteins

**root**   the underground organ in plants that absorbs water and minerals; the part of a tooth that contains nerves and blood vessels

**root cap**   the mass of cells that covers and protects a root tip

**root hair**   a tiny hairlike structure on the outer layer of a root

# S

**saliva**   a liquid in the mouth that helps digestion

**savanna**   a tropical or subtropical biome that is covered with grasses but does not have rich or deep soil

**scientific method**   a set of steps used to find answers to questions

**seed**   a protective covering that surrounds a young plant and its stored food

**selective breeding**   the breeding of parent organisms to produce certain traits in the next generation

**self-pollination**   the movement of pollen from the stamen of a flower to the pistil of the same flower

**semen**   the mixture of fluids in which sperm leave the body

**semipermeable**   having tiny pores or holes so that only certain molecules can pass through

**sensory neuron**   a neuron connected to a sense organ

**sepal**   a special kind of leaf that protects a flower bud

**sex-linked trait**   a trait controlled by genes on the *X* or *Y* chromosome

**skeletal muscle**   a voluntary muscle that is attached to bone

**skeleton**   a group of bones that give structure and support to an organism's body

**smooth muscle**   an involuntary muscle found in the walls of blood vessels, intestines, and other organs

**soil**   a mixture of sand, silt, and clay

**solar energy**   a form of energy from the Sun

**sperm duct**   one of two thin tubes through which sperm travel from the testes to the urethra

**spicule**   a supporting part of marine invertebrates such as sponges

**sponge**   an invertebrate that lives in water and has many pores in its body

**spongy bone**   soft bone tissue filled with many holes or spaces

**spongy layer**   the leaf layer that contains many air sacs

**spore**   the reproductive cell of organisms such as ferns, fungi, and algae

**stamen**   the male part of a flower

**stigma**   the top part of a pistil that receives pollen grains

**stimulus**   a change in the environment that causes a response

**stoma**   a tiny opening in a leaf that allows gases into and out of the leaf

**style**   the tube that supports the stigma

**succession**   the gradual change of the populations in a given community over time

**survival of the fittest**   the process by which organisms that are better suited to their environments survive and reproduce

**symbiosis**   a relationship between two organisms in which at least one organism benefits from the other

**symmetry**   a body plan that forms mirror images along a line

**synapse**   the space between the axon of one neuron and the dendrite of another

# T

**T cells**   a type of lymphocyte that attacks foreign and disease-causing substances directly

**tadpole**   a young undeveloped frog

**taiga**   a biome that has forests of cone-bearing trees and exists between the tundra and temperate deciduous forests

**taproot**   one thick root that grows larger than the other roots

**taste bud**   a group of sensory neurons on the tongue that detects taste

**telophase**   the last stage of mitosis in which the middle of an animal cell pinches together

**temperate forest**   a biome that is composed of trees that lose their leaves in the fall

**temperate grassland**   a biome with a deep layer of soil that is rich in nutrients

**tendon**   a type of tissue that connects muscles to bones or other muscles

**territory**   a space that animals compete for

**testes**   the male organs that make sperm cells and hormones

**testosterone**   the male sex hormone produced in the testes that is responsible for the development of male secondary sex characteristics

**theory**   an idea that explains something based on repeated scientific observations and experiments

**thigmotropism**   a plant's growth in a certain direction in response to touch

**thorax**   the middle segment of an arthropod's body

**threatened species**   organisms that are likely to become endangered in the near future

**tissue**   a group of cells that work together to do a certain job

**tonsil**   a special tissue located on either side of the throat that filters lymph

**trachea**   a tube in the throat through which air moves; the windpipe

**trait**   an inherited characteristic

**transpiration**   the process by which plants lose water through the stomata of their leaves

**trimester**   one of the three periods of time in which a human pregnancy is divided

**tropical rain forest**   a biome located near Earth's equator that has great amounts of rain

**tropism**   the growth of a plant in a certain direction in response to a stimulus

**tuber**   a type of fleshy underground stem often used to store food for the plant

**tundra**   a very cold, dry biome that has very small plants

**tympanic membrane**   a special structure that vibrates in response to sound waves, allowing hearing to occur

## U

**unicellular**   having one cell

**urea**   a waste compound formed when the body uses protein

**ureter**   one of two tubes through which urine leaves the kidneys

**urethra**   the tube through which urine (and also sperm in males) leaves the body

**urinary bladder**   the organ that stores urine

**urine**   a liquid waste in mammals

**uterus**   the female organ in which a fertilized egg develops into a baby

## V

**vaccine**   a substance made from dead or weakened pathogens that stimulates immunity against a disease

**vacuole**   a structure in a cell that stores food, water, or wastes

**vagina**   the canal that leads from a woman's uterus to the outside of her body

**valve**   a flap of tissue in the heart that opens and closes to control the direction of blood flow

**variable**   a factor in an experiment that can change

**variation**   a difference in a trait among individuals in a species

**vascular plant**   a plant that has structures for transporting water

**vegetative propagation**   a kind of asexual reproduction that uses parts of plants to grow new plants

**vein**   a part of a leaf that contains transport tissues; a blood vessel that carries blood toward the heart

**ventricle**   the thick-walled, larger chamber in a heart

**vertebrate**   an animal with a backbone

**villi** tiny finger-shaped structures in the walls of the small intestine that absorb digested food into the blood

**vitamins** organic substances in foods that the body needs to function properly and to remain healthy

**vocal cord** a tissue that vibrates to produce sounds

**voluntary muscle** a muscle that moves because you control it

# W

**warmblooded** having a body temperature that remains about the same

**weathering** the process of wearing away rock

**white blood cell** a type of blood cell that helps to protect the body against disease

**wind energy** a form of energy obtained from wind

**woody stem** a hard stem that is not green

# X

*X* **chromosome** the larger of the two sex chromosomes in humans

**xylem** a type of tissue in roots, stems, and leaves that carries water up a plant to the leaves

# Y

*Y* **chromosome** the smaller sex chromosome in humans that can result in male offspring when paired with an *X* chromosome in a zygote

# Z

**zoology** the study of animals

**zygote** a fertilized egg cell that will grow into an embryo and finally into an adult organism

# Index

## A

Abdomen, 165, 176
Accessory organs, 260
Acid rain, 365, 370, 382
Acquired immunity, 300
Acquired traits, 394
Active acquired immunity, 300
Active transport, 51, 55
Adaptations, 325, 327
Adenine, 73
Adolescence, 317
ADP (adenosine diphosphate), 52
Adrenal gland, 276
Adrenaline, 275
Aggression, 351
AIDS, 300
Air, 376
Air pollution, 382
Alcohol, 304, 305
Algae, 89, 97, 98–99, 103, 125, 329
Algae bloom, 99
Allergic reactions, 153
Alligator, 195, 196
Alternation of generations, 126
Alternative energy resources, 386–387
Alveoli, 246–247
Alzheimer's disease, 300
Amino acids, 74
Amniocentesis, 317
Amoeba, 98
Amoebocytes, 167
Amphibians, 185, 191–194
    body structure of, 191
    body systems of, 191–193
    groups of, 191
    reproduction and
        development of, 193
Anaphase, 43
Anatomy, 10
Andrewin, Kevin, 7
Angiosperms, 131
Animals, 378
    behavior of, 347–357, 358
    energy use by, 19
    extinction of, 194, 385
    kingdom of, 90–92, 166

migration of, 7, 351, 361
organ systems of, 186
role in succession, 335
tracking, 211
Animal trainer, 359
Annual rings, 141
Ant, 177, 349
Antennae, 176
Anther, 152
Antibodies, 297, 299, 300
Anus, 248
Appendages, 165, 175, 204, 208
Appendix, 258
Arachnids, 177
Archaebacteria, 89, 90, 93
Arteries, 239, 242
Arthropods, 165, 175–177
Asexual reproduction, 126, 128, 156
Asthma, 246
Asymmetry, 166
Atoms, 17, 24
ATP (adenosine triphosphate), 51, 52, 56, 57, 58
Atria, 241
Auditory nerves, 283, 291
Axon, 267, 268

## B

Bacteria, 24, 34, 89, 90, 112, 297
    characteristics of, 92–93
    kingdoms of, 93–94
    modern uses of, 96, 388
    role of, 94–95
Ball-and-socket joints, 230, 234
Bats, 210
B cells, 299
Behavior, 347
    innate, 347, 348–349
    instinctive, 350–354
    learned, 355, 358
Bilateral symmetry, 166, 169, 171
Bile, 257, 260
Biodiversity, 375, 385–386
Bio-indicators, 194

Biology, 2–4
    fields of, 10–11
Biomes, 325, 337
    climate and, 340
    land, 337–340
    water, 341
Birds, 204–207
    body structure of, 218
    flight of, 204
    homing instinct in, 207
    migration of, 205
    reproduction of, 204
    types of, 206
Bivalves, 172, 173
Blood, 243
    circulation of, 242
Blood type, 87
Blood vessels, 242
Bones, 41, 225–229, 400
    broken, 228–229
    changes in, 228
    jobs of, 226
    structure of, 227–228
Bony fish, 188
Botany, 3, 10–11
Botulism, 95
Bowman's capsule, 249, 250
Braille, 287
Braille, Louis, 287
Brain, 268, 271–272, 273
Brainstem, 272
Breathing, 56, 245
Bronchi, 246
Budding, 101
Bulbs, 156

## C

Calcium, 226–227
Calories, 61, 303
Cambium, 141
Cancer, 45, 117, 300
Canine, 27, 216, 259
Capillaries, 239, 242, 247
Carbohydrates, 25, 302
Carbon, 24
Carbon cycle, 366–368
Carbon dioxide, 56–57, 114, 144
Carbon monoxide, 382

Cardiac muscle, 225, 232
Carnivores, 332
Carriers, 76, 77
Carrying capacity, 334, 336
Carson, Rachel, 384
Cartilage, 185, 227
Cartilagenous fish, 188
Cataracts, 289
Cavity, 263
Cell, 17, 23–24, 32, 33–34, 52
    animal, 37, 40, 51
    cancer, 45
    diffusion of materials, 53
    discovery of, 34
    kinds of, 34
    needs of, 239, 240
    parts of, 36–39
    plant, 40, 41, 46, 51, 113, 114
    sex, 44
Cell division, 44
Cell membrane, 36, 37, 53, 55
Cell reproduction, 42–43, 44
Cell theory, 35
Cellular respiration, 56, 57, 58, 112
Cellulose, 40
Cell wall, 33, 40, 93
Centipedes, 175
Central nervous system, 268, 270, 271–273
Centrioles, 49
Cephalopods, 172, 173
Cephalothorax, 176
Cerebellum, 267, 272
Cerebrum, 267, 272
Cervix, 313
Chapparal, 340
Chemical energy, 19, 52, 93
Chemosynthesis, 93
Chimpanzees, 214
Chitin, 175
Chlorophyll, 51, 57, 60, 94, 111, 114, 143
Chloroplasts, 33, 40, 57, 99
Chordates, 185, 186
Chromosomes, 33, 42, 43, 74–76
Chyme, 257, 260, 261
Cilia, 97, 285
Circulatory system, 239, 240–243
    in amphibians, 192

Clams, 173, 181
Classical conditioning, 356
Classification, 89, 90
    of animals, 166
    levels of, 91–92
    of plants, 123–124
Clay, 116, 401
Climate, biomes and, 340
Cloning, 81–82, 156
Club fungi, 102
Cnidarians, 165, 168–169
Cnidocysts, 169
Coal, 127, 379
Cochlea, 283, 290
Codominance, 72
Coevolution, 398
Coldblooded vertebrates, 187–197
Collar cells, 167
Colorblindness, 76
Communication, 353
Community, 327
Compact bone, 227
Competition, 347, 350
Complete metamorphosis, 178
Composting, 380
Compounds, 17, 25
Condensation, 364
Cone-bearing plants, 133
    reproduction in, 155
Cones, 130–131
Connective tissue, 41
Conservation, 375, 380–381
    animal, 378
    soil, 384
Consumers, 325, 332
Contagious diseases, 297, 300
Control, 3, 9
Convergent evolution, 398
Coordination, testing, 278
Cornea, 283, 288
Corn smut, 101
Cortex, 137, 139
Cotyledon, 131
Courtship, 350
Crabs, 176, 179
Crayfish, 176
Crickets, 178
Crocodiles, 195, 196
Cro-Magnons, 405
Crossbreeding, 66, 67, 85
Cross-pollination, 154

Crustaceans, 176
Cud, 217
Cuttings, 156, 158
Cytoplasm, 33, 36, 37, 38, 93, 98
Cytosine, 73
Cytotechnologist, 47

**D**

Daphnia, 4
Darwin, Charles, 394–395, 397, 398
Deafness, 291
Decomposers, 325, 332
Decomposition, 95
Deforestation, 386
Dendrites, 267, 268
Depressants, 304
Dermis, 283, 286
Desert, 323, 339
Diagnosis, making, 13
Dialysis, 249
Diaphragm, 245, 247
Dicots, 123, 131, 146
    flowers of, 152
    leaves of, 143
    transport tissue in, 140
Dieting, 301
Dietitian, 307
Diffusion, 51, 53, 276
Digestion, 257–263
    in amphibians, 192
    in fish, 190
    liquid waste from, 249
    in mammals, 216–217
    solid waste from, 248
Digestive system, parts of, 258
Disease, types of, 300
Divergent evolution, 398
DNA, 42, 65, 73, 74, 80, 92–93, 404
    recombinant, 81
DNA cloning, 156
Dolly, 82
Dominant traits, 65, 67, 68
Drought, 371
Drug abuse, 304
Drugs, 304

**E**

Eardrum, 283, 290
Ears, 290–291

Earth, 1
Earthworms, 171
Echinoderms, 165, 174
Ecologist, 343
Ecology, 3, 11
Ecosystem, 325, 326–327
    biodiversity in, 385
    energy flow through, 332
    human impact on, 336
Egg cells, 126, 313, 315
Eldredge, Nile, 399
Elements, 17, 24
Elephants, 16, 211
Embryo, 130, 311, 315
Enamel, tooth, 259, 263
Endangered species, 194, 375, 385
Endocrine glands, 267, 275, 276
Endocrine system, 276–277
Endoplasmic reticulum, 38
Energy, 17, 19–20, 51–52, 55
Environment
    plant responses to, 145
    responding to, 21–22
    traits and, 78
Environmental Protection
    Agency (EPA), 384
Enzyme, 257, 258, 260, 261
Epidermis, 137, 139, 283, 286
Epiglottis, 245, 259
Erosion, 375, 377
Esophagus, 257, 259
Estrogen, 313
Estuary, 341
Euglena, 99, 112
Evaporation, 364
Evolution, 393–405
    evidence of, 404
    human, 405
    types of, 398
Excretory system, 239, 248–250
Exocrine glands, 275
Exoskeleton, 165, 175
Experiments, 3, 6, 9, 12
Extensor, 232
Extinct species, 194, 375, 385, 394
Eyes, 288–289

F

Fallopian tube, 313
Farmer, 119

Fats, 302
Female reproductive system, 313
Ferns, 129
Fertilization, 75, 151, 154–155, 315
Fetus, 310, 311, 315, 316, 317
Fibrous roots, 137, 138
Filament, 152
Fish, 187, 188–191
    body structure of, 189–190
    body systems of, 189–190
    groups of, 188
    keeping, as pets, 199
    reproduction and
        development of, 191
Fission, 93
    nuclear, 387
Flagella, 89, 93, 97, 99
Flatworms, 169, 171
Flexor, 232
Flowers, 130–131
    parts of, 152
Food chains, 98, 103, 330
Food Guide Pyramid, 297, 303
Food poisoning, 95
Food web, 330–331, 344
Fossil fuels, 379
Fossils, 394, 399, 400–402, 404
    determining age of, 401
Fracture, 225, 228–229
Frequency, 291
Frogs, 191, 192, 193
Fruit, 155, 161
Fungi, 89, 90, 101–103, 112, 297, 328
Fusion, nuclear, 387

G

Gallbladder, 257, 260
Gametes, 126
Gastropods, 172
Genes, 36, 65, 67–69
    combinations of, 70, 71
    mutations and, 78
Gene splicing, 80
Genetic code, 73, 78
Genetic drift, 393, 397
Genetic engineering, 65, 80–81, 82
Genetic traits, 77
Genetic variations, 397–398

Genetics, 3, 11, 66 (See also Heredity.)
    history of, 65, 66–67
    modeling, 84
    using, 79
Genome, 83
Genotype, 65, 69
Geologic time scale, 393, 403
Geothermal energy, 386
Germination, 123, 129–130
Gill cover, 189
Gills, 102, 103, 172, 185, 189
Glands, 267, 269
Gliding joints, 230, 234
Global warming, 368
Golgi bodies, 38, 93
Goodall, Jane, 214
Gould, Stephen Jay, 399
Grafting, 156, 157
Grasshoppers, 177, 178
Gravitropism, 145
Gravity, 145
Greenhouse effect, 363, 367
Growth and development, 20–21
Guanine, 73
Guard cells, 144
Gymnosperms, 123, 130

H

Habitat, 325, 327
    loss of, 386
Hair, 208
Half-life, 401
Hallucinogens, 304
Harvey, William, 242
Health, 297–305
Hearing, 290–291
Heart, 241
Hemoglobin, 243
Hemophilia, 77
Herbaceous stems, 137, 141
Herbivores, 332
Heredity, 65, 66 (See also Genetics.)
    patterns of, 72
Heterozygous organisms, 65, 69, 71
Hibernation, 347, 352
Hinge joints, 230, 234
Homeostasis, 17, 24, 99, 248, 277
Homing instinct, 207

*Homo sapiens,* 405
Homozygous organisms, 65, 68, 71
Hooke, Robert, 34, 35
Hookworms, 170
Hormones, 267, 275–277
Horticulturist, 147
Host, 165, 169, 328
Human Genome Project, 83
Humans, 215
    evolution of, 405
    population growth of, 336
    reproduction of, 312–317
    stages of development in, 317
Humus, 111, 116
Hybrid electric vehicle, 389
Hybridization, 79
Hybrids, 85
Hydra, 168
Hydroelectric energy, 387
Hydrogen, 24, 25
Hydroponics, 142
Hydrotropism, 145
Hyphae, 89, 101, 102, 103
Hypothalamus, 276
Hypothesis, 3, 9, 12

**I**

Immovable joints, 229
Immune system, 298–300
Imperfect flowers, 152
Imprinting, 355
Impulses, 269, 274
Inbreeding, 79
Incisor, 216, 259
Incomplete dominance, 72
Incomplete metamorphosis, 178
Index fossils, 401
Inhalants, 304
Innate behavior, 347, 348–349
Insects, 21, 177
    metamorphosis of, 178
    migration of, 346, 351
Insight learning, 357
Instinct, 347, 349, 350–354
Insulin, 81
Interneurons, 269
Invertebrates, 165–179
    body structure of, 166
Involuntary muscles, 225, 232
Iris, 283, 288
Irrigation systems, 119

**J**

Jawless fish, 188
Jellyfish, 168
Johanson, Donald, 405
Joints, 225, 229–230
    types of, 229–230, 234

**K**

Karyotype, 75
Kasperbauer, Michael, 115
Kelp, 98
Kidneys, 248, 249–250
Kingdoms, 89, 90, 91–92, 112

**L**

Lab report, 10, 12
Lactose intolerance, 261
Lamarck, Jean Baptiste, 394
Lancelet, 186
Large-animal veterinarian, 219
Large intestine, 248, 261
Larva, 178
Larynx, 245, 246
Lasik eye surgery, 289
Law, 3, 6
Learned behavior, 347, 355, 358
Leaves, 143–144
    structure of, 144
Leeches, 171
Lens, 283, 288
Lichen, 103, 329
Ligament, 225, 229
Light energy, 19, 20, 52, 93, 112
Limiting factors, 334
Lipid, 25
Liver, 257, 260
Liverwort, 124, 125
Living things, 1, 3, 4, 10
    characteristics of, 17–25, 89
    classifying, 89–90
    need for energy, 52, 55
Lizard, 22, 195, 196
Loam, 111, 116, 117
Lobster, 176, 179
Loop of Henle, 250
Lung, 246–247, 248
Lymphatic system, 239, 244
Lymphocyte, 244, 299
Lysosome, 38, 93

**M**

Male reproductive system, 312
Mammal, 203, 208–214
    characteristics of, 208–209
    classification of, 209–210
    digestion in, 216–217
Mammary gland, 208
Mammoth, 400
Mandible, 176
Mantle, 172
Marrow, 225, 226, 228
Marsupial, 203, 210
Mating, 350
Medulla, 267, 272, 273
Medusa, 168
Meiosis, 33, 44, 75
Mendel, Gregor Johann, 66–67, 68, 69, 72
Menopause, 314
Menstrual cycle, 314
Metamorphosis, 193
Metaphase, 43
Microbiology, 3, 11
Microscope, 34, 35
Migration, 203
    of animals, 7, 351, 361
    of birds, 205, 351
    of insects, 346, 351
Mildew, 102
Millipede, 175
Mineral, 297, 302, 378
Mitochondria, 33, 37, 56, 93
Mitosis, 33, 42, 117
    stages of, 43
Molar, 216, 259
Molds, 102
Molecule, 17, 25, 52, 53
Mollusk, 165, 172–173
Molting, 175
Monocot, 123, 131, 146
    flowers of, 152
    leaves of, 143
    transport tissue in, 140
Monotreme, 203, 209
Moss, 4, 117, 123, 124, 125
Motor neuron, 269
Mouth, 259, 298
Movement, 19
Multicellular organism, 17, 23–24, 41, 98, 112
Multiple allele, 72

Muscles
    kinds of, 231–232
    treating pulled, 235
    work of, 232
Muscle tissue, 41
Muscular dystrophy, 233
Muscular system, 225, 231–233
Mushroom, 102–103
Mutation, 65, 78, 397
Mycology, 10

**N**

Narcotics, 304
Nastic response, 145
Natural gas, 378, 379, 382
Natural immunity, 300
Natural resources, 375–387
Natural selection, 394–395,
    396, 406
Neanderthals, 405
Negative feedback, 277
Negative reinforcement, 356
Nephron, 239, 249, 250
Nerve cell transplant, 274
Nerve tissue, 41
Nervous system, 267, 270
    central, 268, 271–273
    peripheral, 271, 275
Neurons, 267, 268–269
    connections between, 270
Neutral buoyancy, 198
Niche, 325, 327
Night blindness, 76
Nitrate, 363, 368
Nitrogen, 24, 95, 116
Nitrogen cycle, 368–369
Nitrogen-fixation, 95, 363,
    368–369
Nonrenewable resources, 375,
    378–379
Nonvascular plants, 124–126
    life cycle of, 123, 126
Nose, 285, 298
Notochord, 185, 186
Nuclear energy, 387
Nuclear medicine technologist,
    279
Nuclear membrane, 36, 43
Nucleic acids, 25
Nucleus, 33, 36, 37, 42, 93
Nutrients, 24, 53, 116, 301–302

Nutrition, 301–302
Nymph, 178

**O**

Observation, 6, 9, 12
Ocean, 327, 341
Octopus, 22, 173
Offspring, 17, 22
Oil, 378, 379
Oil spills, 96, 388
Omnivores, 332
Operant conditioning, 356
Optic nerve, 283, 288
Optical illusions, 293
Organ system, 41, 165, 166
Organelle, 36, 39
Organic, 116
Organisms, 17, 18
    environment of, 326
    multicellular, 23–24, 41, 98,
        112
    places and jobs for, 327
    relationships between,
        328–329
    responses to environment,
        21–22
    simple, 89–103, 105
    unicellular, 23, 24, 98, 112
Osmosis, 51, 54, 55
Ovary
    in animal, 276, 311, 313
    in seed plant, 151, 152, 154,
        155
Ovulation, 311, 313
Ovules, 152, 154, 155
Oxygen, 24, 25, 112, 114, 144

**P**

Paleobotany, 10
Paleontologist, 402, 407
Palisade layer, 144
Pampas, 323
Pancreas, 81, 257, 260
Paramecium, 97
Parasite, 165, 169
Parasitic worms, 169, 170
Parathyroid gland, 276, 277
Passive acquired immunity, 300
Passive transport, 55
Pathogens, 297, 298, 299
Pedigree, 65, 77

Penis, 312
Peregrine falcon, 51
Perfect flowers, 152
Periosteum, 227
Peripheral nervous system, 270,
    271, 275
Permafrost, 338
Personal trainer, 253
Perspiration, 251
Petal, 151, 152
Petiole, 137, 143
Pharynx, 169, 245
Phenotype, 65, 69, 78
Pheromone, 353, 354
Phloem, 137, 140, 143
Phosphorus, 116, 226
Photosynthesis, 40, 51, 57–58,
    59, 93, 94, 97, 111, 112,
    114, 119, 124, 125, 138,
    139, 140, 143, 144, 327,
    329, 366
Phototropism, 145
Phyla, 92
Physiology, 10
Pine Barrens tree frogs, 194
Pinworms, 170
Pioneer species, 124, 335
Pistil, 151, 152, 153
Pituitary gland, 272, 276
Pivotal joint, 230, 234
Placenta, 311, 315, 316
Placentals, 203, 210–214
Planarian, 169, 179
Plant embryo, 129, 130, 154
Plant researcher, 115
Plants, 378
    careers using, 111
    characteristics of, 112–113
    classifying, 132
    energy use by, 20
    extinction of, 385
    as fuel, 127
    growth of, in soil, 116–117
    importance of, 111–112, 117
    kingdom of, 90, 123
    needs of, 114
    parts of, 113
    responses to environment,
        145
    types of, 123–131
    at work, 137–145

Plaque, 263
Plasma, 239, 243
Plasmids, 80, 81
Platclcts, 239, 243
Polar ice biome, 340
Pollen, 151, 152, 153
Pollen cones, 155
Pollination, 151, 153–154
Pollinators, attracting, to
  garden, 159
Pollution, 375, 382–383
Polygenic traits, 72
Polyp, 168
Population, 325, 327
  changes in, 334
Positive reinforcement, 356
Precipitation, 363, 364, 365
Predator, 328
Pregnancy, 316–317
Premolar, 216, 259
Prenatal care, 317
Prey, 328
Primates, 203, 213–214
Probability, 71, 84
Producers, 325, 332
Progesterone, 313
Prophase, 43
Proteins, 25, 38, 74, 302
Protists, 89, 90, 96–100, 104,
  112, 297
Protozoan, 89, 97–98
Pseudopod, 98
Puberty, 311, 312
Punnett squares, 65, 70, 71
Pupa, 178
Pupil, 283, 288, 292

**R**
Radial symmetry, 166
Radioactive dating, 393, 401
Radiolarian, 96
Receptor cell, 285
Recessive gene, 79
Recessive trait, 65, 67, 68
Recombinant DNA, 81
Rectum, 248, 261
Recycling, 363, 381
Red blood cell, 238, 243, 252,
  298
Redi, Francesco, 23
Redwood tree, 109, 123
Reflex, 270–271, 348

Reflex arc, 271
Regeneration, 165, 174, 179
Relative dating, 393, 401
Renewable resources, 375, 376
Replication, 73
Reproduction, 17, 22
  of amphibians, 193–194
  asexual, 126, 128, 156
  of birds, 204
  cell, 42–43, 44
  in cone-bearing plants, 155
  of fish, 191
  in fungi, 101
  human, 311–317
  of reptiles, 197
  in seed plants, 151–157
  in yeast, 26
Reproductive cells, 74–76
Reproductive isolation, 398
Reptiles, 185, 195–197
  groups of, 195–197
  reproduction and
    development of, 197
Respiratory system, 239,
  245–247
  in amphibians, 192
Restriction enzymes, 80
Retina, 283, 288–289, 292
Rhizomes, 123, 127
Ribosomes, 38
R.I.C.E. treatment method, 235
RNA, 65, 74, 92
Rodents, 210
Root cap, 137, 139
Root hair, 137, 139
Roots, 123, 138–140
  as structure of plant,
    139–140
  of teeth, 259
Rose breeder, 85
Roundworms, 170, 171

**S**
Saliva, 257, 259
Salmonella, 93, 95
Sand, 116, 401
Savanna, 339
Scallops, 173
Science, 5, 6, 12
Scientific law, 6
Scientific method, 3, 8–10, 13
Sea stars, 174, 179

Seaweed, 105
Sedimentary rocks, 401
Seed cones, 155
Seed plants, 129–130
  asexual, 156
  grouping, 130–131
  reproduction in, 151–157
Seeds, scattering of, 155
Segmented worms, 171
Selective breeding, 79, 80
Self-pollination, 153
Semen, 311, 312
Semipermeable membrane, 51,
  53
Sense organs, 268, 283–293
Sensory neuron, 268
Sepal, 151, 152
Sex cells, 44, 74, 75
Sex chromosomes, 75, 76
Sex-linked disorders, 75–76
Sight, 288–289
Silt, 116, 401
Skeletal muscle, 231
Skeletal system, 225, 226
Skeleton, 185, 186, 189
Skin, 41, 251, 286, 298
Slime molds, 100
Small intestine, 248, 261
Smell, 285
Smog, 382
Smoking, dangers of, 247
Smooth muscles, 232
Snakes, 187, 195, 197
Soil, 111, 116–117, 377
  properties of, 118
Soil conservation, 383
Solar energy, 386
Sound waves, 291
Sperm cells, 126, 312, 315
Sperm ducts, 312
Spicules, 167
Spiders, 177
Spinal cord, 268, 272
  treating injuries to, 274
Spleen, 244
Sponges, 165, 167, 179
Spongy bone, 227
Spongy layer, 144
Spontaneous generation, 23
Spore cases, 101, 128
Spores, 101, 103
Squid, 173

Stamen, 152
Starch, digestion of, 262
Stems, 140–141
Stentor, 18
Steppe, 323
Stigma, 152
Stimulants, 304
Stimulus, 17, 21
Stinging cells, 168, 169
Stomach, 217, 260, 298
Stomata, 137, 144
Succession, 325, 335
Sulfur dioxide, 382
Sweat glands, 251
Swim bladder, 190, 198
Symmetry, 165, 166
Synapses, 267, 270

**T**

Tadpoles, 193
Taiga, 338
Tapeworms, 169, 170
Taproot, 137, 138
Target cells, 276
Taste, 285
Taste buds, 283, 285
Taxol, 117
T-bud grafting, 157
T cells, 299
Tears, 288
Teeth, 216, 259, 400
Telophase, 43
Temperate forest, 338
Temperate grassland, 339
Tendons, 225, 231
Tentacles, 168
Territory, 350
Testes, 276, 311, 312
Testosterone, 312
Theory, 3, 6
Thigmotropism, 145
Thorax, 165, 176
Threatened species, 375, 385
Thumb, 213
Thymine, 73
Thymus gland, 244
Thyroid gland, 276, 277
Tissues, 33, 41, 228, 231
Tongue, 285
Tonsils, 244
Tooth decay, preventing, 263
Tortoises, 195, 196, 395

Touch, 286
Trachea, 245–246
Traits, 65, 66, 67, 394
    acquired, 394
    dominant, 67, 68
    environment and, 78
    polygenic, 72
    recessive, 67, 68
    sex-linked, 75–76
Transpiration, 363, 364
Transport tissue, 123, 140
Trees, 136, 141, 143, 378
Tropical rain forest, 338
Tropism, 137, 145
Tubers, 156
Tundra, 338
Turtles, 195, 196
Tympanic membrane, 177

**U**

Ultrasound, 317
Unicellular organism, 23, 24,
    98, 112
Urea, 249
Ureters, 249, 250
Urethra, 249, 312
Urinary bladder, 249
Urine, 249
Uterus, 311, 313, 314, 315,
    316, 317

**V**

Vaccines, 279, 300
Vacuoles, 33, 37, 40, 99
Vagina, 313
Valves, 241
Variables, 9
Vascular plants, 124, 127–128
    complex, 123, 129–130
Vegetative propagation, 151,
    156
Veins, 137, 143, 239, 242
Ventricles, 241
Venus' flytrap, 145
Vertebrates, 165, 166, 185
    body systems of, 186
    coldblooded, 187–197
    warmblooded, 187, 203–217
Victoria, Queen, 77
Villi, 257, 261
Viruses, 24, 297
    classifying, 92

Vitamins, 297, 302
Vocal cords, 193, 245
Voluntary muscles, 225, 231

**W**

Warmblooded vertebrates, 187,
    203–217
Water, 25, 56, 301, 377
Water biomes, 341
Water cycle, 362, 364–365
Water pollution, 382–383
Weathering, 375, 377
Whales, 213
White blood cells, 243, 244,
    252, 297, 298
Wildlife, 386
Wilson, Edward Osborne, 349
Wind energy, 386
Wind farms, 374
Windpipe, 245, 259
Wolves, 64
Woody stems, 137, 141
Worms, 169–171, 297
    comparison of, 180

**X**

X chromosomes, 75, 76
Xylem, 137, 140, 143

**Y**

Y chromosomes, 75, 76
Yeast, 102
Yeast reproduction, 26, 101

**Z**

Zebra, 324
Zoology, 10, 11
Zygote, 315

# Acknowledgments

*All photography copyright Pearson Learning unless otherwise noted.*
Cover: *m.* © Steve Bloom; *t.r.* Art Wolfe/Photo Researchers, Inc.; *m.r.* Siede Preis/Getty Images, Inc.; *b.r.* Siede Preis/Getty Images, Inc. 1: Paolo Curto/Getty Images, Inc. 2: © Stephen Krasemann/Photo Researchers, Inc. 4: © Manfred Kage/Peter Arnold, Inc.; *t.r.* Barry Runk/Grant Heilman Photography, Inc.; *b.* © Manfred Danneger/Peter Arnold, Inc. 5: Sime s.a.s/eStock Photo/PictureQuest. 6: © Robert Browner/PhotoEdit 7: © Keith Sproule, WPTI/Wildlife Trust. 9: Nancy Sheehan Photography. 11: *l.* John Warden/Getty Images, Inc.; *m.l.* Art Wolfe/Tony Stone Images/Getty Images, Inc.; *m.* Jean Claude Revy/Phototake/The Creative Link; *m.r.* PhotoDisc/Getty Images, Inc.; *r.* H. Richard Johnston/Getty Images, Inc. 13: © David Young-Wolff/PhotoEdit. 16: © Stan Osolinski/The Image Bank/Getty Images. 18: *l.* © Kennan Ward/Corbis; *r.* © Manfred Kage/Peter Arnold, Inc. 19: *t.* Renee Lynn/Getty Images, Inc.; *b.* © Fritz Polking/Peter Arnold, Inc. 21: © E.R. Degginger/Color-Pic, Inc. 22: *t.* Belinda Wright/DRK Photo; *b.* © Jeff Foott/DRK Photo. 23: Martha J. Powell/Visuals Unlimited. 26: J. Forsdyke/Gene Cox/SPL/Photo Researchers, Inc. 27: John Giustina/FPG/Getty Images, Inc. 31: Dr. Jeremy Burgess/SPL/Science Source/Photo Researchers, Inc. 32: © M.I. Walker/Science Source/Photo Researchers, Inc. 34: Dr. Jonathan Eisenbach/Phototake/The Creative Link. 35: © Dr. D. Spector/Photo Researchers, Inc. 36: © Ed Reschke/Peter Arnold, Inc. 38: P. Motta & T. Naguro/Science Photo Library/Photo Researchers, Inc. 41: *l.* W.H. Fahrenbach/Visuals Unlimited; *m.l.* Carolina Biological/Visuals Unlimited; *m.r.* © Photo Researchers, Inc.; *r.* © Peter Arnold, Inc.; *b.r.* P. Dayanandan/Photo Researchers, Inc. 45: Jean Claude Revy/Phototake/The Creative Link. 47: *t.* Matthew Borkowski/Index Stock Imagery, Inc.; *b.* M.I. Walker/Dorling Kindersley Ltd. 50: Arthur C. Smith III/Grant Heilman Photography, Inc. 59: SuperStock, Inc. 61: © David Young-Wolff/PhotoEdit. 64: International Stock/ImageState. 66: © Bettmann/Corbis. 75: *t.* L. Willatt, East Anglican Regional Genetics Service/Photo Researchers, Inc.; *m.l.* © Michael Newman/PhotoEdit; *m.r.* Department of Clinical Cytogenetics Addenbrookes Hospital/Science Photo Library/Photo Researchers, Inc.; *b.l.* © John Henley/Corbis; *b.r.* Department of Clinical Cytogenetics Addenbrookes Hospital/Science Photo Library/Photo Researchers, Inc. 78: *t.* © Allen Blake Sheldon/Animals Animals/Earth Scenes; *b.* © Chinch Gryniewicz/ Ecoscene/Corbis. 79: *t.* © Gerard Lacz/Peter Arnold, Inc.; *b.* American Donkey and Mule Society. Used with permission. 81: Volker Steger/Science Photo Library/Photo Researchers, Inc. 82: Roslin Institute. 83: *t.* Sinclair Stammers/ Science Photo Library/Photo Researchers, Inc.; *b.* Klaus Guldbrandsen/Science Photo Library/Photo Researchers, Inc. 85: © Pablo Corral V/Corbis. 88: Jan Hinsch/Science Photo Library/Photo Researchers, Inc.; *b.* Alan L. Detrick/Photo Researchers, Inc. 94: *l.* David M. Phillips/Visuals Unlimited; *m.* G. Murti/Visuals Unlimited; *r.* CNRI/Science Photo Library/Photo Researchers, Inc.; *b.l.* © David M. Dennis/Tom Stack & Associates, Inc.; *b.r.* © Tony Freeman/PhotoEdit; *b.r.* A. B. Dowsett/Photo Researchers, Inc. 96: Manfred Kage/Peter Arnold, Inc. 97: *l.* © Ed Reschke/Peter Arnold, Inc.; *r.* Jerome Paulin/Visuals Unlimited. 98: © Eric V. Grave/Photo Researchers, Inc.; *r.* © Bob Evans/Peter Arnold, Inc. 99: *t.* Daniel Brody/Stock Boston; *b.* R. Kessel-G. Shih/Visuals Unlimited. 100: © Matt Meadows/Peter Arnold, Inc. 101: *t.* © E. R. Degginger/Color-Pic, Inc.; *b.* J. Forsdyke/Gene Cox/SPL/Photo Researchers, Inc. 102: © Hans Pfletschinger/Peter Arnold, Inc. 103: Michael P. Gadomski/Photo Researchers, Inc. 105: Philip Szu/Visuals Unlimited. 109: Philip Schermeister/National Geographic Image Collection. 110: © Gunter Max Photography/Corbis. 111: © James V. Elmore/Peter Arnold, Inc. 112: © Kent Wood/Peter Arnold, Inc. 114: *l.* Ian O'Leary/Dorling Kindersley Ltd.; *r.* John Durham/Photo Researchers, Inc. 115: Courtesy of Michael Kasperbauer. 116: *t.* Dorling Kindersley Ltd.; *b.* © E.R. Degginger/Color-Pic, Inc. 117: © Ray Pfortner/Peter Arnold, Inc. 119: Doug Menuez/Getty Images/PhotoDisc, Inc. 122: © Gunter Ziesler/Peter Arnold, Inc. 125: © Corbis Digital Stock; *b.* Runk/Schoenberger/Grant Heilman Photography, Inc. 127: *t.* Thonig/Premium Stock/PictureQuest; *b.* John Weinstein/The Field Museum of Natural History. 128: © Ed Reschke/Peter Arnold, Inc. 130: © M. & C. Photography/Peter Arnold, Inc. 131: © Charles Gupton/Picturesque Stock Photo. 132: S. Solum/Getty Images, Inc. 133: © Dean Conger/Corbis. 136: © Luiz C. Marigo/Peter Arnold, Inc. 138: *l.* Michael P. Gadomski/Photo Researchers, Inc.; *r.* John D. Cunningham/Visuals Unlimited. 140: © Horst Schafer/Peter Arnold, Inc. 141: *l.* Grace Davies/Omni-Photo Communications, Inc.; *m.* Russel Illiq/PhotoDisc/ Getty Images; *r.* © Manfred Kage/Peter Arnold, Inc. 142: Mike Yamashita/Woodfin Camp/PictureQuest. 144: Dr. Jeremy Burgess/Photo Researchers, Inc. 145: Kim Taylor & Jane Burton/Dorling Kindersley Ltd. 146: *t.* © Ed Reschke/Peter Arnold, Inc.; *b.* © Ed Reschke/Peter Arnold, Inc. 147: Liane Enkelis/Stock Boston. 150: FotoPic/Omni-Photo Communications, Inc. 153: © Hans Pfletschinger/Peter Arnold, Inc. 155: Dr. William M. Harlow/Photo Researchers, Inc. 156: John D. Cunningham/ Visuals Unlimited. 159: Michael Fogden/Animals Animals/Earth Scenes. 163: © Michael & Patricia Fogden/Corbis. 166: Wayne & Karen Brown/Index Stock Imagery, Inc. 167: Andrew J. Martinez/Photo Researchers, Inc. 168: *l.* Biophoto Associates/Photo Researchers, Inc.; *r.* Breck P. Kent/Animals Animals/Earth Scenes. 170: *t.* © Manfred Kage/Peter Arnold, Inc.; *b.* © Sinclair Stammers/Photo Researchers, Inc. 173: *t.* Frank Greenway/Dorling Kindersley Ltd.; *b.* Martin A. Collins, University of Aberdeen, Aberdeen, Scotland, U.K. 174: © E.R. Degginger/Color-Pic, Inc. 175: © Simon D. Pollard/Photo Researchers, Inc. 176: Scott Johnson/Animals Animals/Earth Scenes. 177: *t.* Christoph Burki/Getty Images, Inc.; *b.* © E.R. Degginger/Color-Pic, Inc. 179: Scott Johnson/Animals Animals/Earth Scenes.

181: © Eunice Harris/Photo Researchers, Inc. 184: © Jeffrey L. Rotman/Peter Arnold, Inc. 186: © Pat Lynch/Photo Researchers, Inc. 187: David Hosking/Photo Researchers, Inc. 188: *l.* Oxford Scientific Films/Animals Animals/Earth Scenes; *m.* Steven Frink/Digital Vision; *r.* © E. R. Degginger/Color-Pic, Inc. 191: *t.* Jeff Foott/Alaska Stock Images; *b.* © E. R. Degginger/Color-Pic, Inc. 192: Rauschenbach/Premium Stock/PictureQuest. 193: © Hans Pfletschinger/Peter Arnold, Inc. 194: Breck Kent/Animals Animals/Earth Scenes. 196: *t.* © E.R. Degginger/Photo Researchers, Inc.; *b.* © Joe McDonald/Tom Stack & Associates, Inc. 197: *l.* © Daniel Heuclin–BIOS/Peter Arnold, Inc.; *r.* © Leonard Lessin/Peter Arnold, Inc.; *b.* Frans Lanting/Minden Pictures. 199: © Bios, Yvette Tavernier/Peter Arnold, Inc. 202: © Mark Newman/ SuperStock, Inc. 204: Neil Fletcher/Dorling Kindersley Ltd. 205: Smari/Getty Images, Inc. 206: *l.* Gerard Fuehrer/Visuals Unlimited; *m.l.* David Stuckel/Visuals Unlimited; *m.r.* Shelley Rotner/Omni-Photo Communications, Inc.; *r.* Tui De Roy/Minden Pictures. 208: © Mitch Reardon/Getty Images, Inc. 209: Tom McHugh/Photo Researchers, Inc. 210: © John Warden/Getty Images, Inc. 211: Gerry Ellis/Minden Pictures. 212: © Michael Habicht/Animals Animals/ Earth Scenes. 213: Bruna Stude/Omni-Photo Communications, Inc. 214 © Kennan Ward/Corbis. 219: Stone/Getty Images, Inc. 223: David Job/ Getty Images, Inc. 224: Frank Siteman/Omni-Photo Communications, Inc. 227: Carolina Biological/Visuals Unlimited. 230: The Image Bank/Getty Images. 231: G.W. Willis/Visuals Unlimited. 232: *t.* G.W. Willis/Animals Animals/Earth Scenes; *b.* © Manfred Kage/Peter Arnold, Inc. 233: AP/Wide World Photo. 238: Warren Rosenberg/BPS/Stone/Getty Images. 242: © Science Source/Photo Researchers, Inc. 243: Dorling Kindersley Ltd. 250: Alfred Pasieka/Science Photo Library/Photo Researchers, Inc. 251: Bob Daemmrich/The Image Works Incorporated. 253: Spencer Grant/Stock Boston. 256: CNRI/Photo Researchers, Inc. 261: *t.* Albert Bonniers Forlag AB; *b.* U. S. Department of Agriculture. 263: © David Young-Wolff/PhotoEdit. 266: Mehau Kulyk/Science Photo Library/Photo Researchers, Inc. 269: © Ed Reschke/Peter Arnold, Inc. 274: Josh Mitchell/Index Stock Imagery, Inc. 275: Dann Coffey/The Image Bank/Getty Images. 279: The Image Bank/Getty Images. 282: © Kelly-Mooney Photography/ Corbis. 284 © David Young-Wolff/PhotoEdit. 285: © Omikron/Photo Researchers, Inc. 286: Photo Researchers, Inc. 287: FPG/Getty Images, Inc. 289: Alexander Tsiaras/Photo Researchers, Inc. 296: © Jeff Greenberg/PhotoEdit. 298: Phototake/The Creative Link. 299: © Manfred Kage/Peter Arnold, Inc. 300: *t.* © Michelle D. Bridwell/PhotoEdit; *b.* National Institute for Biological Standards and Control (U.K.)/Science Photo Library/Photo Researchers, Inc. 301: © Jonathan Nourok/PhotoEdit. 305: © Michael Newman/PhotoEdit. 307: © Michael Newman/PhotoEdit. 310: Petit Format/Nestle/Science Source/Photo Researchers, Inc. 313: David M. Phillips/The Population Council/Photo Researchers, Inc. 315: *t.* Dr. Yorgos Nikas/Science Photo Library/Photo Researchers, Inc.; *m.l.* Dr. Yorgos Nikas/Phototake/The Creative Link; *m.r.* Dr. Yorgos Nikas/Phototake/The Creative Link; *r.* Dr. Nikas/Jason Burns/Phototake/The Creative Link. 317: Stone/Getty Images, Inc. 323: © Kennan Ward Photography. 324: Birgit Koch/Animals Animals/Earth Scenes. 326: *t.* Art Wolfe/The Image Bank/Getty Images; *b.* John Anderson/Animals Animals/Earth Scenes. 327: *t.* Lynn Stone/Animals Animals/Earth Scenes; *b.* © Carl Miller/Peter Arnold, Inc. 328: *t.* © ABPL/Lanz Van Horsten/Animals Animals/Earth Scenes; *b.* James H. Robinson/Photo Researchers, Inc. 329: Sharpe, P. OSF/Animals Animals/Earth Scenes. 332: Gerard Lacz/Animals Animals/Earth Scenes. 334: Alan Carey/Photo Researchers, Inc. 336: © Carl R. Sams II/Peter Arnold, Inc. 338: © Johnny Johnson/Animals Animals/Earth Scenes; *b.* Michael Fogden/Animals Animals/ Earth Scenes. 339: © Norbert Wu/Peter Arnold, Inc. 341: *t.* Norbert Wu Wildlife Photographer; *b.* K. Wothe/PictureQuest. 343: William Taufic/Corbis Stock Market. 346: Phyllis Greenberg/Animals Animals/Earth Scenes. 348: Stephen Dalton/Photo Researchers, Inc. 349: Matthew Cavenaugh/AP/Wide World Photo. 350: © Ed Reschke/Peter Arnold, Inc. 351: Phyllis Greenberg/Animals Animals/Earth Scenes. 352: Breck P. Kent/Animals Animals/Earth Scenes. 353: Gerard Lacz/Animals Animals/Earth Scenes. 354: Jack Clark/Animals Animals/Earth Scenes. 355: *t.* © Michael Newman/PhotoEdit; *b.* Steven David Miller/Animals Animals/Earth Scenes. 357: © Michael Newman/PhotoEdit. 359: Richard Nowitz/Photo Researchers, Inc. 362: Martin Bond/Science Photo Library/Photo Researchers, Inc. 365: *t.* Nuridsany& Perennoou/Photo Researchers, Inc.; *b.* Susan Van Etten/Stock Boston. 367: Steve Cole/Stone/Getty Images. 368: NASA Goddard laboratory for Atmospheres. 369: © David M. Dennis/Tom Stack & Associates, Inc. 371: Alan I. Detrick/Photo Researchers, Inc. 374: © Jim Wark/Peter Arnold, Inc. 377: David Schultz/Stone/Getty Images. 378: Dorling Kindersley Ltd. 379: Stone/Getty Images. 382: Stone/Getty Images. 383: Jim Corwin/Photo Researchers, Inc. 384: AP/Wide World Photo. 385: *t.* Gregory G. Dimijian/Photo Researchers, Inc.; *b.* Dominic Johnson/NaturePL. 386: Mark A. Chappell/Animals Animals/Earth Scenes. 387: The Image Bank/Getty Images. 389: © Spencer Grant/PhotoEdit. 392: © John Cancalosi/Peter Arnold, Inc. 394: © Peter Arnold/Peter Arnold, Inc. 395: *t.* Archive/Photo Researchers, Inc.; *b.* Mary Evans Picture Library/Photo Researchers, Inc. 396: E.R. Degginger/Photo Researchers, Inc. 397: Gary Retherford/Photo Researchers, Inc. 398: Michael Fogden/DRK Photo. 399: © Deborah Feingold/Corbis. 400: *t.* Breck P. Kent Photography; *b.* CERPOLEX. 404: *l.* Carolina Biological Supply Company/Phototake/The Creative Link; *r.* Keith Gillett/Animals Animals/Earth Scenes. 405: *t.* John Reader/Science Photo Library/Photo Researchers, Inc.; *b.* John Reader/Science Photo Library/Photo Researchers, Inc. 407: Joseph Nettis/Stock Boston.

438   Acknowledgements